Springer Undergraduate Mathematics Series

Other books in this series

Mícheál Ó Searcóid

Metric Spaces

 Springer

Mícheál Ó Searcóid
UCD School of Mathematical Sciences
University College Dublin
Belfield
Dublin 4
Ireland

Cover illustration elements reproduced by kind permission of:
Aptech Systems, Inc., Publishers of the GAUSS Mathematical and Statistical System, 23804 S.E. Kent-Kangley Road, Maple Valley, WA 98038,
    USA Tel: (206) 432 - 7855 Fax (206) 432 - 7832 email: info@aptech.com URL www.aptech.com
American Statistical Association: Chance Vol 8 No 1, 1995 article by KS and KW Heiner 'Tree Rings of the Northern Shawangunks' page 32 fig 2.
Springer-Verlag: Mathematica in Education and Research Vol 4 Issue 3 1995 article by Roman E Maeder, Beatrice Amrhein and Oliver Gloor
    'Illustrated Mathematics: Visualization of Mathematical Objects' page 9 fig 11, originally published as a CD ROM 'Illustrated Mathematics'
    by TELOS: ISBN 0-387-14222-3, German edition by Birkhauser: ISBN 3-7643-5100-4.
Mathematica in Education and Research Vol 4 Issue 3 1995 article by Richard J Gaylord and Kazume Nishidate 'Traffic Engineering with
    Cellular Aulomata' page 35 fig 2. Mathematica in Education and Research Vol 5 Issue 2 1996 article by Michael Trott 'The Implicitization
    of a Trefoil Knot' page 14.
Mathematica in Education and Research Vol 5 Issue 2 1996 article by Lee de Cola 'Coins, Trees, Bars and Bells: Simulation of the Binomial
    Process' page 19 fig 3. Mathematica in Education and Research Vol 5 Issue 2 1996 article by Richard Gaylord and Kazume Nishidate
    'Contagious Spreading' page 33 fig 1. Mathematica in Education and Research Vol 5 Issue 2 1996 article by Joe Buhler and Stan Wagon
    'Secrets of the Madelung Constant' page 50 fig 1.

Mathematics Subject Classification (2000): 54E35; 54E45; 54E50; 54E52

British Library Cataloguing in Publication Data
A catalogue record for this book is available from the British Library

Library of Congress Control Number: 2006924371

Springer Undergraduate Mathematics Series ISSN 1615-2085
ISBN-10: 1-84628-369-8   Printed on acid-free paper
ISBN-13: 978-1-84628-369-7

Printed in the United States of America   (HAM)

9 8 7 6 5 4 3 2 1

Springer Science + Business Media
springer.com

I gcuimhne ar

Dom Aelred Cousins

Sár-oide

Go gcumhdaí Dia é
02.08.2005

Contents

*There is no philosophy which is not founded upon knowledge of
the phenomena, but to get any profit from this knowledge
it is absolutely necessary
to be a mathematician.   Daniel Bernoulli, 1700–1782*

To the Reader

Mathematics is the most beautiful
and most powerful creation
of the human spirit.   *Stefan Banach, 1892–1945*

A *metric* is a distance function; a *space* is a set with some structure; and a *metric space* is a set with structure determined by a well-defined notion of distance.

I have put together for you in this book an introduction—albeit a fairly thorough introduction—to metrics and metric spaces. The concepts discussed form a foundation for an undergraduate programme in mathematical analysis. They are few in number, but I have treated each of them at some length, and at no point does the book stray far away from the central topic of the metric.

I assume that you are familiar with the formal ideas of convergence of sequences and continuity of functions in the context of the real line. I do not, however, assume that you have mastery of these topics; indeed, much of the analysis of the real line that you have seen is subsumed rather than assumed in the presentation given here. In other words, I am going to present it to you afresh, this time in a more general setting. In some of the examples given in the book, I assume also that you have some practical knowledge of differentiation and integration and that you have studied a little linear algebra.

If you have completed two years of an honours degree in mathematics, then you have probably been introduced to abstract mathematics and should be ready to tackle abstraction at the level of this book. If, on the other hand, you are a lone, but interested, learner of mathematics, then I hope that, by proceeding at a steady pace, by dealing with one concept at a time, and by copiously illustrating the subject, my book will give you, in some measure, the mixture of enjoyment and enlightenment that you seek.

The language and notation of mathematics are designed to make the communication of profound ideas easy; the methods of mathematics are designed to make arguments precise and convincing. I have therefore adopted a rigorous approach to this subject; it is the surest aid to understanding that I can give

you and also the most aesthetically pleasing. The Index will direct you to precise definitions of terms used in the text, and the notation used is indexed in the
List of Symbols. Some pieces of notation are not standard, so it is worth having
a look at the list. Some notes on language and proof are given in Appendix
A, and some preliminary information about sets, set notation, number systems
and algebraic structures is included in Appendix B. In many cases, the List of
Symbols will direct you to these appendices for more accurate descriptions of
notation used in the book. If you are new to rigorous mathematical argument,
you are encouraged to take a quick look at the appendices before proceeding
very far with the main body of the book and to refer to them when you need
to do so.

You are already familiar with the standard function for measuring the distance between any two points in a plane. It is the prototype for all metrics. In
Chapter 1, you will learn that there is a huge variety of metrics even for a single
set of points, and in Chapter 13 you will learn to classify these metrics into
broad categories of equivalence. In between, after four preparatory chapters,
you will have the opportunity to explore the notions of convergence, boundedness, continuity, completeness, connectedness and compactness. You may be
familiar with some or all of these concepts from your study of the real line;
here we tackle them in some depth in the context of a metric space.

Much, though not all, of the material in Chapters 2 to 5 would be covered
rather rapidly in many courses on metric spaces. I prefer to do otherwise. I
am assuming that you are only beginning to develop a taste for abstraction; if
that is the case, it will stand you in good stead later on if you take the time
to digest these early chapters and to learn to juggle mentally with the ideas
they contain. If you are already well seasoned in abstraction, you may wish to
make speedy work of this material; I think you will find much else in the book
to satisfy your intellect.

Each chapter of the book deals with a single concept or with a single collection of related concepts, and all are illustrated by examples. Concepts are
often explored through naturally arising questions. Some of the questions have
non-intuitive answers; by posing them, I hope to emphasize the need for care
in both reading and writing mathematics. Intuition can never take the place of
proof, though it is often a good guide to what is true.

Although the book is designed as a single integrated work, some of the
chapters contain more material than it would normally be possible for a lecturer
to cover in a first undergraduate course on metric spaces. If you are following
such a course, you may therefore expect a few of the sections of this book to
be outside the syllabus adopted by your lecturer. But I am writing primarily
for you, the student. Your interest, motivation and ability, and your desire
to achieve mastery of the subject may not be constrained too tightly by any

particular college curriculum. I have therefore given you a little more than such a curriculum might insist upon.

My book is very heavily cross-referenced. More often than not, when I invoke an earlier theorem in a proof, I attach a reference number. Sometimes you will not need to look back at the earlier result because you understand perfectly what is being asserted; in such cases, please pass over the intruding reference and be aware that some other reader may have need of it. On a grander scale, I have accumulated all the references into a Cumulative Reference Chart that shows which sections of the book you may need to read before embarking on any given section. It shows, at the same time, which sections of the book refer to the given section. I constructed the chart as an aid to course design; it is really more for lecturers than for you.

Each chapter ends with a number of exercises. You should try to do these exercises and then write out the solutions in an ordered and precise manner. Getting solutions is a sign that the material of the chapter has been understood; writing them out is an exercise in communication. If mathematicians were to write down their proofs in the way they actually discover them, most mathematical argument would be unintelligible. Usually we write and rewrite our arguments until we are satisfied that other people will understand them and assent readily to them. Of the 244 exercises, 108 are marked with a dagger and solved at the back of the book. Typeset solutions to all the exercises are available to lecturers at http://www.springer.com/1-84628-369-8, but application must be made to Springer to access them.

You may find mistakes in the book, or you may have suggestions for its improvement. You can get to the book's web site through a link on my home page at the address given below. The site will, in due course, contain comments, corrections and supplementary material, so you may like to look at it from time to time. Please contact me if you want to discuss anything in the book; my email address is also given below.

There are many people who have helped to make this book. First, there are the mathematicians whose pioneering work underlies everything contained here. Some are mentioned in the text or are commemorated by quotations in the chapter headings—the quotations, some in translation, are taken from the wonderful site of John J. O'Connor and Edmund F. Robertson at the University of St. Andrews, Scotland (http://www-groups.dcs.st-and.ac.uk/~history/). Next, I happily acknowledge my debt to Dom Aelred Cousins, monk of Belmont and late of Tororo, Uganda, to whom the book is dedicated. He fostered in me at an early age an appreciation of the beauty and precision of the art of mathematics. I thank Springer's anonymous reviewers and my fellow mathematicians Christopher Boyd, Thomas J. Laffey, Stefan de Wannemacker, Thomas Unger, Richard Moloney, Robin Harte, J. Brendan Quigley, Remo Hügli, Miriam Logan and Patrick Green, who read drafts, commented on the text, alerted me to errors and suggested improvements. I thank my colleagues Wayne Sullivan, Michael Mackey and Alun J. Carr for invaluable technical advice, Colm Ó Searcóid for sharing his graphical expertise and Eoghan Ó Searcóid for his artwork. I typeset the book on a Linux machine using TEX and LATEX, a task made possible only by the provision of free software by Donald Knuth, Leslie Lamport, Linus Torvalds and many others. I am particularly indebted to Niamh and Peter O'Connor, who lent me their house on the West Kerry coast, where a good part of the book was written. Last, I should like to extend my thanks to the Springer teams in Britain and the United States. The copyeditor Hal Henglein saved me from using a lot of inappropriate punctuation, and both Karen Borthwick in London and Herman Makler in New York have been most helpful and cooperative at the various stages of bringing my book to production.

31 May 2006   Mícheál Ó Searcóid
http://maths.ucd.ie/~mos   School of Mathematical Sciences
micheal.osearcoid@ucd.ie   University College Dublin
                                                                          Ireland

Cumulative Reference Chart

I know not what I appear to the world,
but to myself I seem to have been only like
a boy playing on the sea-shore,
and diverting myself in now and then
finding a smoother pebble or a prettier shell,
whilest the great ocean of truth
lay all undiscovered before me.   Sir Isaac Newton, 1643–1727

The Cumulative Reference Chart spread out on the next two pages is intended to give some idea of the interdependencies of the various sections of the book.

The chart uses the 115 sections of the main body of the book as its display units, but the units for its construction are the individual items in those sections, and there are well over a thousand of them. The large numbers at the left and bottom of the chart indicate the chapters of the book; the tiny numbers on the left and on the diagonal indicate the sections in those chapters, as listed in the Contents. Thus each row represents a section, and so does each column.

A mark entered at the cross point of a row and a column indicates that some item in the row section refers to some item in the column section either directly or by way of some intermediate item or items. For example, if a, b, c and d are items that occur in that order in sections A, B, C and D, respectively, and if d refers to c and c refers to b and b refers to a, then marks appear at the various cross points, in particular at that of column A and row D. If we vary this and have, instead of d, some other item d' in section D referring to c, then marks appear at the cross points of column C and row D and of column A and row C, but not necessarily at the cross point of column A and row D. The chart is cumulative item by item but does not exhibit transitivity of sections.

To know what sections include items that depend cumulatively in this sense on a given section A, identify the column associated with A and read off the rows marked in that column. They represent the sections that include items that refer either directly or indirectly to some item in section A. The marks alternate in shape and colour in order to draw the eye along the rows and columns, respectively; nothing intrinsic to the book is signified by this alternation.

Though the chart is not exhaustive, the author hopes that there are few major non-obvious dependencies in the main body of the book that it does not record. Dependence on the appendices is not recorded in the chart.

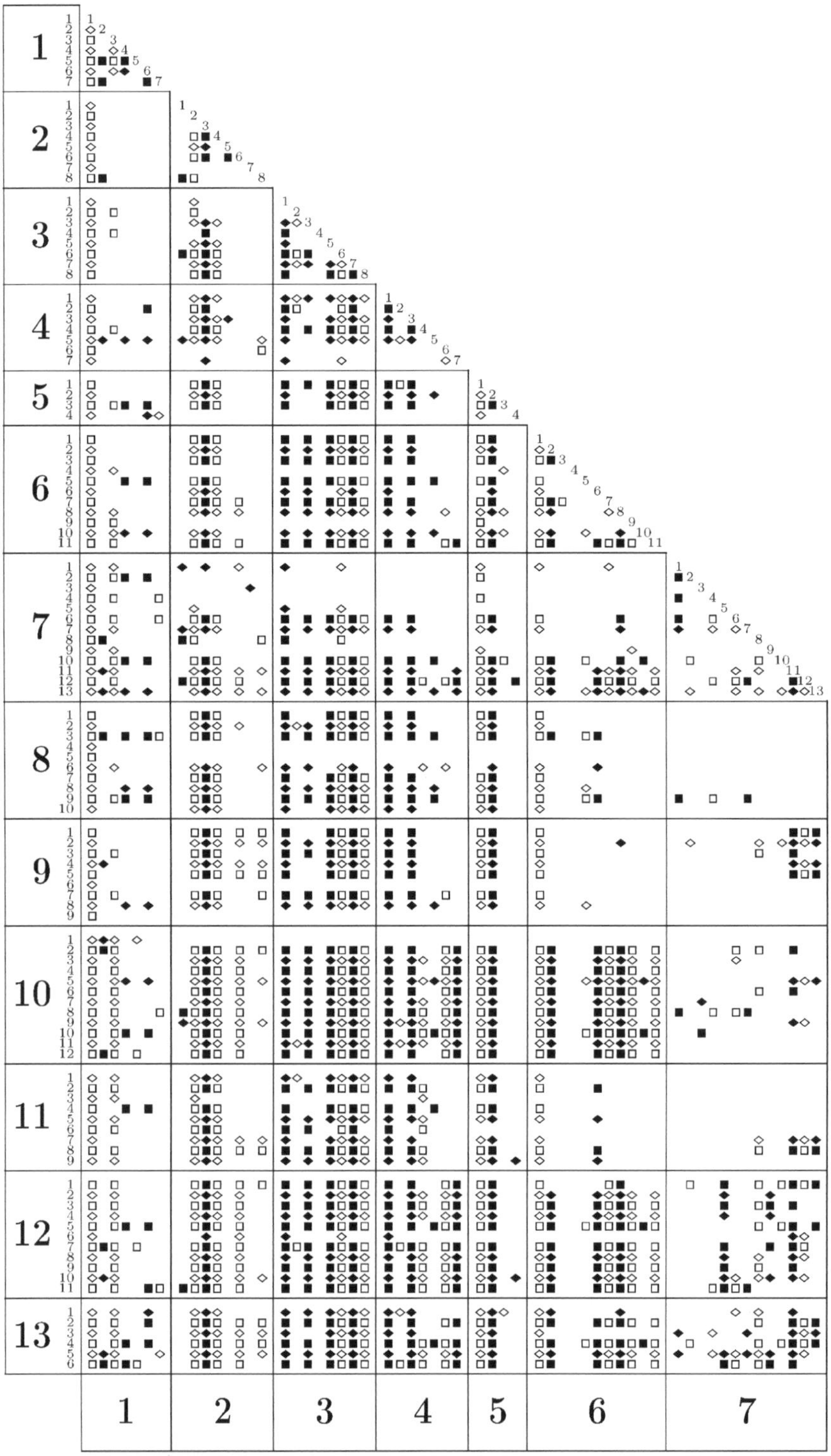
1
2
3
4
5
6
7
8
9
10
11
12
13
1 2 3 4 5 6 7

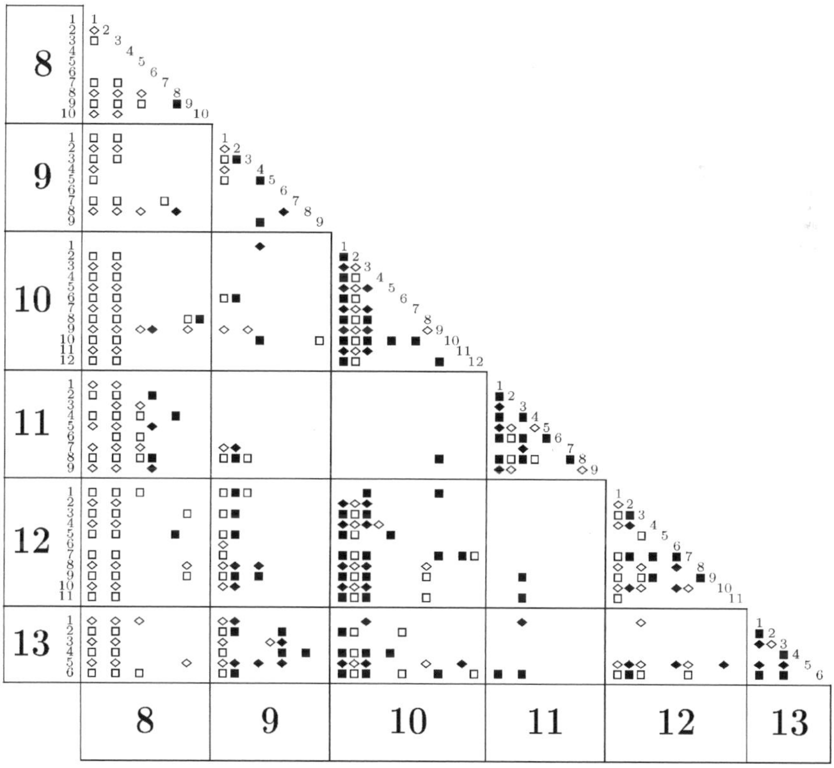

1
Metrics

This book is about functions that measure difference. We shall call this measurement *distance* because our prototype is the distance between two points in a plane along the straight line segment that joins them. The theory we shall develop from the properties associated with this prototype can be applied to a whole range of situations—not just spatial ones—in which we want to measure some particular type of difference between distinct objects. Although *distance* is an appropriate word for the type of measurement we are going to consider, we shall see that there are nonetheless some natural types of distance that do not fall within the scope of our study.

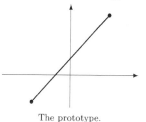

The prototype.

Distance is not always measured in straight lines. Even in our own locality, we do not usually measure as the crow flies, simply because there are more constraints on our motion than on that of the crow; and even the crow meets obstacles. Journeys around the world are measured not by burrowing through the Earth's crust but by following navigable routes on a two-dimensional near-spherical surface in a three-dimensional universe.

We often find ourselves in situations where units of length are not the most informative for measuring distance. In some circumstances, the number of feet or miles is an irrelevant piece of information and phrases such as *'two and a half hours drive'*, *'three days hacking through the jungle'*, *'303 steps to the top'*, *'to the fifth turning on the right'* or *'only two hurdles left to jump'* will tell us exactly what we want to know. Very large distances in space are habitually measured by the amount of time it takes light to travel them; we say, for example, that Sirius, the brightest star in the sky, is 8·6 light years away from the Earth.

Let us, without more ado, examine the fundamental properties of the point-to-point-along-a-straight-line-segment prototype with a view to developing a theory around those properties—a theory that will then be applicable in many situations, some very different from that of the prototype.

We notice first that the prototype deals with a well-defined *set* of objects (Appendix B), namely points in the plane; second, that it measures the distance between pairs of distinct objects in this set by using *positive real numbers* and measures as 0 the distance between each point and itself; and third, that the prototype has symmetry, the distance from a to b being the same as the distance from b to a. Last, we observe that the line segment provides the shortest way possible from one point to another; no legitimate route that passes through a third point is shorter.

Aware of these fundamental properties of the prototype, we shall set the scope of our investigation as follows. Given a set X, we shall study functions d that distinguish between every two points a and b of X by assigning to the ordered pair (a, b), in a symmetric manner, a single positive real number, $d(a, b)$, in such a way that, given any point x of X, $d(a, b)$ does not exceed the sum of $d(a, x)$ and $d(x, b)$, and we always set $d(a, a) = 0$. A function d that has all these properties is called a *metric* on X.

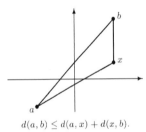

$d(a, b) \leq d(a, x) + d(x, b)$.

1.1 Metric Spaces

Definition 1.1.1

Suppose X is a set and d is a real function defined on the Cartesian product $X \times X$ (see B.5.1). Then d is called a *metric* on X if, and only if, for each $a, b, c \in X$,

- (POSITIVE PROPERTY) $d(a, b) \geq 0$ with equality if, and only if, $a = b$;
- (SYMMETRIC PROPERTY) $d(a, b) = d(b, a)$; and
- (TRIANGLE INEQUALITY) $d(a, b) \leq d(a, c) + d(c, b)$.

In this event, we call the set X endowed with this metric a *metric space* and, for each $a, b \in X$, we call the number $d(a, b)$ the *distance* between a and b with respect to the metric d. Usually, we say simply that X is a metric space; if we need to specify the metric, we say that (X, d) is a metric space.

Theorem 1.1.2 (Rearrangement of the Triangle Inequality)

Suppose X is a metric space and $a, b, c \in X$. Then $|d(a,b) - d(b,c)| \leq d(a,c)$.

Proof

The triangle inequality for d yields first $d(a,b) \leq d(a,c) + d(c,b)$ and second $d(c,b) \leq d(c,a) + d(a,b)$. Using symmetry, rearrangement of the first of these two inequalities gives $d(a,b) - d(b,c) \leq d(a,c)$ and rearrangement of the second gives $d(b,c) - d(a,b) \leq d(a,c)$. The two together prove the theorem.   □

Example 1.1.3

The most familiar metric is that determined by the absolute-value function on $\mathbb{R}$; it is the function $(a,b) \mapsto |a - b|$ defined on $\mathbb{R} \times \mathbb{R}$. This metric is called the *Euclidean metric* on $\mathbb{R}$. It is the usual metric on $\mathbb{R}$ and, unless we state otherwise, we shall generally assume that $\mathbb{R}$ is endowed with this metric. The reader will notice that this usual metric on $\mathbb{R}$ is dependent not simply on the algebraic structure of $\mathbb{R}$ but also on the total ordering (see B.6.1) of $\mathbb{R}$: for $a, b \in \mathbb{R}$, $|a - b|$ is defined to be $a - b$ if $b \leq a$ and $b - a$ otherwise. This relationship between metric and order has many repercussions. For example, the intuitive idea of *betweenness* in the ordering of $\mathbb{R}$ is captured by this particular metric on $\mathbb{R}$. Specifically, if $a, x, b \in \mathbb{R}$ and either $a \leq x \leq b$ or $b \leq x \leq a$, then x lies *between* a and b; once the metric is defined, the same thing can be said in a different, but more obscure, way—x lies between a and b if, and only if, the distances from x to a and from x to b are both less than or equal to the distance from a to b, or, with total precision, if, and only if, $|a - b| = |a - x| + |x - b|$.

Example 1.1.4

Trivially and uselessly, except to confirm that $\varnothing$ is a metric space, the only metric on the empty set is the empty function, namely the function with empty domain (and which therefore does nothing). Not much more interesting is the only metric on a singleton set, which is, of course, the zero function.

Example 1.1.5

The usual metric on $\mathbb{C}$ is the *Euclidean metric* determined by the modulus function, $(z, w) \mapsto |z - w|$. It is, of course, an extension (B.13.1) to $\mathbb{C} \times \mathbb{C}$ of the Euclidean metric on $\mathbb{R}$. We shall assume that $\mathbb{C}$ is endowed with it unless we state otherwise.

Example 1.1.6

Suppose C is a circle and, for each $a, b \in C$, define $d(a, b)$ to be the distance along the line segment from a to b. Then d is a metric on C. The fact that the route from a to b goes outside C is irrelevant; a metric is simply a function defined on ordered pairs and does not take into account any 'route travelled' from one point to another. In this respect our $\mathbb{R}^2$ prototype displays more information than is used for the definition of a metric.

Example 1.1.7

Every set admits a metric. The *discrete metric* on a set X is defined by saying that the distance from each point of X to every other point of X is 1 and, of course, the distance from each point of X to itself is 0 (Q 1.3). If X has more than one member, then there are many metrics on X. Indeed, if d is a metric on X, it is always possible to construct another metric on X that is not equal to d simply by multiplying all the values of d by some positive constant λ. A less obvious way of defining a new metric e in terms of d is to set $e(a, b) = d(a, b)/(1 + d(a, b))$ for each $a, b \in X$. The reader should check that, for $x, y, z \in \mathbb{R}^{\oplus}$, $z \le x + y \Rightarrow z/(1 + z) \le x/(1 + x) + y/(1 + y)$ and use this result to verify the triangle inequality for e (Q 1.11). Note that all the values of the metric e lie in the interval $[0, 1)$ irrespective of the values of d.

Example 1.1.8

The *Euclidean metric* on $\mathbb{R}^2$ is the familiar distance function we used as our prototype. It is the usual metric assigned to $\mathbb{R}^2$, though we shall later be investigating many others. Under this metric, the distance between $a = (a_1, a_2)$ and $b = (b_1, b_2)$ in $\mathbb{R}^2$, namely $\sqrt{(b_1 - a_1)^2 + (b_2 - a_2)^2}$, is obtained by using Pythagoras's Theorem. The reader will notice that, by using that theorem twice, we get a formula for the usual distance in $\mathbb{R}^3$. This is the *Euclidean metric* on $\mathbb{R}^3$; it is the function defined on $\mathbb{R}^3 \times \mathbb{R}^3$ by setting the distance from a to b to be $\sqrt{(a_1 - b_1)^2 + (a_2 - b_2)^2 + (a_3 - b_3)^2}$, where it is to be understood that a is (a_1, a_2, a_3) and b is (b_1, b_2, b_3).

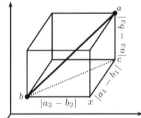

Use Pythagoras's Theorem first in triangle cbx to get the length of bc and then in triangle abc to get the length of ab.

Example 1.1.9

Suppose C is a circle and, for each $a, b \in C$, define $d(a, b)$ to be the shorter distance along the circle from a to b. Then d is a metric on C. Similarly, if S is a sphere and $a, b \in S$, we can define $d(a, b)$ to be the shortest distance along a great circle joining a and b; this is well defined and thus determines a metric on S because the great circle is unique unless a and b are equal or antipodal.

Example 1.1.10

Example 1.1.8 suggests that the function $(a, b) \mapsto \sqrt{\sum_{i=1}^{n}(b_i - a_i)^2}$ defined on $\mathbb{R}^n \times \mathbb{R}^n$, where a is the n-tuple $(a_1, \ldots, a_n)$ and b is the n-tuple $(b_1, \ldots, b_n)$, might be a metric on $\mathbb{R}^n$. It is indeed the case. This metric is called the *Euclidean metric* on $\mathbb{R}^n$. It is the usual metric on $\mathbb{R}^n$, and we shall assume that $\mathbb{R}^n$ is endowed with it unless we state otherwise. For the time being, we leave it to the reader to provide a proof that this function satisfies the triangle inequality and is therefore a metric (Q 1.4). We shall give a proof of a much more general result later in the book (12.11.3).

Example 1.1.11

There are many metrics on $\mathbb{R}$ itself. It may be appropriate, for example, to use metrics that stretch or shrink sections of the real line when we want to model situations in which weighting is required; distances over rough terrain or through liquid or solid media might be stretched to incorporate in measurement the difficulty of passing through them. This can always be done; in fact, every injective function $f: \mathbb{R} \to \mathbb{R}$ generates a metric $(a, b) \mapsto |f(a) - f(b)|$ on $\mathbb{R}$—and a more general result applies to injective functions into any metric space (Q 1.12). The injectivity is necessary to ensure that distances between distinct points are nonzero, symmetry is built into the definition and the triangle inequality is a consequence of the triangle inequality for the absolute-value function. If, for example, f is the exponential function, then this new metric measures the dis-

$\mathbb{R}$ with the exponential metric.

tance between real numbers a and b as $|e^a - e^b|$; it stretches the positive part of the real line and shrinks the negative part, all negative numbers being less than 1 apart.

Example 1.1.12

Since the inverse function $x \mapsto x^{-1}$ is injective, the function $(a, b) \mapsto |a^{-1} - b^{-1}|$ determines a metric on $\mathbb{R}^+$ (Q 1.12), called the *inverse metric*. A somewhat similar metric can be put on $\tilde{\mathbb{N}} = \mathbb{N} \cup \{\infty\}$ (see B.7.1) by defining $d(m, n) = |m^{-1} - n^{-1}|$ and $d(n, \infty) = d(\infty, n) = n^{-1}$ for all $m, n \in \mathbb{N}$ and $d(\infty, \infty) = 0$ (Q 1.14). We shall call this the *inverse metric* on $\tilde{\mathbb{N}}$.

Example 1.1.13

Much of New York City's road network consists of two sets of parallel roads that intersect at right angles. The distance from one place to another using this network is calculated by adding the distance in the direction parallel to one of the sets of roads to the distance in the direction parallel to the other. An idealized model of this situation is obtained by endowing $\mathbb{R}^2$ with the function defined on $\mathbb{R}^2 \times \mathbb{R}^2$ by $(a, b) \mapsto |b_1 - a_1| + |b_2 - a_2|$, where $a = (a_1, a_2)$ and $b = (b_1, b_2)$. It is easy to check that this function is a metric; it is known as the *taxicab metric*. More generally, the function μ_1 defined on $\mathbb{R}^n \times \mathbb{R}^n$ by $(a, b) \mapsto \sum_{i=1}^{n} |a_i - b_i|$ is a metric on $\mathbb{R}^n$. We leave the verification to the reader (Q 1.5).

Example 1.1.14

Blood pressure is measured using two numbers. The higher is referred to as the systolic reading and the lower the diastolic. Let us suppose that a patient's blood pressure is fluctuating wildly and that we want to measure the difference between two systolic readings or the difference between the two associated diastolic readings, whichever difference is greater. This difference is represented by the function $(a, b) \mapsto \max\{|b_1 - a_1|, |b_2 - a_2|\}$ defined on $\mathbb{R}^2 \times \mathbb{R}^2$, and it is easy to check that this is a metric on $\mathbb{R}^2$. If $a = (a_1, a_2)$ and $b = (b_1, b_2)$ are two blood pressure readings, this metric gives the number we want. More generally, the function μ_∞ defined on $\mathbb{R}^n \times \mathbb{R}^n$ by $(a, b) \mapsto \max\{|a_i - b_i| \mid i \in \mathbb{N}_n\}$ is a metric on $\mathbb{R}^n$; we leave the verification to the reader (Q 1.5).

Example 1.1.15

Define d on $\mathbb{R}^2 \times \mathbb{R}^2$ as follows: $d(a, a) = 0$ for all $a \in \mathbb{R}^2$ and, for $a, b \in \mathbb{R}^2$ with $a \neq b$, $d(a, b) = \sqrt{(b_1 - a_1)^2 + (b_2 - a_2)^2}$ if neither a nor b is the origin $(0, 0)$ of $\mathbb{R}^2$ and $d(a, b) = 1 + \sqrt{(b_1 - a_1)^2 + (b_2 - a_2)^2}$ otherwise. Then d has the positive property and is symmetric. Moreover, d coincides with the Euclidean metric on $\mathbb{R}^2$ except when exactly one of a and b is the origin. For all $a, b, c \in \mathbb{R}^2$, we have $d(a, (0, 0)) \leq d(a, c) + d(c, (0, 0))$ and $d(a, b) \leq d(a, (0, 0)) + d((0, 0), b)$, by the triangle inequality for the Euclidean metric on $\mathbb{R}^2$, so that d also satisfies the triangle inequality and is consequently a metric on $\mathbb{R}^2$.

Example 1.1.16

Let p be a prime number. Each non-zero rational number x can be expressed as $p^k r/s$ for a unique value of $k \in \mathbb{Z}$, where $r \in \mathbb{Z}$ and $s \in \mathbb{N}$ and neither r nor s is divisible by p; we define $|x|_p$ to be p^{-k}. Also, we set $|0|_p$ to be 0. The *p-adic metric* on $\mathbb{Q}$ is then defined to be $(a, b) \mapsto |a - b|_p$. This is clearly a non-negative symmetric function and is 0 only when $a = b$. For the triangle inequality, it is easily verified that, if $|a - c|_p = p^{-m}$ and $|c - b|_p = p^{-n}$, then $|a - b|_p \leq \max\{p^{-m}, p^{-n}\}$.

Example 1.1.17

Consider the set $\mathcal{F}$ of functions from $[0, 1]$ to $[0, 1]$. For each $f, g \in \mathcal{F}$, define

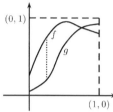

$s(f, g) = \sup\{|f(x) - g(x)| \mid x \in [0, 1]\}$. We show that s is a metric on $\mathcal{F}$. Certainly, s is symmetric and its values are in $\mathbb{R}^{\oplus}$. Also, $s(f, f) = 0$ for each $f \in \mathcal{F}$. For $f, g \in \mathcal{F}$ with $s(f, g) = 0$, we have $|f(x) - g(x)| \leq 0$ for every $x \in [0, 1]$, which forces $f(x) = g(x)$ for every $x \in [0, 1]$ and therefore $f = g$. For the trian-

$s(f, g) = 0\cdot5525$ as shown by the dotted line.

gle inequality, if $f, g, h \in \mathcal{F}$, then, for all $x \in [0, 1]$, we have $|f(x) - g(x)| \leq |f(x) - h(x)| + |h(x) - g(x)|$; therefore $|f(x) - g(x)| \leq s(f, h) + s(h, g)$ and, since this is true for all $x \in [0, 1]$, it follows that $s(f, g) \leq s(f, h) + s(h, g)$ (B.6.6).

Example 1.1.18

Let $\mathcal{I}$ denote the collection of intervals of $\mathbb{R}$ of the type $[a, b]$, where $a, b \in \mathbb{R}$ with $a \leq b$ (B.8.2). For each $I = [a, b]$ and $J = [r, s]$ in $\mathcal{I}$, define $d(I, J) = \max\{|r - a|, |s - b|\}$. Then $d(I, J) = d(J, I) \geq 0$ and, moreover, if $d(I, J) = 0$, then $r = a$ and $s = b$, so that $J = I$. To show that

$d([a, b], [r, s]) = 1\cdot5 - 0\cdot5 = 1.$

d is a metric on $\mathcal{I}$, we therefore need only establish that the triangle inequality holds. Towards this, let $K = [u, v]$ also be a member of $\mathcal{I}$. Then, using the triangle inequality for the absolute-value function, we have

$$
\begin{aligned}
d(I, K) + d(K, J) &= \max\{|u - a|, |v - b|\} + \max\{|r - u|, |s - v|\} \\
&\geq \max\{|u - a| + |r - u|, |v - b| + |s - v|\} \\
&\geq \max\{|r - a|, |s - b|\} \\
&= d(I, J).
\end{aligned}
$$

It follows that d is a metric on $\mathcal{I}$. The restriction to intervals of the type $[a, b]$ in this example is necessary. We avoided intervals of the types (a, ∞), $[a, \infty)$, $(-\infty, b)$, $(-\infty, b]$ and $(-\infty, \infty)$ in order to ensure real values for d, and we avoided intervals of the types (a, b), $[a, b)$ and $(a, b]$ in order to ensure that $d(I, J) = 0 \Rightarrow I = J$. Nonetheless, as we shall see later (7.3.1), there are possibilities for extending this metric to an even larger collection of subsets of $\mathbb{R}$ than $\mathcal{I}$.

Example 1.1.19

Although there are many real situations in which distances are determined
by metrics, there are other familiar situations that are not encompassed by
this theory. For example, many cities have one-way traffic systems, so that the
appropriate distance function may not be symmetric even though it does satisfy
the triangle inequality.

Example 1.1.20

We can measure the distance between cousins in a family tree. One way to do
this is to say that, for persons a and b, the distance from a to b is 0 if a and b
are the same person and is obtained otherwise by counting the smallest number
of generations from a to b up and down the family tree, peaking once. Care is
needed as the shortest route may not go through the common ancestor nearest
to a or through the one nearest to b. The resulting function is symmetric, but
only in very strange human circumstances will it satisfy the triangle inequality.
For example, the distance from Elizabeth Tudor to her lifelong friend and ally
Black Tom Butler is 12 and the distance from Elizabeth Tudor to her arch-rival
Mary Stuart is 5, whereas the distance between the Irish earl and the Scottish
queen, whatever it is, surely exceeds 17.

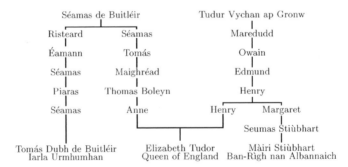

Butlers, stewards and courtly kinfolk.

1.2   Point Functions and Pointlike Functions

A metric on a set X is defined on $X \times X$, not on X. Nonetheless, some metrics,
such as the Euclidean metric on $\mathbb{R}$, are determined by a function that is defined
on X itself. We do not expect to have such a function for every metric on every
set, but we always have *point functions*, which help to repair the deficiency.

Definition 1.2.1

Suppose (X, d) is a metric space and $z \in X$. We shall call the non-negative real function $x \mapsto d(z, x)$ defined on X the *point function* at z and denote it by δ_z. The set $\{\delta_z \mid z \in X\}$ that consists of all point functions on X will be denoted by $\delta(X)$.

Theorem 1.2.2

Suppose (X, d) is a metric space. The function $z \mapsto \delta_z$ is a bijective function from X onto $\delta(X)$.

Proof

Certainly the function is surjective, by definition. We must show that it is injective. Suppose $z, w \in X$ and $\delta_z = \delta_w$. We have the succession of equalities $d(z, w) = \delta_z(w) = \delta_w(w) = d(w, w) = 0$. So $z = w$ by the positive property of the metric d. $\qquad\square$

Example 1.2.3

Let $n \in \mathbb{N}$. In $\mathbb{R}^n$ with the Euclidean metric d, we can recover the metric entirely by knowing the function δ_0 and using the equation $d(a, b) = \delta_0(b - a)$. Indeed,

d can be recovered by knowing δ_z for any one z in $\mathbb{R}^n$ because $d(a, b) = \delta_z(z - b + a)$ for all $a, b \in \mathbb{R}^n$. This sort of thing is not possible in an arbitrary metric space because we do not have the algebraic operations of addition and scalar multiplication. We can, however, easily recover the metric by knowing all, or a good many, of the point functions because $\delta_b(a) = d(a, b) = \delta_a(b)$ for all $a, b \in X$.

Point functions mimic the metric in a way that is made precise now in the two properties listed in 1.2.4. The first of these properties is shared by many other functions; these we shall call *pointlike functions* in 1.2.5.

Theorem 1.2.4

Suppose (X, d) is a non-empty metric space and $z \in X$. Then

(i) $\delta_z(b) - \delta_z(a) \leq d(a, b) \leq \delta_z(b) + \delta_z(a)$ for all $a, b \in X$; and

(ii) $\delta_z(z) = 0$.

Proof

The inequalities of (i) can be rewritten as $|d(a,b) - \delta_z(b)| \le \delta_z(a)$, which is true by 1.1.2. And we certainly have $\delta_z(z) = d(z,z) = 0$. $\qquad\square$

Definition 1.2.5

Suppose (X,d) is a non-empty metric space and $u: X \to \mathbb{R}^{\oplus}$. We shall call u a *pointlike function* on X if, and only if, $u(a) - u(b) \le d(a,b) \le u(a) + u(b)$ for all $a, b \in X$.

Theorem 1.2.6

Suppose (X,d) is a metric space and $u: X \to \mathbb{R}^{\oplus}$. Then $u \in \delta(X)$ if, and only if, u is a pointlike function on X and $0 \in u(X)$. In that case, there is a unique $w \in X$ such that $u = \delta_w$.

Proof

If $u \in \delta(X)$, u is a pointlike function and $0 \in u(X)$ by 1.2.4. On the other hand, if u is a pointlike function and there exists $w \in X$ with $u(w) = 0$, then $u(b) = u(b) - u(w) \le d(b,w) \le u(b) + u(w) = u(b)$ for all $b \in X$. Therefore $u(b) = d(b,w) = \delta_w(b)$, yielding $u = \delta_w$, and w is unique by 1.2.2. $\qquad\square$

1.3 Metric Subspaces and Metric Superspaces

Each set X has a collection of subsets, its power set $\mathcal{P}(X)$ (B.2.2). Each metric defined on X not only makes X into a metric space, but determines a metric on each member of $\mathcal{P}(X)$ and makes it into *metric subspace* of X.

Definition 1.3.1

Suppose (X,d) and (Y,e) are metric spaces. We say that X is a *metric subspace* of Y and that Y is a *metric superspace* of X if, and only if, X is a subset of Y and d is a restriction of e. We shall often, provided it leads to no confusion, use the same letter to designate the metric on a subspace and the metric on its superspace.

Example 1.3.2

$\mathbb{R}$ with its usual metric is a metric subspace of $\mathbb{C}$ since the absolute-value function on $\mathbb{R}$ is a restriction of the modulus function on $\mathbb{C}$; similarly $\mathbb{Q}$ and $\mathbb{N}$ are metric subspaces of $\mathbb{R}$. In fact, any subset of $\mathbb{R}$ may be regarded as a metric subspace of $\mathbb{R}$ simply by using on it the appropriate restriction of the usual distance function. The subset $(0,1) \cup (4,6)$ of $\mathbb{R}$, for example, is a metric space when endowed with the usual distance function inherited from $\mathbb{R}$.

Question 1.3.3

If Y is a metric space and X is a subset of Y, then X is made into a metric subspace of Y in one way only, by restricting the metric of Y to $X \times X$. Let us turn the tables. Suppose (X, d) is a non-empty metric space and Y is a proper superset of X. Is it always possible to define a metric on Y that is an extension of d? There are many ways to do it; here is one. Extend d to $(X \times X) \cup (Y \backslash X \times Y \backslash X)$ by endowing $Y \backslash X$ with any metric, labelling it d also. Fix two points, $a \in X$ and $b \in Y \backslash X$. Then, for each $y \in Y \backslash X$ and $x \in X$, set $d(y, x) = d(x, y) = d(x, a) + 1 + d(b, y)$, thus extending d to $Y \times Y$. Then d is a metric on Y (Q 1.19).

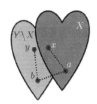

The measured route from x to y.

1.4 Isometries

Despite the differences in algebraic structure that exist between $\mathbb{C}$ and $\mathbb{R}^2$—in particular that complex numbers have a unique multiplication, whereas $\mathbb{R}^2$ may or may not be endowed with one of several different multiplicative operations—there is no detectable difference, other than labelling, between $\mathbb{C}$ and $\mathbb{R}^2$ when they are viewed simply as metric spaces, each with its usual metric. The technical way to describe this similarity is to say that the spaces are *isometric*.

Definition 1.4.1

Suppose (X, d) and (Y, e) are metric spaces and $\phi \colon X \to Y$. Then ϕ is called an *isometry* or an *isometric map* if, and only if, $e(\phi(a), \phi(b)) = d(a, b)$ for all $a, b \in X$. If ϕ is an isometry, we say that the metric subspace $(\phi(X), e)$ of (Y, e) (1.3.1) is an *isometric copy* of the space (X, d).

Suppose (X, d) and (Y, e) are metric spaces and $\phi \colon X \to Y$ is an isometry. Then ϕ is necessarily injective (Q 1.21) and its inverse $\phi^{-1} \colon \phi(X) \to X$ is also an isometry. As metric spaces, X and $\phi(X)$ are indistinguishable; $\phi(X)$ is merely a relabelling of X with $\phi(a)$ in place of a for each $a \in X$, all distances in X being preserved in $\phi(X)$ by ϕ. Therefore, we shall often, when we are concerned only with metric properties, suppress the function ϕ and treat such spaces as if they were identical. Notwithstanding this identification, let it be remembered that X and $\phi(X)$ may have differing non-metric structure, as is the case with $\mathbb{C}$ and $\mathbb{R}^2$.

Example 1.4.2

Each $(a, b) \in \mathbb{R}^2$ is associated with a unique complex number $z = a + ib$, where $a = \Re z$ and $b = \Im z$ (B.4.1); the function $(a, b) \mapsto a + ib$ from $\mathbb{R}^2$ onto $\mathbb{C}$ is a bijective map that preserves the metric and so is an isometry.

Example 1.4.3

We like to think of $\mathbb{R}$ and $\mathbb{R}^2$ as metric subspaces of $\mathbb{R}^3$, although, strictly speaking, they are not even subsets of $\mathbb{R}^3$. However, the isomorphic copy (B.21.1) $\{a \in \mathbb{R}^3 \mid a_2 = a_3 = 0\}$ of $\mathbb{R}$ in $\mathbb{R}^3$, popularly known as the x-axis, is isometric to $\mathbb{R}$, and the isomorphic copy $\{a \in \mathbb{R}^3 \mid a_3 = 0\}$ of $\mathbb{R}^2$ in $\mathbb{R}^3$, popularly known as the xy-plane, is isometric to $\mathbb{R}^2$, all spaces being endowed with their Euclidean metrics. Of course, the various other lines and planes of $\mathbb{R}^3$ are also isometric to $\mathbb{R}$ and $\mathbb{R}^2$, respectively.

Example 1.4.4

Suppose Z is a set, (X, d) is a metric space and $f \colon Z \to X$ is an injective function. Then f and d induce a metric on Z, namely $(a, b) \mapsto d(f(a), f(b))$ (Q 1.12). This metric makes Z an isometric copy of the metric subspace $(f(Z), d)$ of (X, d) and f an isometry.

1.5   Extending a Metric Space

Here we show how to build a natural extension X' of an arbitrary metric space X. Just as $\mathbb{R}^2$ and $\mathbb{R}^3$ are not strictly metric superspaces of $\mathbb{R}$ but are thought of as such because they include naturally occurring isometric copies of $\mathbb{R}$ (1.4.3), neither is our extension strictly a metric superspace of X, but it may be thought of as such because it includes a naturally occurring isometric copy of X. We shall make use of it later (10.12.2).

Theorem 1.5.1

Suppose (X, d) is a non-empty metric space, and let X' denote the set of all pointlike functions on X (1.2.5). Define s on $X' \times X'$ to be the function

$$(u, v) \mapsto \sup\{|u(x) - v(x)| \mid x \in X\}.$$

Then s is a metric on X'. Moreover, $\delta(X) \subseteq X'$ and the map $z \mapsto \delta_z$ from (X, d) onto the subspace $(\delta(X), s)$ of (X', s) is an isometry, so $(\delta(X), s)$ is an isometric copy of (X, d).

Proof

For $u, v \in X'$, we have $u(x) - u(b) \leq d(x, b) \leq v(x) + v(b)$ and therefore, by rearrangement, $u(x) - v(x) \leq u(b) + v(b)$ for all $x, b \in X$. It follows that $\sup\{|u(x) - v(x)| \mid x \in X\}$ is real, so that s is a real function. Moreover, s is certainly non-negative and symmetric and satisfies $s(u, v) = 0 \Rightarrow u = v$. The triangle inequality is obtained as follows. Suppose $u, v, w \in X'$. Then, for all $x \in X$, we have

$$|u(x) - v(x)| \leq |u(x) - w(x)| + |w(x) - v(x)| \leq s(u, w) + s(w, v),$$

so that $s(u, v) = \sup\{|u(x) - v(x)| \mid x \in X\} \leq s(u, w) + s(w, v)$. Therefore s is a metric on X'.

That $\delta(X) \subseteq X'$ was proved in 1.2.4. By 1.1.2, $|\delta_p(a) - \delta_y(a)| \leq d(p, y)$ for all $a, p, y \in X$. This becomes an equality when $a = p$ or $a = y$ and so gives $s(\delta_p, \delta_y) = d(p, y)$ for all $p, y \in X$. So the map $z \mapsto \delta_z$ is an isometry from (X, d) onto $(\delta(X), s)$. $\qquad \square$

1.6   Metrics on Products

There are usually many ways in which a product of metric spaces can be endowed with a metric. There is no reason why an arbitrary metric on a product should bear any clear relationship to the metrics on the individual spaces, but we are particularly interested in those that do. What clear relationship is desirable may be open to question; for the moment, we shall fix our attention on what we shall call *conserving metrics*, the name being suggested by 1.6.4.

Theorem 1.6.1

Suppose $n \in \mathbb{N}$ and, for each $i \in \mathbb{N}_n$, (X_i, τ_i) is a metric space. Then the following three functions are metrics on $\prod_{i=1}^n X_i$:

- $\mu_1 : (a, b) \mapsto \sum_{i=1}^{n} \tau_i(a_i, b_i)$;
- $\mu_2 : (a, b) \mapsto \sqrt{\sum_{i=1}^{n} (\tau_i(a_i, b_i))^2}$;
- $\mu_\infty : (a, b) \mapsto \max\{\tau_i(a_i, b_i) \mid i \in \mathbb{N}_n\}$.

Moreover, for each $a, b \in \prod_{i=1}^{n} X_i$, we have $\mu_\infty(a, b) \leq \mu_2(a, b) \leq \mu_1(a, b)$. The metric μ_2 will be called the *Euclidean product metric*.

Proof

These functions are all symmetric and non-negative and are zero only when $a = b$. It is easy to check, as in Q 1.5 and Q 1.4, that μ_1, μ_2 and μ_∞ satisfy the triangle inequality. Towards the inequalities, we have, for each $a, b \in \prod_{i=1}^{n} X_i$,

$$\max\{\tau_i(a_i, b_i) \mid i \in \mathbb{N}_n\} = \tau_j(a_j, b_j) = \sqrt{(\tau_j(a_j, b_j))^2} \leq \sqrt{\sum_{i=1}^{n} (\tau_i(a_i, b_i))^2}$$

for some $j \in \mathbb{N}_n$, which gives $\mu_\infty(a, b) \leq \mu_2(a, b)$; that $\mu_2(a, b) \leq \mu_1(a, b)$ is obtained by noticing that $\sum_{i=1}^{n} (\tau_i(a_i, b_i))^2 \leq (\sum_{i=1}^{n} \tau_i(a_i, b_i))^2$. $\qquad\square$

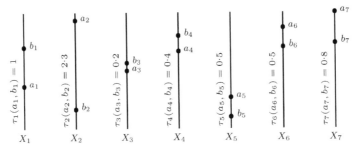

A graphic representation of two points a and b in a product space $\prod_{i=1}^{7} X_i$.
$\mu_1(a, b) = 5\cdot 7$, $\mu_2(a, b) \approx 2\cdot 76$, $\mu_\infty(a, b) = 2\cdot 3$.

Definition 1.6.2

Suppose $n \in \mathbb{N}$ and, for each $i \in \mathbb{N}_n$, (X_i, τ_i) is a metric space. Suppose e is a metric on the product $\prod_{i=1}^{n} X_i$. We shall call e a *conserving metric* on the product $\prod_{i=1}^{n} X_i$ with respect to the given metrics τ_i if, and only if, for all $a, b \in \prod_{i=1}^{n} X_i$, we have $\mu_\infty(a, b) \leq e(a, b) \leq \mu_1(a, b)$. Usually, we shall call a metric e that satisfies this condition simply a conserving metric, the metrics on the spaces X_i being understood from the context.

Example 1.6.3

μ_1, μ_2 and μ_∞ are themselves conserving metrics. The functions defined in 1.1.13, 1.1.14 and Q 1.4 are the metrics μ_1, μ_∞ and μ_2, respectively, on $\mathbb{R}^n$, μ_2

being the usual Euclidean metric. So, assuming each copy of $\mathbb{R}$ has its usual metric, the Euclidean metric on $\mathbb{R}^n$ is conserving. In contrast, the metric d defined on $\mathbb{R}^2$ in 1.1.15 is not conserving (Q 1.20).

What do conserving metrics conserve? Although Definition 1.6.2 is designed to deal with non-trivial products, it also covers the case when $n = 1$; but when $n = 1$, the only conserving metric is the given metric. More generally, if $n \in \mathbb{N}$ and, for each $i \in \mathbb{N}_n$, (X_i, τ_i) is a non-empty metric space and the finite product $\prod_{i=1}^{n} X_i$ is given a conserving metric, then, just as $\mathbb{R}^3$ includes isometric copies of $\mathbb{R}$ (1.4.3), the product $\prod_{i=1}^{n} X_i$ includes various naturally occurring isomorphic copies of each coordinate space X_j (1.6.4).

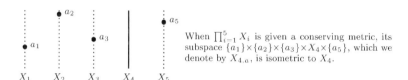

When $\prod_{i=1}^{5} X_i$ is given a conserving metric, its subspace $\{a_1\} \times \{a_2\} \times \{a_3\} \times X_4 \times \{a_5\}$, which we denote by $X_{4,a}$, is isometric to X_4.

Theorem 1.6.4

Suppose $n \in \mathbb{N}$ and, for each $i \in \mathbb{N}_n$, (X_i, τ_i) is a non-empty metric space. Let e be a conserving metric on $P = \prod_{i=1}^{n} X_i$. For each $j \in \mathbb{N}_n$ and $a \in P$, let $X_{j,a} = \{x \in P \mid x_i = a_i \text{ for all } i \in \mathbb{N}_n \setminus \{j\}\}$. Then the map $x \mapsto x_j$ is an isometry from $X_{j,a}$ onto X_j.

Proof

Suppose $j \in \mathbb{N}_n$ and $a \in P$. Then the map $x \mapsto x_j$ from $X_{j,a}$ to X_j is surjective because the coordinate spaces are all non-empty (see B.13). Consider elements $x, y \in X_{j,a}$. Then $x_i = y_i = a_i$ for all $i \in \mathbb{N}_n$ for which $i \neq j$, which yields $\tau_j(x_j, y_j) = \sup\{\tau_i(x_i, y_i) \mid i \in \mathbb{N}_n\} \leq e(x, y) \leq \sum_{i=1}^{n} \tau_i(x_i, y_i) = \tau_j(x_j, y_j)$, because e is a conserving metric. This in turn gives $e(x, y) = \tau_j(x_j, y_j)$. $\qquad\square$

Example 1.6.5

In the notation we have introduced in 1.6.4, the subspace $\mathbb{R}_{1,(0,0,0)}$ of $\mathbb{R}^3$ is the axis $\{a \in \mathbb{R}^3 \mid a_2 = a_3 = 0\}$, the so-called x-axis. Theorem 1.6.4 shows that certain lines of $\mathbb{R}^3$ parallel to an axis are isometric to $\mathbb{R}$. But, in this case, more can be said: in fact, all the lines of $\mathbb{R}^3$ are isometric to $\mathbb{R}$ because, for $a, b \in \mathbb{R}^3$ with $b \neq 0$, the map $t \mapsto a + tb/\delta_{(0,0,0)}(b)$ from $\mathbb{R}$ onto $\{a + tb \mid t \in \mathbb{R}\}$ is an isometry.

1.7 Metrics and Norms on Linear Spaces

The concept of a metric space is very broad indeed. An arbitrary metric space need not have the algebraic properties that we associate with the most familiar examples such as $\mathbb{R}$, $\mathbb{R}^2$ and $\mathbb{C}$; it need not admit an operation of addition; there may be no special element designated as zero; and there may be no concept corresponding to that of the absolute value in $\mathbb{R}$, the modulus in $\mathbb{C}$ or the length of a vector in $\mathbb{R}^2$, which in those spaces is used to define the usual metric.

When we define a metric on an arbitrary linear space, we usually follow the pattern established in those familiar spaces, requiring that the metric interact with the algebraic operations. Specifically, suppose V is a linear space over $\mathbb{R}$ or $\mathbb{C}$. We want to define a metric d on V. For each $a, b \in V$, we should like the *length* $d(b-a, 0)$ of the vector $b-a$ to be the same as the distance $d(a, b)$ from a to b. We should also like the length $d(\lambda a, 0)$ of the vector λa to be $|\lambda| \, d(a, 0)$ for each scalar λ. If d has these two properties, it is a metric determined by a length function (1.7.8); such length functions are called *norms*.

Definition 1.7.1

Suppose V is a linear space over $\mathbb{R}$ or $\mathbb{C}$ (B.20.3). Suppose $\|\cdot\|$ is a real function defined on V such that, for each $x, y \in V$ and each scalar α, we have

- $\|x\| \geq 0$ with equality if, and only if, $x = 0$;
- $\|\alpha x\| = |\alpha| \, \|x\|$; and
- (TRIANGLE INEQUALITY) $\|x + y\| \leq \|x\| + \|y\|$.

Then $\|\cdot\|$ is called a *norm* on V, and V equipped with this norm is called a *normed linear space*. If we need to specify the norm, we say that $(V, \|\cdot\|)$ is a normed linear space. For each $x \in V$, the number $\|x\|$ is called the *length* of x with respect to the norm $\|\cdot\|$. If W is a linear subspace of V (B.20.5), the restriction of $\|\cdot\|$ to W is clearly a norm on W; W equipped with this norm is called a *normed linear subspace* of V.

Theorem 1.7.2

Suppose V is a normed linear space. Then the function d defined on $V \times V$ by $(a, b) \mapsto \|a - b\|$ is a metric on V.

Proof

d is a non-negative real symmetric function, and $d(a, b) = 0$ if, and only if, $a = b$. Also, $d(a, b) = \|a - b\| \leq \|a - c\| + \|c - b\| = d(a, c) + d(c, b)$ for all $a, b, c \in V$. Therefore d is a metric on V. $\qquad \square$

Definition 1.7.3

Given a normed linear space V, the metric of 1.7.2 is styled the *metric determined by the norm*. We shall always, without comment, regard a given normed linear space as a metric space with the metric determined by the norm.

Example 1.7.4

The absolute-value function on $\mathbb{R}$ and the modulus function on $\mathbb{C}$ are the prototypes for norms; they determine the Euclidean metrics of 1.1.3 and 1.1.5, respectively. The Euclidean metric on $\mathbb{R}^2$ given in 1.1.8 is determined by the norm $x \mapsto \sqrt{x_1{}^2 + x_2{}^2}$; similarly, the Euclidean metric on $\mathbb{R}^3$ given in 1.1.8 is determined by the norm $x \mapsto \sqrt{x_1{}^2 + x_2{}^2 + x_3{}^2}$. In general, for each $n \in \mathbb{N}$, the function $\|\cdot\|_2$ given by $\|x\|_2 = \sqrt{\sum_{i=1}^{n} x_i^2}$ for each $x \in \mathbb{R}^n$ is the norm that determines the Euclidean metric on $\mathbb{R}^n$ (1.1.10). These norms belong to an even more general class of norm in which square roots and squares are replaced by pth roots and pth powers: for each $p \in [1, \infty)$, the function $\|\cdot\|_p$ defined on $\mathbb{R}^n$ by $\|x\|_p = \left(\sum_{i=1}^{n} |x_i|^p\right)^{1/p}$ is a norm on $\mathbb{R}^n$, but we shall not prove this until 12.11.3.

Example 1.7.5

The metric $(a, b) \mapsto |b_1 - a_1| + |b_2 - a_2|$ on $\mathbb{R}^2$ mentioned in 1.1.13 is determined by the norm $x \mapsto |x_1| + |x_2|$. More generally, for each $n \in \mathbb{N}$, the function $x \mapsto \sum_{i=1}^{n} |x_i|$ is the norm $\|\cdot\|_p$ for $p = 1$ of 1.7.4, but the proof that this function, $\|\cdot\|_1$, is a norm (Q 1.24) is much easier than the general result for all $p \in [1, \infty)$.

Example 1.7.6

Suppose $n \in \mathbb{N}$. The norms $\|\cdot\|_1$ and $\|\cdot\|_2$ of 1.7.4 are clearly the norms that determine the metrics μ_1 and μ_2, respectively, on $\mathbb{R}^n$ (see 1.6.1). The metric μ_∞ on $\mathbb{R}^n$ is also determined by a norm; this norm is denoted by $\|\cdot\|_\infty$ and is defined by setting $\|x\|_\infty = \max\{|x_i| \mid i \in \mathbb{N}_n\}$ for each $x \in \mathbb{R}^n$. It is not difficult to show that this function is a norm (Q 1.24).

Example 1.7.7

We have defined norms on linear spaces over $\mathbb{R}$ or $\mathbb{C}$, but exactly the same definition can be used on vector spaces over the field $\mathbb{Q}$. In particular, $\mathbb{Q}$ is a vector space over itself with its usual norm inherited from $\mathbb{R}$. For each prime number p, the function $x \mapsto |x|_p$ of 1.1.16 is another norm on $\mathbb{Q}$.

Example 1.7.8

Suppose V is a linear space. Suppose d is a metric on V that respects the algebraic operations on V in the way we discussed in the introduction to this section. Then d is a metric determined by a norm. Specifically, suppose d has the two properties listed there. Consider the point function δ_0 on (V, d) (1.2.1). For all $x, y \in V$ and scalar α, we have

- $\delta_0(x) = d(x, 0) \geq 0$ with equality if, and only if, $x = 0$;
- $\delta_0(\alpha x) = d(\alpha x, 0) = |\alpha| \, d(x, 0) = |\alpha| \, \delta_0(x)$; and
- $d(x + y, 0) \leq d(x + y, x) + d(x, 0) = d(y, 0) + d(x, 0)$, which says that $\delta_0(x + y) \leq \delta_0(y) + \delta_0(x)$.

So δ_0 is a norm on V and, clearly, d is the metric determined by δ_0.

Example 1.7.9

Linear spaces admit metrics that are not determined by norms. Suppose V is any non-trivial normed linear space (that is, $V \neq \{0\}$), and let d denote the metric determined by the norm. Suppose v is a any non-zero vector of V. Then $d(v, 0) = \|v\|$ and $d(2v, 0) = \|2v\| = 2\|v\|$. Since $\|v\| \neq 0$, at least one of $d(v, 0)$ and $d(2v, 0)$ is neither 1 nor 0, so that d is not the discrete metric on V.

Question 1.7.10

Every set can be endowed with a metric. Can every linear space be endowed with a norm? We give only a partial answer to this question here, namely that existence of a basis (B.22.1) implies existence of a norm. Suppose V is a linear space and S is a basis for V. Then each vector of V is a linear combination of members of S (B.22.1). The minimality of S as a spanning set for V ensures that the representation of each vector v as a sum $\sum_{s \in S} \lambda_{v,s} s$, where only a finite number of the $\lambda_{v,s}$ are non-zero, is unique. Set $\|v\| = \max\{|\lambda_{v,s}| \mid s \in S\}$. Using the triangle inequality for the modulus function, it is easy to show that this yields a norm on V.

Summary

In this chapter, we have defined the terms *metric* and *metric space* and given various examples. We have defined *norms* on linear spaces and shown that they determine metrics. We have introduced metric subspaces and metric superspaces and metrics on products of metric spaces. We have explained the concept of isometric spaces and have shown that an isometric copy can be made of any given metric space simply be endowing its set of point functions with a suitable metric; we have also shown that this isometric copy of the given space sits inside a naturally defined metric superspace.

EXERCISES

Q1.1 Suppose X is a set and $d: X \times X \to \mathbb{R}$. Show that d is a metric on X if, and only if, for all $a, b, z \in X$, the two conditions $d(a, b) = 0 \Leftrightarrow a = b$ and $d(a, b) \leq d(z, a) + d(z, b)$ are both satisfied.

Q1.2 Suppose that d is a metric on a set X. Prove that the inequality $|d(x, y) - d(z, w)| \leq d(x, z) + d(y, w)$ holds for all $w, x, y, z \in X$.

Q1.3 Suppose X is a non-empty set and $d(a, a) = 0$ for all $a \in X$ and $d(a, b) = 1$ for all $a, b \in X$ with $a \neq b$. Show that d is a metric on X.

†Q1.4 Use induction to verify that the function μ_2 defined on $\mathbb{R}^n \times \mathbb{R}^n$ by $(a, b) \mapsto \sqrt{\sum_{i=1}^{n}(b_i - a_i)^2}$ satisfies the triangle inequality and is therefore a metric on $\mathbb{R}^n$.

Q1.5 Verify that the functions μ_1 and μ_∞ of 1.1.13 and 1.1.14 are metrics.

Q1.6 Suppose d and e are metrics on a set X. Let g be the function $(x, y) \mapsto \min\{d(x, y), e(x, y)\}$ defined on $X \times X$. Show that g need not be a metric on X and find a condition under which it is a metric.

†Q1.7 Devise a metric that compares words and puts words of similar spelling, such as *complement* and *compliment*, close together. Your set of words may be written in the Roman alphabet and involve no diacritical marks.

Q1.8 Let $\mathcal{F}(S)$ be the set of all finite subsets of a set S. For all $A, B \in \mathcal{F}(S)$, let $\Delta(A, B) = (A \backslash B) \cup (B \backslash A)$ be the *symmetric difference* between A and B. Let $d(A, B)$ be the cardinality (B.17) of $\Delta(A, B)$. Is d a metric?

Q1.9 A metric on a set X that satisfies $d(a, c) \leq \max\{d(a, b), d(b, c)\}$ for all $a, b, c \in X$ is called an *ultrametric* on X. Identify an ultrametric amongst the examples of metrics given in this chapter.

Q1.10 Consider the collection poly$(\mathbb{R})$ of polynomial functions from $\mathbb{R}$ to $\mathbb{R}$ (B.20.10). Each member of poly$(\mathbb{R})$ can be represented as a sum $\sum_{i=0}^{\infty} \alpha_i x^i$, where the α_i are real numbers, all except a finite number of which are zero, and the x^i are the power functions $x \mapsto x^i$. For each $p, q \in$ poly$(\mathbb{R})$ with $p = \sum_{i=0}^{\infty} \alpha_i x^i$ and $q = \sum_{i=0}^{\infty} \beta_i x^i$, define $d(p, q)$ to be $\sup\{|\alpha_i - \beta_i| \mid i \in \mathbb{N} \cup \{0\}\}$. Explain why this supremum must be real, and show that d defines a metric on poly$(\mathbb{R})$.

Q 1.11 Suppose (X, d) is a metric space and $e(x, y) = d(x, y)/(1 + d(x, y))$ for each $x, y \in X$ (see 1.1.7). Show that e is a metric on X.

†Q 1.12 Suppose Z is a set, (X, d) is a metric space and $f: Z \to X$ is an injective function. Show that $(a, b) \mapsto d(f(a), f(b))$ is a metric on Z.

Q 1.13 Let X be the collection of all *continuous* real functions defined on $[0, 1]$. For each $f, g \in X$, set $e(f, g)$ to be $\int_0^1 |f(x) - g(x)| \, dx$. Show that e is a metric on X. Can this metric be extended, using the same formula, to all functions defined on $[0, 1]$ for which the integral is well defined?

†Q 1.14 Set $d(m, n) = \left| m^{-1} - n^{-1} \right|$, $d(n, \infty) = d(\infty, n) = n^{-1}$ and $d(\infty, \infty) = 0$ for all $m, n \in \mathbb{N}$ (1.1.12). Show that d is a metric on $\tilde{\mathbb{N}}$.

†Q 1.15 Define a function $d: \mathbb{R} \times \mathbb{R} \to \mathbb{R}$ by setting $d(a, b) = |b - a|$ if $a, b \in \mathbb{R}^-$, $d(a, b) = b^2 - a$ if $a \in \mathbb{R}^-$ and $b \in \mathbb{R}^\oplus$, $d(a, b) = a^2 - b$ if $a \in \mathbb{R}^\oplus$ and $b \in \mathbb{R}^-$ and $d(a, b) = \left| b^2 - a^2 \right|$ if $a, b \in \mathbb{R}^\oplus$. Is d a metric on $\mathbb{R}$?

†Q 1.16 Suppose (X, d) is a metric space and $f: X \to \mathbb{R}$. Show that the function $(a, b) \mapsto d(a, b) + |f(a) - f(b)|$ is a metric on X.

†Q 1.17 Suppose (X, d) is a metric space and $z \in X$. Suppose $k \in \mathbb{R}^+$. Show that $v : x \mapsto \delta_z(x) + k$ is pointlike.

Q 1.18 Find a pointlike function that does not attain its minimum value; deduce that it is not of the type described in Q 1.17.

†Q 1.19 Check that the extension of d given in 1.3.3 is a metric on Y.

†Q 1.20 Show that the metric defined on $\mathbb{R}^2$ in 1.1.15 is not conserving.

Q 1.21 Show that every isometry is injective.

†Q 1.22 Suppose $(X, \|\cdot\|_X)$ and $(Y, \|\cdot\|_Y)$ are normed linear spaces and $\phi: X \to Y$ is a bijective function that preserves the norm in the sense that $\|\phi(x)\|_Y = \|x\|_X$ for all $x \in X$. Need ϕ be an isometry?

Q 1.23 If (X, d) and (Y, e) are metric spaces, write $(X, d) \simeq (Y, e)$ if there exists an isometry from (X, d) onto (Y, e). Show that, for all metric spaces (X, d), (Y, e) and (Z, m),

(i) $(X, d) \simeq (X, d)$;

(ii) $(X, d) \simeq (Y, e) \Rightarrow (Y, e) \simeq (X, d)$; and

(iii) if $(X, d) \simeq (Y, e)$ and $(Y, e) \simeq (Z, m)$, then $(X, d) \simeq (Z, m)$.

†Q 1.24 Let $n \in \mathbb{N}$. Verify the triangle inequality for $\|\cdot\|_\infty$ and $\|\cdot\|_1$ on $\mathbb{R}^n$ (see 1.7.6 and 1.7.5). Deduce that these functions are norms on $\mathbb{R}^n$.

2
Distance

Metrics are designed to measure distances between points. They permit us to define *diameters* for sets and also to measure distances between points and sets and distances between sets and other sets. Moreover, they prompt us to consider the notions of *isolated point*, *accumulation point* and *nearest point*.

2.1 Diameter

The diameter of a circle in a plane is the maximum distance between its points. We extend this idea to subsets of arbitrary metric spaces. In the general case, however, there may not be a maximum distance, so we have to settle for the supremum of the various distances between points of the set.

Definition 2.1.1

Suppose (X, d) is a metric space and A is a subset of X. We define the *diameter* of A to be $\sup\{d(r, s) \mid r, s \in A\}$. We denote this quantity by $\mathrm{diam}(A)$. The diameter is dependent on the metric; if we need to avoid any ambiguity, we may augment our notation with a subscript, as in $\mathrm{diam}_d(A)$.

Example 2.1.2

The diameter of the empty set is of only marginal interest to us; it is $\sup \varnothing$, which is defined to be $-\infty$ (B.7.2). Singleton subsets of a metric space all have diameter 0; every larger set has diameter in $\mathbb{R}^+ \cup \{\infty\}$.

Theorem 2.1.3

Suppose X is a metric space and A and B are subsets of X for which $A \subseteq B$.
Then $\operatorname{diam}(A) \le \operatorname{diam}(B)$.

Proof

 Since $A \subseteq B$, we have $\{d(c,a) \mid c, a \in A\} \subseteq \{d(e,b) \mid e, b \in B\}$ and
so, by B.7.4, $\sup\{d(c,a) \mid c, a \in A\} \le \sup\{d(e,b) \mid e, b \in B\}$, which
is precisely the inequality that is required.   □

Theorem 2.1.4

Suppose S is a subset of $\mathbb{R}$. Then $\operatorname{diam}(S) = \sup S - \inf S$.

Proof

First, if $S = \varnothing$, then $\inf S = \infty$, $\sup S = -\infty$ and $\operatorname{diam}(S) = \sup \varnothing = -\infty$,
so the proposition is true by definition (B.7.1). Second, if $\sup S = \infty$, pick
$b \in S$. Then, for each $p \in \mathbb{R}^+$, there exists some $a \in S$ with $a > b + p$, so that
$\operatorname{diam}(S) > p$. Since p is arbitrary in $\mathbb{R}^+$, we have $\operatorname{diam}(S) = \infty = \sup S - \inf S$.
By a similar argument, the proposition is true if $\inf S = -\infty$.

Finally, suppose that $\inf S$ and $\sup S$ are both real. Let $r \in \mathbb{R}^+$. Then there
exist $a, b \in S$ such that $a - r/2 < \inf S \le a \le b \le \sup S \le b + r/2$. So
$\sup S - \inf S \le b - a + r \le \operatorname{diam}(S) + r$. Since r is arbitrary in $\mathbb{R}^+$, we then
have $\sup S - \inf S \le \operatorname{diam}(S)$. But, for all $x, y \in S$, we have $\inf S \le x \le \sup S$
and $\inf S \le y \le \sup S$, so that $|y - x| \le \sup S - \inf S$. Since x and y are
arbitrary in S, this yields $\operatorname{diam}(S) \le \sup S - \inf S$. The two inequalities then
lead to the desired conclusion.   □

Example 2.1.5

Suppose $a, b \in \mathbb{R}$ and $a < b$. The intervals $(a\,,b)$, $[a\,,b)$, $(a\,,b]$ and $[a\,,b]$ all have
diameter $b - a$. The intervals $(a\,,\infty)$, $[a\,,\infty)$, $(-\infty\,,b)$, $(-\infty\,,b]$ and $(-\infty\,,\infty)$
all have diameter ∞ (see B.7.1). Degenerate intervals $[a\,,a]$ have diameter 0.

2.2 Distances from Points to Sets

We should like to say that the distance from a point x to a non-empty set A in
a metric space is the distance from x to the *nearest point* of A. Unfortunately,
there may not be a nearest point of A to x. So, in order to capture the idea

of the distance from x to A, we have to settle for the infimum of the distances from x to the various points of A.

Definition 2.2.1

Suppose (X, d) is a metric space, A is a subset of X and $x \in X$. We define the *distance* from x to A to be $\text{dist}(x, A) = \inf\{d(x, a) \mid a \in A\}$. It is, of course, dependent on the metric; if it is necessary to specify which metric is being used in order to avoid ambiguity, we may use a subscript, as in $\text{dist}_d(x, A)$.

Example 2.2.2

Suppose X is a metric space. The distance from any point of X to a non-empty subset of X is a non-negative real number. However, since we define $\inf \varnothing$ to be ∞ (see B.7.2), it follows that $\text{dist}(x, \varnothing) = \infty$ for all $x \in X$.

Example 2.2.3

Every point of $\mathbb{R}$ is zero distance from $\mathbb{Q}$. To see this, suppose $x \in \mathbb{R}$ and $r \in \mathbb{R}^+$. By B.6.11, $\mathbb{Q} \cap (x, x + r) \neq \varnothing$. So $\text{dist}(x, \mathbb{Q}) < r$. Since r is arbitrary in $\mathbb{R}^+$, it follows that $\text{dist}(x, \mathbb{Q}) = 0$. A similar argument shows that $\text{dist}(x, \mathbb{R}\backslash\mathbb{Q}) = 0$ also.

Example 2.2.4

It is not easy in general to calculate the distance between a point and a set. There are, however, some familiar cases in which formulae are available. A line in $\mathbb{R}^2$ is a particular set of points and the reader will no doubt recall that the distance from a point $z \in \mathbb{R}^2$ to a line $\{x \in \mathbb{R}^2 \mid ax_1 + bx_2 + c = 0\}$ is given by the formula $|az_1 + bz_2 + c| / \sqrt{a^2 + b^2}$.

Is the supremum of a subset of $\mathbb{R}$ necessarily zero distance from the subset itself? Recall that the supremum is defined in terms of the order, not of the metric. So, if we had no other knowledge of $\mathbb{R}$, there would be no immediate reason to suppose such a pleasing concurrence (see Q 2.3). But the strong relationship between metric and order in $\mathbb{R}$ (1.1.3) leads to an affirmative answer.

Theorem 2.2.5

Suppose S is a subset of $\mathbb{R}$ and $z \in \mathbb{R}$. Then

(i) $\text{dist}(z, S) \leq |z - \sup S|$ with equality if $z \geq \sup S$;

(ii) $\text{dist}(z, S) \le |z - \inf S|$ with equality if $z \le \inf S$;

(iii) if $\sup S \in \mathbb{R}$, then $\text{dist}(\sup S, S) = 0$; and

(iv) if $\inf S \in \mathbb{R}$, then $\text{dist}(\inf S, S) = 0$.

Proof

First, if $S = \varnothing$, then $\inf S = \infty$, $\sup S = -\infty$ and $\text{dist}(z, S) = \inf \varnothing = \infty$, so the assertions are true by definition (B.7.1). Second, if $\sup S = \infty$, then (i) is clearly true. Third, if $\inf S = -\infty$, then (ii) is clearly true.

We now consider the case when $\inf S$ and $\sup S$ are both real. Let $s = \sup S$ and $r = \text{dist}(z, S)$. As $s \in \mathbb{R}$, certainly $S \ne \varnothing$ and r is real. If $z < s$, then, since $s = \sup S$, there exists some $x \in S$ for which $z < x \le s$, yielding $\text{dist}(z, S) \le x - z \le s - z$, as required. If, on the other hand, $s \le z$, we let $t \in \mathbb{R}^+$. Then, since $s = \sup S$, $(s - t, s] \cap S \ne \varnothing$, so that $\text{dist}(z, S) \le z - s + t$ and, since t is arbitrary in $\mathbb{R}^+$, it follows that $\text{dist}(z, S) \le z - s$. Moreover, in this case, since $s = \sup S$, we have $S \cap (s, \infty) = \varnothing$, so that $d(z, x) \ge z - s$ for all $x \in S$, whence $\text{dist}(z, S) \ge z - s$, and we have $\text{dist}(z, S) = z - s$, as required. This proves (i). There is a similar proof for (ii). Then (iii) follows from (i) by putting $z = \sup S$, and (iv) follows from (ii) by putting $z = \inf S$. So all the statements are true.   □

2.3 Inequalities for Distances

Metrics satisfy the triangle inequality and its variants. In this section, we identify some similarly useful inequalities concerning distances between points and sets.

Theorem 2.3.1

Suppose X is a metric space, $x \in X$ and A and B are non-empty subsets of X for which $A \subseteq B$. Then $\text{dist}(x, B) \le \text{dist}(x, A) \le \text{dist}(x, B) + \text{diam}(B)$.

Proof

Since A is a subset of B, it follows that $\{d(x, a) \mid a \in A\} \subseteq \{d(x, b) \mid b \in B\}$. So, by B.7.4, $\inf\{d(x, b) \mid b \in B\}$ cannot exceed $\inf\{d(x, a) \mid a \in A\}$, a fact that is written more succinctly as $\text{dist}(x, B) \le \text{dist}(x, A)$. For the second inequality, consider $a \in A$ and $b \in B$. We certainly have $\text{dist}(x, A) \le d(x, a)$ and, since both a and b are in B, it follows that $d(b,a) \leq \mathrm{diam}(B)$, so $\mathrm{dist}(x,A) \leq d(x,a) \leq d(x,b)+d(b,a) \leq d(x,b)+\mathrm{diam}(B)$. This is true for all $b \in B$, so B.6.6 yields the inequality $\mathrm{dist}(x,A) \leq \mathrm{dist}(x,B)+\mathrm{diam}(B)$.   □

Theorem 2.3.2

Suppose (X,d) is a metric space, $a,b \in X$, $S \subseteq X$ and $S \neq \varnothing$. Then
(i) $\mathrm{dist}(a,S) \leq d(a,b)+\mathrm{dist}(b,S)$; and
(ii) $|\mathrm{dist}(a,S)-\mathrm{dist}(b,S)| \leq d(a,b) \leq \mathrm{dist}(a,S)+\mathrm{diam}(S)+\mathrm{dist}(b,S)$.

Proof

Suppose $x \in S$. Then $\mathrm{dist}(a,S) \leq d(a,x) \leq d(a,b)+d(b,x)$. Since this is true for all $x \in S$, B.6.6 gives $\mathrm{dist}(a,S) \leq d(a,b)+\mathrm{dist}(b,S)$, as required. By reversing the roles of a and b, we have also $\mathrm{dist}(b,S) \leq d(a,b)+\mathrm{dist}(a,S)$, and the two inequalities yield $|\mathrm{dist}(a,S)-\mathrm{dist}(b,S)| \leq d(a,b)$, as required. Now, for $p,q \in S$, we have $d(a,b) \leq d(a,p)+d(p,q)+d(q,b) \leq d(a,p)+\mathrm{diam}(S)+d(q,b)$. Then B.6.6 gives $d(a,b) \leq \inf\{d(a,p) \mid p \in S\}+\mathrm{diam}(S)+\inf\{d(q,b) \mid q \in S\}$, which is rewritten as $d(a,b) \leq \mathrm{dist}(a,S)+\mathrm{diam}(S)+\mathrm{dist}(b,S)$.   □

2.4 Distances to Unions and Intersections

It is possible to calculate the distance from a point to a union of subsets of a metric space in terms of the distances to the individual sets (2.4.1). The reader who has little experience with arbitrary unions and intersections is urged to look at Section B.11 in Appendix B before embarking on this section.

Theorem 2.4.1

Suppose X is a metric space, $x \in X$ and $\mathcal{C}$ is a collection of subsets of X. Then $\mathrm{dist}(x,\bigcup\mathcal{C}) = \inf\{\mathrm{dist}(x,A) \mid A \in \mathcal{C}\}$.

Proof

If $\bigcup\mathcal{C} = \varnothing$, the result follows from 2.2.2, so we assume otherwise. Since the members of $\mathcal{C}$ are subsets of $\bigcup\mathcal{C}$, 2.3.1 gives $\mathrm{dist}(x,\bigcup\mathcal{C}) \leq \mathrm{dist}(x,A)$ for all $A \in \mathcal{C}$ and therefore also $\mathrm{dist}(x,\bigcup\mathcal{C}) \leq \inf\{\mathrm{dist}(x,A) \mid A \in \mathcal{C}\}$. We prove the reverse inequality. Let $r \in \mathbb{R}^+$. Then there exists $z \in \bigcup \mathcal{C}$ such that $d(x, z) \leq \mathrm{dist}(x, \bigcup \mathcal{C}) + r$, and there exists $S \in \mathcal{C}$ with $z \in S$. So $\inf\{\mathrm{dist}(x, A) \mid A \in \mathcal{C}\} \leq \mathrm{dist}(x, S) \leq d(x, z) \leq \mathrm{dist}(x, \bigcup \mathcal{C}) + r$. Because r is arbitrary in $\mathbb{R}^+$, it follows that $\inf\{\mathrm{dist}(x, A) \mid A \in \mathcal{C}\} \leq \mathrm{dist}(x, \bigcup \mathcal{C})$. $\qquad\square$

Example 2.4.2

Unfortunately, there is no theorem like 2.4.1 for intersection. There is only an inequality, which we give below in 2.4.3. For a counterexample to anything like 2.4.1, consider the disc $A = \{z \in \mathbb{C} \mid |z - 1| \leq 1\}$ and its companion $B = \{z \in \mathbb{C} \mid |z + 1| \leq 1\}$. The intersection $A \cap B$ is the singleton set $\{0\}$, and the distance from the complex number i to this intersection is 1. But the distance from i to A is $\sqrt{2} - 1$, and so is the distance from i to B.

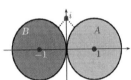

The lightest of kisses.

Theorem 2.4.3

Suppose X is a metric space, $x \in X$ and $\mathcal{C}$ is a non-empty collection of subsets of X. Then $\sup\{\mathrm{dist}(x, A) \mid A \in \mathcal{C}\} \leq \mathrm{dist}(x, \bigcap \mathcal{C})$.

Proof

If $\bigcap \mathcal{C} = \varnothing$, the result follows from 2.2.2. Otherwise, for each $A \in \mathcal{C}$, we have $\bigcap \mathcal{C} \subseteq A$ and the assertion follows from 2.3.1. $\qquad\square$

2.5 Isolated Points

A point a of a subset S of a metric space is necessarily zero distance from S. It may or may not, however, be zero distance from the rest of S; that is, from $S \backslash \{a\}$. Points that are not zero distance from the rest of S are called *isolated* points of S.

Definition 2.5.1

Suppose X is a metric space, S is a subset of X and $z \in S$. Then z is called an *isolated point* of S if, and only if, $\mathrm{dist}(z, S \backslash \{z\}) \neq 0$. In this case, we say that z is *isolated* in S. The collection of all isolated points of S will be denoted by $\mathrm{iso}(S)$.

Example 2.5.2

No point of a space with the discrete metric (1.1.7) is of distance zero from the rest of the space. It follows that every point of the space is isolated.

Example 2.5.3

Consider the subset $\mathbb{N}$ of $\mathbb{R}$ with its usual metric. For each $n \in \mathbb{N}$, we have $\mathrm{dist}(n, \mathbb{N}\backslash\{n\}) = 1$, so that n is isolated in $\mathbb{N}$. It should be clear, however, that n is not an isolated point of $\mathbb{R}$. Indeed, $\mathbb{R}$ has no isolated points.

Example 2.5.4

Endow $S = \{1/n \mid n \in \mathbb{N}\} \cup \{0\}$ with the usual metric inherited from $\mathbb{R}$. For each $n \in \mathbb{N}$, $\mathrm{dist}(1/n, S\backslash\{1/n\}) = 1/(n(n+1)) > 0$, so that $1/n$ is an isolated point of S. But $\mathrm{dist}(0, S\backslash\{0\}) = \inf\{1/n \mid n \in \mathbb{N}\} = 0$ (B.6.12), so that 0 is not isolated in S.

Example 2.5.5

Suppose $r \in \mathbb{Q}$. Then $\mathrm{dist}(r, \mathbb{Q}\backslash\{r\}) = 0$ (2.2.3), so that r is not an isolated point of $\mathbb{Q}$. Since r is arbitrary in $\mathbb{Q}$, it follows that $\mathbb{Q}$ has no isolated points.

Theorem 2.5.6

Suppose X is a metric space and F is a non-empty finite subset of X. Then every point of F is isolated in F.

Proof

If F is a singleton set $\{a\}$, then $\mathrm{dist}(a, F\backslash\{a\}) = \infty$ (B.7.2). Otherwise $\{d(a,b) \mid a, b \in F,\ a \neq b\}$ is a finite subset of $\mathbb{R}^+$ and so has a minimum in $\mathbb{R}^+$ (B.6.4), and $\mathrm{dist}(x, F\backslash\{x\})$ is not less than this minimum for any $x \in F$. $\quad\square$

Theorem 2.5.7

Suppose X is a metric space and A and B are subsets of X with $A \subseteq B$. Then $A \cap \mathrm{iso}(B) \subseteq \mathrm{iso}(A)$.

Proof

Suppose $z \in A \cap \mathrm{iso}(B)$. Then $\mathrm{dist}(z, B\backslash\{z\}) \neq 0$. Since $A\backslash\{z\} \subseteq B\backslash\{z\}$, 2.3.1 then gives $\mathrm{dist}(z, A\backslash\{z\}) \neq 0$, so that $z \in \mathrm{iso}(A)$. $\quad\square$

2.6 Accumulation Points

If X is a metric space and $S \subseteq X$, then points of X that are zero distance from the rest of S are called accumulation points of S. Such points need not be members of S, but the members of S that are accumulation points of S are precisely those that are not isolated points of S.

Definition 2.6.1

Suppose X is a metric space, $z \in X$ and S is a subset of X. Then z is called an *accumulation point* or a *limit point* of S in X if, and only if, $\operatorname{dist}(z, S\backslash\{z\}) = 0$. The collection of all accumulation points of S in X will be denoted by $\operatorname{acc}(S)$, or by $\operatorname{acc}_X(S)$ if it is deemed necessary to specify the space.

Example 2.6.2

By 2.2.3, every point of $\mathbb{R}$ is an accumulation point of $\mathbb{Q}$ and is also an accumulation point of $\mathbb{R}$ itself.

Example 2.6.3

The only accumulation point of the set $I = \{1/n \mid n \in \mathbb{N}\}$ in $\mathbb{R}$ is 0, which is not a member of I. Every member of I is isolated in I (see 2.5.4).

Theorem 2.6.4

Suppose X is a metric space, $z \in X$ and S is a subset of X.
(i)   If $z \notin S$, then $z \in \operatorname{acc}(S)$ if, and only if, $\operatorname{dist}(z, S) = 0$.
(ii) If $z \in S$, then $z \in \operatorname{acc}(S)$ if, and only if, $z \notin \operatorname{iso}(S)$.
(iii) $z \in \operatorname{acc}(S)$ if, and only if, $z \notin \operatorname{iso}(S)$ and $\operatorname{dist}(z, S) = 0$.

Proof

If $z \notin S$, then $S\backslash\{z\} = S$, so that the definition of an accumulation point yields (i). If, on the other hand, $z \in S$, then the definitions of an isolated point and an accumulation point yield (ii). Since every isolated point of S is a member of S and every point of S is zero distance from S, (i) and (ii) imply (iii). □

Theorem 2.6.5

Suppose X is a metric space and $A \subseteq B \subseteq X$. Then $\operatorname{acc}(A) \subseteq \operatorname{acc}(B)$.

Proof

Suppose $z \in \mathrm{acc}(A)$. Then $\mathrm{dist}(z, A\backslash\{z\}) = 0$. Since $A\backslash\{z\} \subseteq B\backslash\{z\}$, 2.3.1 gives $\mathrm{dist}(z, B\backslash\{z\}) = 0$, whence $z \in \mathrm{acc}(B)$.   □

Theorem 2.6.6

Suppose X is a metric space and $S \subseteq Z \subseteq X$. Then $\mathrm{acc}_Z(S) = Z \cap \mathrm{acc}_X(S)$.

Proof

For each $z \in Z$, $\mathrm{dist}_Z(z, S\backslash\{z\}) = \mathrm{dist}_X(z, S\backslash\{z\})$.   □

2.7   Distances from Sets to Sets

There is more than one concept of distance between subsets of a metric space. The most straightforward is given in 2.7.1, but we shall explore another in 2.7.4.

Definition 2.7.1

Suppose (X, d) is a metric space and A and B are subsets of X. We define the *distance* from A to B to be $\inf\{d(a,b) \mid a \in A, b \in B\}$ and denote it by $\mathrm{dist}(A, B)$. This distance is dependent on the metric. To avoid ambiguity, we may sometimes use an index in the notation, as in $\mathrm{dist}_d(A, B)$ or $\mathrm{dist}_X(A, B)$.

Theorem 2.7.2

Suppose (X, d) is a metric space, $x \in X$ and A and B are subsets of X. Then $\mathrm{dist}(A, B) \leq \mathrm{dist}(x, A) + \mathrm{dist}(x, B)$.

Proof

If A or B is empty, the result is clear; we suppose otherwise. Let $r \in \mathbb{R}^+$, let $a \in A$ be such that $d(x, a) \leq \mathrm{dist}(x, A) + r/2$ and let $b \in B$ be such that $d(x, b) \leq \mathrm{dist}(x, B) + r/2$. Then $\mathrm{dist}(A, B) \leq d(a, b) \leq d(a, x) + d(x, b) \leq \mathrm{dist}(x, A) + \mathrm{dist}(x, B) + r$ and, because r is an arbitrary member of $\mathbb{R}^+$, there follows the required inequality, $\mathrm{dist}(A, B) \leq \mathrm{dist}(x, A) + \mathrm{dist}(x, B)$.   □

Example 2.7.3

Suppose X is a metric space. The distance between any two non-empty subsets of X is a non-negative real number. In many cases, this distance is zero; it is certainly so when the two sets have non-empty intersection but may still be so even when the sets are disjoint—the subsets $(0,1)$ and $(1,2)$ of $\mathbb{R}$, for example, are zero distance apart but do not intersect. Note, however, that since $\inf \varnothing = \infty$, we have $\mathrm{dist}(A,\varnothing) = \infty$ for every subset A of X.

Example 2.7.4

Suppose (X,d) is a metric space. The function $(A,B) \mapsto \mathrm{dist}(A,B)$ is a symmetric real non-negative function on $(\mathcal{P}(X) \setminus \{\varnothing\}) \times (\mathcal{P}(X) \setminus \{\varnothing\})$ that mimics d on $\{\{x\} \mid x \in X\}$ in the sense that $\mathrm{dist}(\{a\},\{b\}) = d(a,b)$ for all $a,b \in X$. But it is not a metric on $\mathcal{P}(X) \setminus \{\varnothing\}$ because it does not satisfy the triangle inequality and because distinct subsets of X may be zero distance apart. If we want to define a metric with the mimicking property on as large a collection of subsets of X as possible, we need to look elsewhere. The function $(A,B) \mapsto \max\{\sup\{\mathrm{dist}(x,A) \mid x \in B\}, \sup\{\mathrm{dist}(x,B) \mid x \in A\}\}$ looks more promising; it, too, is non-negative and symmetric—indeed, it is the formula that was used, in a different guise, to define a metric on a subset of $\mathcal{P}(\mathbb{R})$ in 1.1.18. In general, it does not give a metric on $\mathcal{P}(X) \setminus \{\varnothing\}$, but we shall show in due course that there is a well-defined subset of $\mathcal{P}(X)$ on which it does do so (7.3.1).

2.8 Nearest Points

The distance from a point x to a subset S of a given metric space is $\mathrm{dist}(x,S)$. Under what conditions is there an element of S that is distant exactly $\mathrm{dist}(x,S)$ from x? We are interested particularly in knowing what property of S will ensure that such a *nearest point* of S exists irrespective of what metric superspace X enfolds S and what point x of X we are considering. We can relate this property immediately (2.8.3) to the pointlike functions of 1.2.5, but we need to do a lot more work before we can characterize it more fully, which we shall do in 7.11.1 and 12.6.1.

Definition 2.8.1

Suppose (X,d) is a metric space, S is a subset of X, and $z \in X$. A member s of S is called a *nearest point* of S to z in X if, and only if, $d(z,s) = \mathrm{dist}(z,S)$.

Example 2.8.2

Nearest points need not exist: there is no nearest point of the interval $(7,8)$ to 5. Nearest points need not be unique: there are two nearest points of the union of intervals $[0,1] \cup [3,4]$ to 2. Nearest points may be useless: every point of the circle $\mathbb{T} = \{z \in \mathbb{C} \mid |z| = 1\}$ is nearest to 0 in $\mathbb{C}$. That all of these possibilities may occur in a general setting is clear when we realize that a nearest point of a subset S to a point z is a member of S at which the point function δ_z attains its minimum value on S. There is nothing new here; indeed, a great deal of time is spent in calculus courses identifying conditions under which functions attain minima and exploring methods for evaluating them and the points at which they occur.

Theorem 2.8.3

Suppose (X, d) is a metric space. The following statements are equivalent:

(i) $X = \varnothing$ or X admits a nearest point to each point in every metric super-space of X.

(ii) Every pointlike function on X attains its minimum value on X.

Proof

No pointlike functions are defined on the empty metric space (1.2.5), so we suppose $X \neq \varnothing$. Suppose (Y, d) is a metric superspace of X and $z \in Y$. Then $x \mapsto d(z, x)$ is a pointlike function on X. If this attains its minimum value on X, then there exists $w \in X$ such that $d(z, w) = \inf\{d(z, x) \mid x \in X\} = \mathrm{dist}(z, X)$, so that w is a nearest point of X to z in Y. For the converse, suppose that u is a pointlike function on X. We want to show that if X admits a nearest point to each point in every metric superspace of X, then u attains its minimum value on X. If 0 is in the range of u, then certainly u attains its minimum value, so we suppose otherwise. Pick $w \notin X$, let $X' = X \cup \{w\}$ and extend d to $X' \times X'$ by setting $d(w, w) = 0$ and $d(x, w) = d(w, x) = u(x) \neq 0$ for all $x \in X$. It is easily verified that d is then a metric on X'. If there is a nearest point p of X to w in X', then $u(p) = d(w, p) = \mathrm{dist}_{X'}(w, X) = \inf\{u(x) \mid x \in X\}$, so that u attains its minimum value at p. $\square$

The next theorem is a variant form of the Bolzano–Weierstrass Theorem from real analysis, which states that every bounded infinite subset of $\mathbb{R}$ has an accumulation point. The reader may like to try using the Bolzano–Weierstrass Theorem to prove 2.8.4. The presentation here is the other way around, for in 7.11.3 we present the Bolzano–Weierstrass Theorem as a consequence of 2.8.4.

Theorem 2.8.4

Suppose (X, d) is a metric superspace of $\mathbb{R}$ with its usual metric and $z \in X$. Then there is a nearest point of $\mathbb{R}$ to z in X.

Proof

Let $t = \text{dist}(z, \mathbb{R})$. For each $r \in \mathbb{R}^+$, let $S_r = \{x \in \mathbb{R} \mid d(x, z) \leq t + r\}$; the sets S_r are all non-empty because $\text{dist}(z, \mathbb{R}) = t$, so that $\inf S_r \leq \sup S_r$ (B.7.4). Set $a_r = \inf S_r$ and $b_r = \sup S_r$. By the triangle inequality, $\text{diam}(S_r) \leq 2(t + r)$. Then 2.1.4 ensures that a_r and b_r are real and 2.2.5 yields $\text{dist}(a_r, S_r) = 0$. The

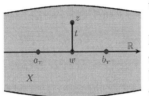

triangle inequality then gives $d(a_r, z) \leq t + r$.

Let $\epsilon \in \mathbb{R}^+$ and $w = \sup\{a_r \mid r \in \mathbb{R}^+\}$. By B.7.4, $a_s \leq a_r \leq b_1$ for all $r, s \in \mathbb{R}^+$ with $r \leq s$. It follows that $w \in \mathbb{R}$ and that w can be expressed as $\sup\{a_r \mid r \in (0, \epsilon/2)\}$. By 2.2.5, there exists $p \in (0, \epsilon/2)$ with $w - a_p < \epsilon/2$. Then $t \leq d(w, z) \leq d(z, a_p) + d(a_p, w) < t + p + \epsilon/2 < t + \epsilon$. Since this is true for all $\epsilon \in \mathbb{R}^+$, we have $d(w, z) = t$, as required. $\square$

Example 2.8.5

Suppose $a, b \in \mathbb{R}^2$ with $b \neq (0, 0)$. Let $L = \{a + tb \mid t \in \mathbb{R}\}$ be the line in $\mathbb{R}^2$ through a in the direction determined by b. Endow $\mathbb{R}^2$ with any metric d that has the property that $d(a + ub, a + vb) = |u - v|$ for all $u, v \in \mathbb{R}$. Then (L, d) is an isometric copy of $(\mathbb{R}, |\cdot|)$ in $(\mathbb{R}^2, d)$, where $|\cdot|$ denotes the usual metric on $\mathbb{R}$. So (L, d) has metric properties identical to those of $(\mathbb{R}, |\cdot|)$. In particular, if $z \in \mathbb{R}^2$, there exists $w \in L$ with $d(z, w) = \text{dist}_d(z, L)$. But beware: whether or not w is uniquely determined depends on the metric on $\mathbb{R}^2$ (2.8.6).

Example 2.8.6

The nearest point of 2.8.4 need not be unique. We know that $\mathbb{C}$ is a superset of $\mathbb{R}$. If we endow $\mathbb{C}$ with the metric $(z, w) \mapsto \max\{|\Re z - \Re w|, |\Im z - \Im w|\}$ (compare 1.1.14), then its subspace $\mathbb{R}$ has its usual metric. But $\text{dist}(i, \mathbb{R}) = 1$ and every point of the interval $[-1, 1]$ is distance 1 from i.

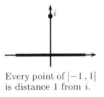

Every point of $[-1, 1]$ is distance 1 from i.

Summary

In this chapter, we have explained what we mean by the distance from a point to a set, the distance between two sets and the diameter of a set. We have introduced the concepts of *nearest point*, *isolated point* and *accumulation point*. We have learnt also the seminal fact that $\mathbb{R}$ contains at least one nearest point to any given point in any superspace.

EXERCISES

Q 2.1 Suppose (X, d) is a metric space with more than one point. Find subsets A and B of X such that $\text{diam}(A \cup B) \geq \text{diam}(A) + \text{diam}(B)$.

Q 2.2 With reference to 2.1.3, find a condition on a metric space (X, d) that ensures that there exist subsets A and B of X with $A \subset B$ such that $\text{diam}(A) = \text{diam}(B)$.

†Q 2.3 Consider the metric subspace $X = I \cup J$ of $\mathbb{R}$, where I is the interval $(0, 1)$ and J is the interval $[4, 7)$. Show that $\text{dist}(\sup_X I, I) \neq 0$.

Q 2.4 With reference to 2.3.1, find a metric space X, an element x of X and non-empty subsets A and B of X with $A \subseteq B$ such that $\text{dist}(x, A) > \text{dist}(x, B) + \text{diam}(B \backslash A)$.

†Q 2.5 Suppose that X is a metric space and that S is a subset of X. Show that $\text{iso}(S) = S \backslash \text{acc}(S)$.

Q 2.6 Suppose (X, d) is a metric space and F is a finite subset of X. Show that $\text{acc}(F) = \varnothing$.

Q 2.7 Suppose $\mathcal{C}$ is a collection of subsets of a metric space X and $S \in \mathcal{C}$. Show that every isolated point of S that belongs to $\bigcap \mathcal{C}$ is an isolated point of $\bigcap \mathcal{C}$. Show also that every isolated point of $\bigcup \mathcal{C}$ is isolated in every member of $\mathcal{C}$ to which it belongs.

†Q 2.8 Suppose $\mathcal{C}$ is a non-empty collection of subsets of a metric space X. Show that there are inclusions $\text{acc}(\bigcap \mathcal{C}) \subseteq \bigcap \{\text{acc}(S) \mid S \in \mathcal{C}\}$ and $\bigcup \{\text{acc}(S) \mid S \in \mathcal{C}\} \subseteq \text{acc}(\bigcup \mathcal{C})$ and that each of them may be proper.

†Q 2.9 Suppose $n \in \mathbb{N}$ and, for each $i \in \mathbb{N}_n$, (X_i, τ_i) is a metric space. Suppose d is a conserving metric on $P = \prod_{i=1}^{n} X_i$. Suppose $S \subseteq P$ and $a \in S$. Is it true that $a \in \text{iso}(S)$ if, and only if, $a_i \in \text{iso}(\pi_i(S))$ for all $i \in \mathbb{N}_n$, where π_i denotes the natural projection of P onto X_i (B.13.3)?

†Q 2.10 Suppose $n \in \mathbb{N}$ and, for each $i \in \mathbb{N}_n$, (X_i, τ_i) is a metric space. Let d be a conserving metric on $P = \prod_{i=1}^{n} X_i$. Suppose $S \subseteq P$ and $a \in S$. Is it true that $a \in \text{acc}(S)$ if, and only if, $a_j \in \text{acc}(\pi_j(S))$ for some $j \in \mathbb{N}_n$?

Q 2.11 In the proof of 2.8.4, why did we not define w to be $\sup\{a_r \mid r \in (0, \epsilon/2)\}$ immediately after picking ϵ?

Q 2.12 Suppose S is a subset of $\mathbb{R}$ and $x \in \mathbb{R}$. Show that there are at most two
nearest points of S to x.

3
Boundary

The mathematical term *boundary* is intended to correspond to our intuitive idea of *boundary*, *frontier* or *border*. How well it does so the reader might judge after looking through some examples.

3.1 Boundary Points

The boundary points of a subset S of a metric space X are those points of X, whether in S or not, that are zero distance both from S and from its complement S^c in X. Each point of X is in either S or S^c and is, of course, zero distance from the one it belongs to. What distinguishes a boundary point of S, therefore, is that it is zero distance from the one of S and S^c to which it does not belong.

Definition 3.1.1

Suppose X is a metric space, S is a subset of X and $a \in X$. Then a is called a *boundary point* of S in X if, and only if, $\text{dist}(a, S) = 0 = \text{dist}(a, S^c)$. The collection of boundary points of S in X is called the *boundary* of S in X and will be denoted by ∂S, or by $\partial_X S$ if it is not clear from the context that the metric space under consideration is X.

Theorem 3.1.2

Suppose X is a metric space and S is a subset of X. Then $\partial S = \partial(S^c)$.

Proof

Since $(S^c)^c = S$, we have, for each $a \in X$, $\mathrm{dist}(a, (S^c)^c) = \mathrm{dist}(a, S)$. Thus $\mathrm{dist}(a, S) = 0 = \mathrm{dist}(a, S^c)$ if, and only if, $\mathrm{dist}(a, S^c) = 0 = \mathrm{dist}(a, (S^c)^c)$. In other words, $a \in \partial S$ if, and only if, $a \in \partial(S^c)$. $\square$

Example 3.1.3

Let us look at boundary points of intervals. Suppose $a, b \in \mathbb{R}$ with $a < b$. Then each of the intervals (a, b), $[a, b)$, $(a, b]$ and $[a, b]$ has just the two boundary points, a and b, in $\mathbb{R}$ (Q 3.1). Each of the intervals (a, ∞), $[a, \infty)$, $(-\infty, b)$ and $(-\infty, b]$ has only one boundary point in $\mathbb{R}$, a in the first two cases and b in the others. The interval $(-\infty, \infty)$ has no boundary point in $\mathbb{R}$. The degenerate interval $[a, a]$ has a as its sole boundary point.

Example 3.1.4

Consider the closed unit disc $\mathbb{D} = \{z \in \mathbb{C} \mid |z| \leq 1\}$ of the complex plane. Its boundary points are those points lying on the unit circle $\mathbb{T} = \{z \in \mathbb{C} \mid |z| = 1\}$.
The points of $\mathbb{T}$ are the boundary points of the open unit disc $\{z \in \mathbb{C} \mid |z| < 1\}$ also; indeed, these same points are the boundary points of every disc S of $\mathbb{C}$ that includes the open unit disc and is included in the closed unit disc. Whether the points of $\mathbb{T}$ are members of S or not is irrelevant. What is important is that each point of $\mathbb{T}$ is on the frontier of S with its complement $\mathbb{C} \backslash S$.

Part of the boundary
$\mathbb{T}$ of S is not in S.

Example 3.1.5

Our intuition about boundary points is usually going to be correct. However, lest we get carried away by an incomplete understanding, let us look at the interval $(7, 17]$ of the real line. Examination of this interval of the real line gives us its two boundary points, 7 and 17, their nature differing only in that the first boundary point is not in the set, whereas the second is. That is all there is to say, provided we are considering the interval $(7, 17]$ as a subset of the real line—in other words, if the metric space in question is $\mathbb{R}$. If, however, we are looking at $(7, 17]$ as a subset of the larger metric space $\mathbb{C}$, then every point of the interval $[7, 17]$ is a boundary point of $(7, 17]$ since each of these points is on the frontier of $(7, 17]$ with $\mathbb{C} \backslash (7, 17]$. An important lesson is to be learnt from this example: the boundary points of a set are always calculated relative to the metric space of which it is being considered a subset.

Accumulation points and boundary points are both determined by zero distances. But accumulation points need not be boundary points: the points of the disc $\{z \in \mathbb{C} \mid |z| < 1\}$ are all accumulation points, and none of them is a boundary point. So we need to examine the relationship between accumulation points and boundary points. Fortunately, it is very easy to state, as we see now in 3.1.6.

Theorem 3.1.6

Suppose X is a metric space, S is a subset of X and $a \in X$.

(i)   If $a \notin S$, then $a \in \partial S$ if, and only if, $a \in \mathrm{acc}(S)$.

(ii) If $a \in S$, then $a \in \partial S$ if, and only if, $a \in \mathrm{acc}(S^c)$.

Proof

If $a \in S^c$, then certainly $\mathrm{dist}(a, S^c) = 0$, so that $a \in \partial S$ if, and only if, $\mathrm{dist}(a, S) = 0$, which, since $a \notin S$, is equivalent to $\mathrm{dist}(a, S \backslash \{a\}) = 0$, or equivalently, $a \in \mathrm{acc}(S)$. This proves (i). Reversing the roles of S and S^c, we deduce that, if $a \in S$, then $a \in \partial(S^c)$ if, and only if, $a \in \mathrm{acc}(S^c)$. Then 3.1.2 clinches the proof of (ii). □

3.2 Sets with Empty Boundary

The empty subset has empty boundary in every metric space, as does the space itself. Metric spaces that have no proper non-trivial subset with empty boundary are said to be *connected*, and we shall study them in Chapter 11.

Theorem 3.2.1

Suppose X is a metric space. Then $\partial \varnothing = \varnothing$ and $\partial X = \varnothing$.

Proof

Every point of X is of distance ∞ from the empty set, so no point of X is zero distance from $\varnothing$ or from X^c, which is the same thing. The first assertion yields $\partial \varnothing = \varnothing$; the second yields $\partial X = \varnothing$. □

Question 3.2.2

Does X have any subset other than X and $\varnothing$ that has empty boundary? This seemingly innocuous question leads to a whole new area of enquiry. The answer depends on the space. When X is $\mathbb{R}$ or $\mathbb{C}$ or $\mathbb{R}^3$, it is relatively easy to show that the answer is *no*. Such spaces X are said to be *connected*. We can appreciate why the word *connected* is used by examining spaces that do not have the property. Consider $X = (0,1) \cup (7,8)$ as a metric subspace of $\mathbb{R}$ (1.3.1). It is easy to establish that the subsets $(0,1)$ and $(7,8)$ of X both have empty boundary in X. The disconnection of X is clear in the diagram.

The space $(0,1) \cup (7,8)$ is disconnected.

Question 3.2.3

Consider the metric space $X = [0,1] \cup [7,8]$ with the usual metric induced from $\mathbb{R}$. This is very similar to the space discussed in 3.2.2 above. Do the subsets $[0,1]$ and $[7,8]$ have non-empty boundary in X? There is a temptation to respond without thinking, saying that 0 and 1 are boundary points of the first and that 7 and 8 are boundary points of the second. But this is not so. Both sets, like their counterparts in 3.2.2, have empty boundary. Indeed, all the points of $[7,8]$ are of distance at least 6 from its complement $[0,1]$; and all the points of $[0,1]$ are of distance at least 6 from its complement $[7,8]$. It is true, however, that 0 and 1 are boundary points of $(0,1)$ in X and that 7 and 8 are boundary points of $(7,8)$ in X.

Example 3.2.4

The definition of a boundary does not lead us to expect that inclusion of one set in another implies inclusion of the boundary of the smaller set in that of the larger. Indeed, the reverse may be the case. There are many sets with empty boundary that have subsets with as large a boundary as possible. To appreciate this, consider $\mathbb{R}$ as a subset of itself. It has empty boundary (3.2.1). But its subsets $\mathbb{Q}$ and $\mathbb{R}\backslash\mathbb{Q}$ both have $\mathbb{R}$ as their boundary in $\mathbb{R}$ This is so by 2.2.3 since, for all $x \in \mathbb{R}$, $\operatorname{dist}(x,\mathbb{Q}) = 0 = \operatorname{dist}(x,\mathbb{R}\backslash\mathbb{Q})$.

3.3 Boundary Inclusion

Every set that has empty boundary includes its own boundary. But there are many other sets that have this nice property; they are said to be *closed* and will be studied in more detail in Chapter 4. Here we look at a few such sets.

Theorem 3.3.1

Suppose X is a metric space and F is a finite subset of X. Then F includes its own boundary in X.

Proof

If $F = X$, we invoke 3.2.1. Otherwise $F \neq X$. Suppose $x \in F^c$. By B.6.4, every non-empty finite subset of $\mathbb{R}$ has a minimum element, so there exists $w \in F$ such that $d(x, w) = \inf\{d(x, a) \mid a \in F\} = \text{dist}(x, F)$. But $x \in F^c$ and $w \in F$, so that $x \neq w$ and $d(x, w) > 0$. Therefore $\text{dist}(x, F) > 0$, whence $x \notin \partial F$. Since x is arbitrary in F^c, this yields $\partial F \subseteq F$. $\qquad\square$

Example 3.3.2

The inclusion of 3.3.1 may be proper. Consider, for example, the subset $\{0\}$ of the metric subspace $X = [1, 2] \cup \{0\}$ of $\mathbb{R}$; it has no boundary point in X.

Example 3.3.3

We define the *Cantor set* as follows. From the interval $I_0 = [0, 1]$, delete the middle third $\left(\frac{1}{3}, \frac{2}{3}\right)$. Call the resulting set I_1; it is the union of the two intervals $\left[0, \frac{1}{3}\right]$ and $\left[\frac{2}{3}, 1\right]$, each of length $1/3$ and a distance $1/3$ apart. Delete the middle third from each of these two intervals to get a new set, called I_2, which is the union of the four intervals $\left[0, \frac{1}{9}\right]$, $\left[\frac{2}{9}, \frac{1}{3}\right]$, $\left[\frac{2}{3}, \frac{7}{9}\right]$ and $\left[\frac{8}{9}, 1\right]$, each of length $1/9$ and each pair a distance at least $1/9$ apart. Continue this process, recursively defining I_n for each $n \in \mathbb{N}$ to be the union of the 2^n intervals remaining after the nth round of deletion (B.19); each of the intervals will be of length $1/3^n$, and each pair will be a distance at least $1/3^n$ apart. The Cantor set, which we label $\mathcal{K}$, is defined to be the set of points common to all I_n, namely $\bigcap\{I_n \mid n \in \mathbb{N}\}$.

What is the boundary of $\mathcal{K}$ in $\mathbb{R}$? To answer this question, we shall look separately at real numbers that are in $\mathcal{K}$ and at those that are not in $\mathcal{K}$. We note here that our question would be valid but worthless if $\mathcal{K}$ were empty; $\mathcal{K}$ is, in fact, a highly structured non-empty set. Its members are all those real numbers that can be written as $\sum_{n=1}^{\infty} x_n/3^n$, where, for each $n \in \mathbb{N}$, either $x_n = 0$ or $x_n = 2$. We leave it to the reader to prove this (Q 3.8).

Suppose $x \in \mathcal{K}$. Then, for each $n \in \mathbb{N}$, we have $x \in I_n$. Since I_n is a union of intervals of length $1/3^n$ and any two are a distance at least $1/3^n$ apart, it follows that $\text{dist}(x, (I_n)^c) < 1/3^n$. Since $\mathcal{K} \subseteq I_n$, we have $(I_n)^c \subseteq \mathcal{K}^c$ by B.3.3 and $\text{dist}(x, \mathcal{K}^c) < 1/3^n < 1/n$ by 2.3.1. This is true for all $n \in \mathbb{N}$, so B.6.12 yields $\text{dist}(x, \mathcal{K}^c) = 0$. Therefore $x \in \partial \mathcal{K}$.

Now suppose that $z \in \mathcal{K}^c$. Then there exists $m \in \mathbb{N}$ such that $z \notin I_m$.

Since I_m is a union of intervals of $\mathbb{R}$, it follows that z is not in any of them. Since each of the intervals includes its boundary, the distance from z to each of them is not zero. There is only a finite number of such distances and, by 2.4.1 and B.6.4, the least of those distances is the distance from z to the union I_m. This distance is not zero, and since $\mathcal{K} \subseteq I_m$, 2.3.1 gives $\mathrm{dist}(z, \mathcal{K}) \neq 0$ also. So $z \notin \partial \mathcal{K}$. This and the foregoing calculation yield $\partial \mathcal{K} = \mathcal{K}$.

Example 3.3.4

Let $\Gamma = \{(x, 1/x) \in \mathbb{R}^2 \mid x \in \mathbb{R}, x \neq 0\}$. Then Γ is the graph of the function $x \mapsto 1/x$ defined on $\mathbb{R}\backslash\{0\}$. We show that $\partial_{\mathbb{R}^2}\Gamma = \Gamma$, where $\mathbb{R}^2$ is assumed to have its Euclidean metric. For each $(a, 1/a)$ in Γ and $s \in \mathbb{R}^+$, the point $(a + s, 1/a)$ is in $\mathbb{R}^2\backslash\Gamma$ and is of distance s from $(a, 1/a)$. Since s is arbitrary in $\mathbb{R}^+$, it follows that $(a, 1/a)$ is zero distance from $\mathbb{R}^2\backslash\Gamma$ and is therefore in $\partial\Gamma$. Now consider a point $(b, c) \in \mathbb{R}^2\backslash\Gamma$. Then $bc \neq 1$. Let $r = \min\{1, |1 - bc|/(1 + |b| + |c|)\}$. Since $bc \neq 1$, we have $r \in \mathbb{R}^+$. We claim that no point of Γ is of distance less than r from (b, c). If there were x in $\mathbb{R}\backslash\{0\}$ such that the distance from $(x, 1/x)$ to (b, c) were less than r, we should have, in

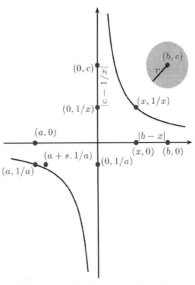

The graph of $1/x$ is its own boundary.

particular, $|b - x| < r$ and $|c - 1/x| < r$, which yield $|x| < r + |b|$ and subsequently $|1 - bc| \leq |1 - cx| + |cx - cb| = |c - 1/x|\,|x| + |b - x|\,|c| < r^2 + r|b| + r|c|$, which, since $r \leq 1$, does not exceed $r(1 + |b| + |c|)$, giving $|1 - bc| < r(1 + |b| + |c|)$ and contradicting the definition of r. We conclude that $(b, c) \notin \partial\Gamma$.

Example 3.3.5

To provide a contrast to Example 3.3.4, we consider the graph of the function $f : x \mapsto \sin(1/x)$ defined on $\mathbb{R}^+$. An argument similar to the one given above shows that each point of the graph is a boundary point of the graph in $\mathbb{R}^2$ with the Euclidean metric. In this case, however, there are points not in the graph that are boundary points. Indeed, it is easy to check that each point in $\{(0, y) \mid y \in [-1, 1]\}$ is also zero distance from the graph (see Q 3.4). But the observant reader will note that none of these extra boundary points has its first coordinate in the domain of f; if we consider the graph as a subset of $\mathrm{dom}(f) \times \mathbb{R}$ rather than of $\mathbb{R}^2$, we see that it does indeed contain all of its boundary points, which is what gen-

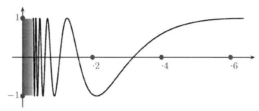

In the region shaded grey, $\sin(1/x)$ fluctuates with ever-increasing intensity between the values -1 and 1.

erally happens for a continuous function (see Q 8.13).

We have seen in 3.3.3 and 3.3.4 that the Cantor set and the graph of $1/x$ both have the property of being equal to their respective boundaries. This phenomenon, though of interest, is not unusual. For example, it is easy to show that, in the complex plane, $\partial\mathbb{R} = \mathbb{R}$ and $\partial\mathbb{T} = \mathbb{T}$ (3.1.4). This last equation can be rewritten $\partial(\partial\mathbb{D}) = \partial\mathbb{D}$ (3.1.4) and prompts us to ask whether or not $\partial(\partial S) = \partial S$ for every subset S of a metric space. The answer is *no*, and $\mathbb{Q}$ is a counterexample when regarded as a subset of $\mathbb{R}$: its boundary is $\mathbb{R}$, and $\mathbb{R}$ has empty boundary. We salvage something, however, in the inclusion of 3.3.6.

Theorem 3.3.6

Suppose X is a metric space and S is a subset of X. Then $\partial(\partial S) \subseteq \partial S$.

Proof

If $\partial(\partial S) = \varnothing$, the result is clearly true, so we suppose otherwise. Suppose $x \in \partial(\partial S)$. Then $\mathrm{dist}(x, \partial S) = 0$. So, for each $r \in \mathbb{R}^+$, there exists $z \in \partial S$ such that $d(x, z) < r/2$. Since $z \in \partial S$, there exist $a \in S$ and $b \in S^c$ such that $d(z, a) < r/2$ and $d(z, b) < r/2$. The triangle inequality then yields $d(x, a) < r$ and $d(x, b) < r$, so that $\mathrm{dist}(x, S) < r$ and $\mathrm{dist}(x, S^c) < r$. Since r is arbitrary in $\mathbb{R}^+$, it follows that $\mathrm{dist}(x, S) = 0 = \mathrm{dist}(x, S^c)$, whence $x \in \partial S$. $\qquad\square$

3.4 Boundaries in Subspaces and Superspaces

Boundary is a relative term; the boundary of a set with a fixed metric depends on which superspace is deemed to envelop it. If Y is a metric superspace of a metric space X and $S \subseteq X$, how does the boundary of S in X relate to the boundary of S in Y? Example 3.1.5 teaches us that there may be inclusion $\partial_X S \subseteq \partial_Y S$, and we show in 3.4.1 that this always occurs.

Theorem 3.4.1

Suppose X is a subspace of a metric space Y and $S \subseteq X$. Then $\partial_X S \subseteq \partial_Y S$.

Proof

Suppose $z \in \partial_X S$. Then $\mathrm{dist}(z, S) = 0$ and $\mathrm{dist}(z, X\backslash S) = 0$. Then, because $X\backslash S \subseteq Y\backslash S$, it follows from 2.3.1 that $\mathrm{dist}(z, Y\backslash S) = 0$, so that $z \in \partial_Y S$. $\square$

Example 3.4.2

Suppose Y is a metric superspace of a metric space X and $S \subseteq X$. We know from 3.4.1 that $\partial_X S \subseteq \partial_Y S$. When this inclusion is proper, the points of $\partial_Y S \backslash \partial_X S$ may all be in X, as is the case in 3.1.5. But some may be in X and some in $Y\backslash X$; compare, for example, $\partial_{\mathbb{Q}} \mathbb{Q}$, which is empty, with $\partial_{\mathbb{R}} \mathbb{Q}$, which is $\mathbb{R}$. It may also happen that all are in $Y\backslash X$; consider the interval $(0, 1)$ and compare $\partial_{(0,1)}(0, 1)$, which is empty, with $\partial_{\mathbb{R}}(0, 1)$, which is $\{0, 1\}$.

3.5 Boundaries of Unions and Intersections

Is the boundary of a union or an intersection related to the boundary of the individual sets that are being united or intersected? The answer to this question is not straightforward. We shall see in 3.5.1 that, if the collection of sets is finite, then both boundaries are included in the union of the boundaries of the individual sets. This need not be the case for infinite collections (3.5.2).

Theorem 3.5.1

Suppose X is a metric space and $\mathcal{C}$ is a non-empty finite collection of subsets of X. Then

(i) $\partial(\bigcup \mathcal{C}) \subseteq \bigcup \{\partial A \mid A \in \mathcal{C}\}$; and

(ii) $\partial(\bigcap \mathcal{C}) \subseteq \bigcup \{\partial A \mid A \in \mathcal{C}\}$.

Proof

Suppose $x \in \partial(\bigcup \mathcal{C})$. Then $\mathrm{dist}(x, \bigcup \mathcal{C}) = 0$ and $\mathrm{dist}(x, X\backslash \bigcup \mathcal{C}) = 0$. From the first equation and 2.4.1, we get $\inf\{\mathrm{dist}(x, A) \mid A \in \mathcal{C}\} = 0$. Since $\mathcal{C}$ is finite, B.6.4 ensures that there exists $S \in \mathcal{C}$ such that $\mathrm{dist}(x, S) = 0$. The second equation, with De Morgan's Theorem (B.11.2), gives $\mathrm{dist}(x, \bigcap \{A^c \mid A \in \mathcal{C}\}) = 0$, which yields $\mathrm{dist}(x, A^c) = 0$ for all $A \in \mathcal{C}$, by 2.4.3. In particular, we have $\text{dist}(x, S^c) = 0$. This, together with the fact, already established, that $\text{dist}(x, S) = 0$, yields $x \in \partial S$. Since x is arbitrary in $\partial(\bigcup \mathcal{C})$, (i) is proven.

In order to prove (ii), we make use of (i) and 3.1.2 together with De Morgan's Theorem (B.11.2) to justify the following sequence of assertions:
$$\partial(\bigcap \mathcal{C}) = \partial((\bigcap \mathcal{C})^c) = \partial(\bigcup\{A^c \mid A \in \mathcal{C}\}) \subseteq \bigcup\{\partial(A^c) \mid A \in \mathcal{C}\} = \bigcup\{\partial A \mid A \in \mathcal{C}\},$$
giving exactly what is required. $\square$

Example 3.5.2

Finiteness is necessary in 3.5.1. Zero, for example, is in the boundary of $\bigcup\{(r,1) \mid r \in (0,1)\}$, although it is not in the boundary of any of the intervals $(r,1)$. Zero is also in the boundary of $\bigcap\{(-r,r) \mid r \in \mathbb{R}^+\}$ but is not in the boundary of any of the intervals $(-r,r)$.

Question 3.5.3

Given a subset S of a metric space X, is it always the case that if we form a new set from S by removing or appending some boundary points, then the boundary of the new set will be the same as that of S? This was exactly the situation we observed in 3.1.4. A quick check reveals, however, that the outcome is not always so neat. The boundary of $\mathbb{Q}$ in $\mathbb{R}$ is $\mathbb{R}$, but the boundary of $\mathbb{Q} \cup [0,1]$ in $\mathbb{R}$ is $\mathbb{R}\backslash(0,1)$, and the boundary of $\mathbb{Q}\backslash[0,1]$ in $\mathbb{R}$ is also $\mathbb{R}\backslash(0,1)$. What does remain unaltered is, in the case of appending boundary points, the *closure* of the set and, in the case of removing boundary points, the *interior* of the set—as we shall see in 3.7.3.

3.6   Closure and Interior

The process of removing from a set all of its boundary points leaves us with its *interior*. The process of appending to a set all of its boundary points yields its *closure*. The concepts of *interior* and *closure* are relative ones: just as the boundary of a set depends on the metric space in which the set is considered to reside, so also do its closure and interior.

Definition 3.6.1

Suppose X is a metric space and S is a subset of X. We define

- the *closure* of S in X to be the union $S \cup \partial S$;
- the *interior* of S in X to be the difference $S \backslash \partial S$; and

- the *exterior* of S in X to be the complement of the closure of S in X.

The members of the interior of S are called *interior points* of S, and the members of the exterior of S are called *exterior points* of S. We denote the closure of S in X by $\overline{S}$ or $\mathrm{Cl}(S)$ and the interior of S in X by S° or $\mathrm{Int}(S)$. If it is thought necessary to specify the metric space, we may write $\mathrm{Cl}_X(S)$ and $\mathrm{Int}_X(S)$, respectively.

Example 3.6.2

For each metric space X, we have $\partial X = \varnothing = \partial\varnothing$, so $\mathrm{Cl}_X(\varnothing) = \varnothing = \mathrm{Int}_X(\varnothing)$ and $\mathrm{Cl}_X(X) = X = \mathrm{Int}_X(X)$.

Example 3.6.3

By 3.2.4, the boundaries of $\mathbb{Q}$ and $\mathbb{R}\backslash\mathbb{Q}$ in $\mathbb{R}$ are $\mathbb{R}$, so we have $\overline{\mathbb{Q}} = \mathbb{R} = \overline{\mathbb{R}\backslash\mathbb{Q}}$ and $\mathbb{Q}^\circ = \varnothing = (\mathbb{R}\backslash\mathbb{Q})^\circ$.

Example 3.6.4

Each of the discs examined in 3.1.4 has the unit circle $\mathbb{T}$ for its boundary, the closed unit disc $\mathbb{D}$ for its closure and the open unit disc $\{z \in \mathbb{C} \mid |z| < 1\}$ for its interior.

Example 3.6.5

In 3.1.3, we listed the boundary points of intervals. Using the information from 3.1.3, we now list their closures and their interiors. Suppose $a, b \in \mathbb{R}$ and $a < b$. Then the intervals (a,b), $[a,b)$, $(a,b]$ and $[a,b]$ all have closure $[a,b]$ and interior (a,b). The intervals (a,∞) and $[a,\infty)$ have closure $[a,\infty)$ and interior (a,∞), and the intervals $(-\infty,b)$ and $(-\infty,b]$ have closure $(-\infty,b]$ and interior $(-\infty,b)$. The degenerate interval $[a,a]$ is its own closure and has empty interior. The interval $(-\infty,\infty)$ is its own closure and its own interior.

Example 3.6.6

The Cantor set $\mathcal{K}$ of 3.3.3 satisfies $\partial\mathcal{K} = \mathcal{K}$. It follows that $\overline{\mathcal{K}} = \mathcal{K}$ and $\mathcal{K}^\circ = \varnothing$.

Question 3.6.7

Suppose X is a metric space and $S \subseteq X$. It follows immediately from the definition of closure and 3.3.6 that $\overline{\partial S} = \partial S$. Our intuition might be that boundaries have empty interior. Such intuition would be backed up by the examples of 3.1.4; they all have $\mathbb{T}$ as their boundary and $\mathbb{T}$ certainly has empty interior in $\mathbb{C}$. Is the same true of all boundaries? Indeed it is not. An immediate counterexample is given by the fact that the interior of the boundary of $\mathbb{Q}$ in $\mathbb{R}$ is $\mathbb{R}$ itself.

Theorem 3.6.8

Suppose X is a metric space and $S \subseteq X$. Then

(i) $\overline{S} = S \cup \mathrm{acc}(S)$; and

(ii) $S^\circ = S \backslash \mathrm{acc}(S^c)$.

Proof

These two results follow immediately from 3.1.6 and Definition 3.6.1.   □

Theorem 3.6.9

Suppose X is a metric space and $S \subseteq X$. Then

(i) $(S^\circ)^c = \overline{S^c}$; and

(ii) $(\overline{S})^c = (S^c)^\circ$—that is, the exterior of S in X is the interior of its complement in X.

Proof

Elementary set theory gives $S^\circ = S \backslash \partial S = S \cap (\partial S)^c$, so that, using De Morgan's Theorem (B.11.2), we get $(S^\circ)^c = S^c \cup \partial S$. Since $\partial S = \partial(S^c)$ by 3.1.2, we then have $(S^\circ)^c = S^c \cup \partial(S^c) = \overline{S^c}$. This proves (i). Then (ii) follows by reversing the roles of S and S^c; indeed, the argument yields $((S^c)^\circ)^c = \overline{(S^c)^c} = \overline{S}$, which implies $(S^c)^\circ = (\overline{S})^c$.   □

Theorem 3.6.10

Suppose X is a metric space and $S \subseteq X$. Then

(i) $\overline{S} = \{x \in X \mid \mathrm{dist}(x, S) = 0\}$;

(ii) the exterior of S is $\{x \in X \mid \mathrm{dist}(x, S) > 0\}$; and

(iii) $S^\circ = \{x \in X \mid \mathrm{dist}(x, S^c) > 0\}$.

Proof

For each $x \in \overline{S}$, either $x \in S$ or $x \in \partial S$ and, in either case, $\mathrm{dist}(x, S) = 0$. Conversely, for each $x \in X \backslash S$ with $\mathrm{dist}(x, S) = 0$, we have $x \in \partial S$ by definition, so that $x \in \overline{S}$. Thus (i) is proved, and (ii) follows by definition. Then (iii) is obtained by replacing S by S^c in (ii) and noting that the exterior of S^c is the interior of S (3.6.9). □

Corollary 3.6.11

Suppose (X, d) is a metric space, $w \in X$ and $A \subseteq X$. Then

(i) $\operatorname{diam}(\overline{A}) = \operatorname{diam}(A)$;

(ii) $\operatorname{dist}(w, A) \leq \operatorname{dist}(w, \partial A)$; and

(iii) $\operatorname{dist}(w, \overline{A}) = \operatorname{dist}(w, A)$.

Proof

Suppose $r \in \mathbb{R}^+$. For each $a, b \in \overline{A}$, there are $x, y \in A$ such that $d(a, x) < r/2$ and $d(b, y) < r/2$ by 3.6.10. So $d(a, b) \leq d(a, x) + d(x, y) + d(y, b) \leq \operatorname{diam}(A) + r$. Since a and b are arbitrary in $\overline{A}$, we then have $\operatorname{diam}(\overline{A}) \leq \operatorname{diam}(A) + r$ and, because r is arbitrary in $\mathbb{R}^+$, $\operatorname{diam}(\overline{A}) \leq \operatorname{diam}(A)$. The reverse inequality comes from 2.1.3. So (i) is proved.

For (ii), we proceed as follows. If $\partial A = \varnothing$, then $\operatorname{dist}(w, \partial A) = \infty$ and the result holds. Otherwise we consider arbitrary $z \in \partial A$. By definition, $\operatorname{dist}(z, A) = 0$, so that, by 2.3.2, $\operatorname{dist}(w, A) \leq d(w, z) + \operatorname{dist}(z, A) = d(w, z)$. Since this is true for all z in ∂A, it follows that $\operatorname{dist}(w, A) \leq \operatorname{dist}(w, \partial A)$, as required.

For (iii), we recall first that $\overline{A} = A \cup \partial A$ and then invoke 2.4.1 to get $\operatorname{dist}(w, \overline{A}) = \min\{\operatorname{dist}(w, A), \operatorname{dist}(w, \partial A)\}$. Then (ii) clinches the matter. □

Example 3.6.12

Interiors do not behave in the same way as closures in relation to distance (3.6.11). If X is a metric space and $A \subseteq X$, there is no guarantee that $\operatorname{diam}(\operatorname{Int}(A)) = \operatorname{diam}(A)$ or that, for $x \in X$, $\operatorname{dist}(x, \operatorname{Int}(A)) = \operatorname{dist}(x, A)$. Consider, for example, $X = \mathbb{R}$ and $A = \mathbb{Q}$. Because $\operatorname{Int}(\mathbb{Q}) = \varnothing$, we have $\operatorname{diam}(\operatorname{Int}(\mathbb{Q})) = -\infty$, whereas $\operatorname{diam}(\mathbb{Q}) = \infty$, and, for every $x \in \mathbb{R}$, we have $\operatorname{dist}(x, \mathbb{Q}) = 0$, whereas $\operatorname{dist}(x, \operatorname{Int}(\mathbb{Q})) = \infty$.

Corollary 3.6.13

Suppose X is a metric space and S is a subset of X. Then the interior, the boundary and the exterior of S are mutually disjoint and their union is equal to X.

Proof

This follows easily from the definition of boundary and the second and third parts of 3.6.10.   □

 A hole-in-heart subset of a square metric space. Its boundary is shown in black, its interior is shown in dark grey, and its exterior is shown in light grey; together they make up the whole space.

3.7 Inclusion of Closures and Interiors

The inclusion of one set in another does not imply the inclusion of the boundary of the one in the boundary of the other. Indeed, we have learnt in 3.2.4 that such a proposition may fail spectacularly to be true. Nonetheless, we ask hopefully whether or not the inclusion of one set in another need imply the inclusion of the closure of the first in the closure of the second and perhaps also the inclusion of the interior of the first in the interior of the second. In such hopes we are not disappointed (3.7.1).

Theorem 3.7.1

Suppose X is a metric space, A and B are subsets of X and $A \subseteq B$. Then

(i) $\overline{A} \subseteq \overline{B}$; and

(ii) $A^{\circ} \subseteq B^{\circ}$.

Proof

Since $A \subseteq B$, 2.3.1 yields $\{x \in X \mid \operatorname{dist}(x, A) = 0\} \subseteq \{x \in X \mid \operatorname{dist}(x, B) = 0\}$, which 3.6.10 translates to $\overline{A} \subseteq \overline{B}$. Since $A \subseteq B$, B.3.3 gives $B^c \subseteq A^c$, so that, by 2.3.1, $\{x \in X \mid \operatorname{dist}(x, B^c) = 0\} \subseteq \{x \in X \mid \operatorname{dist}(x, A^c) = 0\}$. Now B.3.3 gives $\{x \in X \mid \operatorname{dist}(x, A^c) \neq 0\} \subseteq \{x \in X \mid \operatorname{dist}(x, B^c) \neq 0\}$, which 3.6.10 translates to $A^{\circ} \subseteq B^{\circ}$.   □

The closure of a set S in a metric space X is constructed by appending to it all of its boundary points; the interior of S is constructed by removing all of its boundary points. The intention is to produce the smallest superset of S that includes its own boundary and the largest subset of S that is disjoint from its own boundary. Is this intention realized? In particular, does the closure of S include its own boundary, which may, as we have learnt from 3.5.3, differ from the boundary of S, and is the interior of S disjoint from its own boundary, which may differ from the boundary of S? The answer to this question is *yes*, as we see in 3.7.2. Later, in 4.1.14, we shall realize our whole stated intention.

Theorem 3.7.2

Suppose X is a metric space and $S \subseteq X$. Then

(i) $\partial \overline{S} \subseteq \overline{S}$;

(ii) $\partial(S^\circ) \cap S^\circ = \varnothing$;

(iii) $\overline{\overline{S}} = \overline{S}$; and

(iv) $(S^\circ)^\circ = S^\circ$.

Proof

For (i), suppose $x \in \partial \overline{S}$. Then $\operatorname{dist}(x, S \cup \partial S) = 0$, so that, by 2.4.1, either $\operatorname{dist}(x, S) = 0$ or $\operatorname{dist}(x, \partial S) = 0$. So, by 3.6.10, $x \in \overline{S}$ or $x \in \overline{\partial S}$. Since, by 3.3.6, $\overline{\partial S} = \partial S \subseteq \overline{S}$, we have $x \in \overline{S}$, as required. For (ii), we have, by 3.6.9, $(S^\circ)^c = \overline{S^c}$. We have also $\partial(\overline{S^c}) \subseteq \overline{S^c}$ by applying (i) to S^c. So, using 3.1.2, we get $\partial(S^\circ) = \partial((S^\circ)^c) = \partial(\overline{S^c}) \subseteq \overline{S^c} = (S^\circ)^c$, which is what we require. Then (iii) and (iv) follow immediately from (i) and (ii), respectively.   □

Corollary 3.7.3

Suppose X is a metric space and S and A are subsets of X.

(i)   If $S \subseteq A \subseteq \overline{S}$, then $\overline{A} = \overline{S}$.

(ii) If $S^\circ \subseteq A \subseteq S$, then $A^\circ = S^\circ$.

Proof

Using 3.7.1 and 3.7.2, we have, for (i), $S \subseteq A \subseteq \overline{S} \Rightarrow \overline{S} \subseteq \overline{A} \subseteq \overline{\overline{S}} = \overline{S}$, and, for (ii), $S^\circ \subseteq A \subseteq S \Rightarrow S^\circ = (S^\circ)^\circ \subseteq A^\circ \subseteq S^\circ$.   □

Question 3.7.4

Can 3.7.3 be extended to sets A outside the ranges stated? Can we, for example, get similar results if $S^\circ \subseteq A \subseteq \overline{S}$? A little reflection shows that we cannot. If $S^\circ \subseteq A \subseteq \overline{S}$, it does not necessarily follow that $\overline{A} = \overline{S}$ or that $A^\circ = S^\circ$. Indeed, every subset A of $\mathbb{R}$ satisfies the inclusions $\mathbb{Q}^\circ \subseteq A \subseteq \overline{\mathbb{Q}}$ because $\mathbb{Q}^\circ = \varnothing$ and $\overline{\mathbb{Q}} = \mathbb{R}$, but, provided only that A include an interval and $\mathbb{R} \backslash A$ include another, A has neither empty interior nor closure equal to $\mathbb{R}$.

3.8   Closure and Interior of Unions and Intersections

How well do closure and interior behave under unions and intersections? Is a union of closures the same as the closure of the union? Is an intersection of closures the same as the closure of the intersection? If we do not have equality, do we get inclusion in either case? We ask similar questions about interiors. The full answer to these questions (3.8.1) holds a surprise that is unlikely to be guessed by any but the most astute reader, despite the simplicity of the examples that show that things must be so (3.8.2).

Theorem 3.8.1

Suppose X is a metric space and $\mathcal{C}$ is a non-empty set of subsets of X. Then

(i) $\overline{\bigcap \mathcal{C}} \subseteq \bigcap \{\overline{A} \mid A \in \mathcal{C}\}$;

(ii) $\bigcup \{\overline{A} \mid A \in \mathcal{C}\} \subseteq \overline{\bigcup \mathcal{C}}$ with equality if $\mathcal{C}$ is finite;

(iii) $\bigcup \{A^\circ \mid A \in \mathcal{C}\} \subseteq (\bigcup \mathcal{C})^\circ$; and

(iv) $(\bigcap \mathcal{C})^\circ \subseteq \bigcap \{A^\circ \mid A \in \mathcal{C}\}$ with equality if $\mathcal{C}$ is finite.

Proof

For each $A \in \mathcal{C}$, we have $\bigcap \mathcal{C} \subseteq A \subseteq \bigcup \mathcal{C}$, so that $\overline{\bigcap \mathcal{C}} \subseteq \overline{A} \subseteq \overline{\bigcup \mathcal{C}}$ and $(\bigcap \mathcal{C})^\circ \subseteq A^\circ \subseteq (\bigcup \mathcal{C})^\circ$ by 3.7.1. Since this is true for all $A \in \mathcal{C}$, the inclusions of (i), (ii), (iii) and (iv) all hold. To complete the proof of the theorem, we need therefore consider only the cases when $\mathcal{C}$ is finite.

Suppose now that $\mathcal{C}$ is finite. Then 3.5.1 gives $\partial(\bigcup \mathcal{C}) \subseteq \bigcup \{\partial A \mid A \in \mathcal{C}\}$. It follows that $\overline{\bigcup \mathcal{C}} = \bigcup \mathcal{C} \cup \partial(\bigcup \mathcal{C}) \subseteq \bigcup \{A \cup \partial A \mid A \in \mathcal{C}\} = \bigcup \{\overline{A} \mid A \in \mathcal{C}\}$, which completes the proof of (ii). For (iv), we note that, by De Morgan's Theorem (B.11.2) and 3.6.9, $\bigcap \{A^\circ \mid A \in \mathcal{C}\} = (\bigcup \{(A^\circ)^c \mid A \in \mathcal{C}\})^c = (\bigcup \{\overline{A^c} \mid A \in \mathcal{C}\})^c$ and $(\bigcap \mathcal{C})^\circ = ((\bigcup \{A^c \mid A \in \mathcal{C}\})^c)^\circ = (\overline{\bigcup \{A^c \mid A \in \mathcal{C}\}})^c$. Then, since $\mathcal{C}$ is finite, (ii) gives $\bigcup \{\overline{A^c} \mid A \in \mathcal{C}\} = \overline{\bigcup \{A^c \mid A \in \mathcal{C}\}}$, so that $(\bigcap \mathcal{C})^\circ = \bigcap \{A^\circ \mid A \in \mathcal{C}\}$, as required. $\square$

Example 3.8.2

To appreciate that finiteness of the collection $\mathcal{C}$ is needed to ensure equality in parts (ii) and (iv) of 3.8.1, we return to the examples of 3.5.2. Zero is in the closure of $\bigcup \{(r, 1) \mid r \in (0, 1)\}$, even though it is not in the closure of any of the intervals $(r, 1)$. On the other hand, $\bigcap \{(-r, r) \mid r \in (0, 1)\} = \{0\}$, which has empty interior, despite the fact that 0 is in the interior of every one of the intervals $(-r, r)$. To convince ourselves that the inclusions in the two other parts are not generally reversible even for finite collections of sets, we consider the intervals $I = (0, 1)$ and $J = [1, 2]$ of $\mathbb{R}$. We have $\overline{I \cap J} = \varnothing$, whereas $\overline{I} \cap \overline{J} = \{1\}$, yielding the proper inclusion $\overline{I \cap J} \subset \overline{I} \cap \overline{J}$. Looking at the interiors, we have $I^\circ \cup J^\circ = (0,2) \setminus \{1\}$, whereas $(I \cup J)^\circ = (0,2)$, giving the proper inclusion $I^\circ \cup J^\circ \subset (I \cup J)^\circ$.

Summary

This chapter opened with a discussion about boundary points of subsets of a metric space. We then examined how boundary points relate to isolated points and accumulation points. We have talked about sets with empty boundary and how they relate to *connectedness*, and sets that include their boundaries (*closed sets*). We have explored boundaries of unions and intersections of sets. We have defined the Cantor set, more of which we shall see later. We have defined the closure and interior of subsets in terms of their boundaries. We have looked at the relationships that exist between closure and interior and have examined how they behave under the basic set-theoretic operations.

EXERCISES

†Q 3.1   Suppose $a, b \in \mathbb{R}$ and $a < b$. Show that the boundary points of the interval (a, b) are a and b.

Q 3.2   Suppose X is a metric space and A and B are subsets of X for which $\partial B \subseteq A \subseteq B$. Show that $\partial B \subseteq \partial A$.

Q 3.3   Suppose X is a metric space, S is a subset of X and a is an isolated point of S. Show that a is a boundary point of S in X if, and only if, $a \notin \mathrm{iso}(X)$.

†Q 3.4   Verify that the graph $\Gamma = \{(x, \sin(1/x)) \mid x \in \mathbb{R}^+\}$ of $x \mapsto \sin(1/x)$ defined on $\mathbb{R} \setminus \{0\}$ has boundary $\Gamma \cup \{(0, y) \mid y \in [-1, 1]\}$ in $\mathbb{R}^2$.

Q 3.5   With reference to 3.5.1, show that the intersection of the boundaries of a finite number of subsets of a metric space need not be included in the boundary of their intersection or in the boundary of their union.

Q 3.6   Find a countable subset A of $\mathbb{R}$ (see B.17) such that $(\partial_{\mathbb{R}} A) \setminus A$ is a singleton set.

Q 3.7   Consider the set $\mathcal{F}$ of functions from $[0, 1]$ to $[0, 1]$ with the metric $(f, g) \mapsto \sup\{|f(x) - g(x)| \mid x \in [0, 1]\}$ discussed in 1.1.17. Let $\mathcal{C}$ denote the collection of constant functions in $\mathcal{F}$. Show that $\partial \mathcal{C} = \mathcal{C}$.

†Q 3.8 Each number in $[0\,,1]$ has a decimal expansion of the type $\sum_{n=1}^{\infty} x_n/10^n$, where each x_n is a non-negative integer less than 10. In a similar way, it has a *ternary* expansion $\sum_{n=1}^{\infty} x_n/3^n$, where each x_n is a non-negative integer less than 3. Some numbers have two expansions of each type, one terminating and the other ending in a recurring sequence of nines for decimal expansion or of twos for ternary expansion. For example, the decimal number $0{\cdot}12$ can be written as $0{\cdot}11\dot{9}$, where the dot indicates that the nine recurs, and the ternary number $0{\cdot}12$ can be written as $0{\cdot}11\dot{2}$, where the two recurs. Show that the Cantor set consists of all the numbers in $[0\,,1]$ that have a ternary expansion consisting only of zeroes and twos.

†Q 3.9 Let $S = \mathcal{K} \cap \{(a+b)/2 \mid a,b \in \mathcal{K},\ a \neq b\}$, where $\mathcal{K}$ denotes the Cantor set. Show that $S = \mathcal{K} \cap \{k/3^n \mid n \in \mathbb{N},\ k \in \mathbb{N}_{3^n},\ k/3 \notin \mathbb{N}\}$.

Q 3.10 Is it possible for an uncountable subset S of $\mathbb{R}$ (see B.17) to satisfy $\partial S = S$?

†Q 3.11 Suppose $n \in \mathbb{N}$ and, for each $i \in \mathbb{N}_n$, (X_i, τ_i) is a non-empty metric space. Suppose d is a conserving metric on $P = \prod_{i=1}^{n} X_i$. Suppose $S \subseteq P$. Explore the relationship between $\partial_P S$ and $\partial_{X_i} \pi_i(S)$ for $i \in \mathbb{N}_n$.

Q 3.12 Show that every countable subset of $\mathbb{R}$ (B.17.3) has empty interior in $\mathbb{R}$ and is therefore included in its own boundary in $\mathbb{R}$.

Q 3.13 Find a metric space in which no non-empty countable subset has empty interior.

Q 3.14 Suppose (X,d) is a metric space and A is a subset of X. Show that $A^\circ \cup \partial A = \overline{A}$ and $\partial A = \overline{A} \cap \overline{A^c}$.

†Q 3.15 Suppose X is a metric space and A is a subset of X. Is it necessarily the case that ∂A and $\partial(\overline{A})$ are identical?

Q 3.16 Suppose X is a metric space and S is a subset of X. Show that $\mathrm{diam}(S^\circ)$ need not be the same as $\mathrm{diam}(S)$.

Q 3.17 Suppose X is a metric space and $S \subseteq X$. Show that $\overline{S} = \mathrm{acc}(S) \cup \mathrm{iso}(S)$.

Q 3.18 Suppose (X,d) is a metric space and A and B are subsets of X. Show that $\mathrm{dist}\big(\overline{A}\,,\overline{B}\big) = \mathrm{dist}(A\,,B)$.

†Q 3.19 Suppose X is a metric space and $A \subseteq X$. Must $(A^\circ)^c$ be equal to $\mathrm{Cl}\big((\overline{A})^c\big)$?

4
Open, Closed and Dense Subsets

*How thoroughly it is ingrained in mathematical science
that every real advance goes hand in hand with the invention of
sharper tools and simpler methods which, at the same time,
assist in understanding earlier theories and in casting aside
some more complicated developments.*   *David Hilbert, 1862–1943*

The interval $[0,1]$ includes its boundary; it is called a *closed* interval. The interval $(0,1)$ is disjoint from its boundary; it is called an *open* interval. Every real number is zero distance from the set of rational numbers; we say that $\mathbb{Q}$ is *dense* in $\mathbb{R}$. In this chapter, we shall extend these ideas of open, closed and dense subsets to all metric spaces.

4.1 Open and Closed Subsets

Here are the definitions, not to be forgotten: a subset of a metric space that includes all of its boundary is closed; a subset that contains no point of its boundary is open; and all other subsets are neither open nor closed. Most subsets, like the interval $[0,1)$, are neither open nor closed. There are, however, subsets of metric spaces that, unlike honey pots, are both open and closed.

As the analysis of metric spaces develops, we shall see that one of the most important questions that can be asked about a subset of a metric space is whether or not it includes its boundary; in other words, whether or not it is closed. The importance of open subsets comes about in another way: it rests on the fact that the theories of *convergence* (Chapter 6), *continuity* (Chapter 8), *connectedness* (Chapter 11) and *compactness* (Chapter 12) depend on the open subsets of the space rather than on the metric that produces them.

We begin in 4.1.1 with some equivalent criteria for subsets to be open and in 4.1.2 with some equivalent criteria for subsets to be closed. In 4.1.4, we tie together the two ideas.

Theorem 4.1.1 (Criteria for Being Open)

Suppose X is a metric space and $S \subseteq X$. These statements are equivalent:

(i) $\partial S \cap S = \varnothing$.

(ii) $S = S^\circ$.

(iii) $\mathrm{acc}(S^c) \cap S = \varnothing$.

Proof

Statements (i) and (ii) are equivalent because $S^\circ = S \backslash \partial S$; (ii) and (iii) are equivalent by 3.6.8.   □

Theorem 4.1.2 (Criteria for Being Closed)

Suppose X is a metric space and $S \subseteq X$. These statements are equivalent:

(i) $\partial S \subseteq S$.

(ii) $S = \overline{S}$.

(iii) $\mathrm{acc}(S) \subseteq S$.

Proof

Statements (i) and (ii) are equivalent because $\overline{S} = S \cup \partial S$; (ii) and (iii) are equivalent by 3.6.8.   □

Definition 4.1.3

Suppose X is a metric space and S is a subset of X. Then S is said to be

- an *open* subset of X, or *open in* X, if, and only if, $S \cap \partial S = \varnothing$; and
- a *closed* subset of X, or *closed in* X if, and only if, $\partial S \subseteq S$.

Now 4.1.1 allows us to use any one of the criteria listed there as a criterion for openness. Equally, a subset of a metric space is closed if, and only if, it satisfies any one of the criteria listed in 4.1.2. Moreover, as we see now in 4.1.4, a subset of a metric space is open if, and only if, its complement is closed.

Theorem 4.1.4

Suppose X is a metric space and S is a subset of X. The following statements are equivalent:

(i) S is closed in X.

(ii) S^c is open in X.

Proof

S is closed in X if, and only if, $\partial S \subseteq S$, but, since $\partial S = \partial(S^c)$ (3.1.2), this is the same as saying that $\partial(S^c) \cap S^c = \varnothing$ or, in other words, that S^c is open in X. □

Example 4.1.5

There are many sets that are neither open nor closed in their enveloping metric spaces. Since $\overline{\mathbb{Q}} = \mathbb{R}$ and $\mathbb{Q}^\circ = \varnothing$, $\mathbb{Q}$ is neither open nor closed in $\mathbb{R}$.

Example 4.1.6

Suppose (X, d) is a metric space, $a \in X$ and $r \in \mathbb{R}^+$. Then the *ball* $B = \{x \in X \mid d(x, a) < r\}$ is open. Specifically, if $z \in B$, then there exists $s \in (0, r)$ such that $d(a, z) < r - s$ and, for $w \in X \backslash B$, we have $d(w, a) \geq r$, so that $d(z, w) \geq d(w, a) - d(a, z) > s$ (1.1.2). Since w is arbitrary in $X \backslash B$, we have $\mathrm{dist}(z, X \backslash B) \geq s$ and $z \in B^\circ$ by 3.6.10. Since z is arbitrary in B, it follows

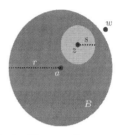

that $B = B^\circ$ and B is open by 4.1.1. We shall study balls in Chapter 5.

Example 4.1.7

All boundaries are closed. Specifically, if X is a metric space and S is a subset of X, then $\partial(\partial S) \subseteq \partial S$ by 3.3.6, so that ∂S is closed in X by definition.

Example 4.1.8

Suppose X is a metric space. Every finite subset of X includes its boundary in X (3.3.1) and so is closed in X. In particular, all singleton subsets of X are closed in X. A singleton set is open in X if, and only if, it is an isolated point of X: for $z \in X$, $\{z\} \cap \partial\{z\} = \varnothing \Leftrightarrow z \notin \partial\{z\} \Leftrightarrow \mathrm{dist}(z, X \backslash \{z\}) \neq 0 \Leftrightarrow z \in \mathrm{iso}(X)$.

Example 4.1.9

Let us examine the intervals of $\mathbb{R}$. In 3.1.3, we looked at their boundaries. From the information gathered there, the following facts can be gleaned: for $a, b \in \mathbb{R}$ with $a < b$, the intervals $[a, b]$, $[a, \infty)$ and $(-\infty, b]$ are all closed in $\mathbb{R}$; the intervals (a, b), (a, ∞) and $(-\infty, b)$ are all open in $\mathbb{R}$; the degenerate interval $[a, a]$ is closed in $\mathbb{R}$; and the interval $(-\infty, \infty)$ is both open and closed in $\mathbb{R}$. The intervals $[a, b)$ and $(a, b]$ are neither open nor closed in $\mathbb{R}$; we might call them *half-open* or, perhaps during bouts of agoraphobia, *half-closed*.

Example 4.1.10

The *open upper half-plane* of $\mathbb{C}$, $\{z \in \mathbb{C} \mid \Im(z) > 0\}$, is open in $\mathbb{C}$ since it contains no point of its boundary $\mathbb{R}$. On the other hand, the *closed upper half-plane*, $\{z \in \mathbb{C} \mid \Im(z) \geq 0\}$, includes its boundary and so is closed.

Example 4.1.11

We have seen in 3.3.3 that the Cantor set not only includes its boundary but actually equals its boundary. Certainly the Cantor set is closed in $\mathbb{R}$.

Question 4.1.12

From 3.3.4, the graph $\Gamma = \{(x, 1/x) \mid x \in \mathbb{R}, x \neq 0\}$ of the function $1/x$ is equal to its boundary in $\mathbb{R}^2$ and is therefore closed in $\mathbb{R}^2$. It is closed in $(\mathbb{R}\backslash\{0\}) \times \mathbb{R}$ also. In fact, it is true in general that a continuous function $f \colon X \to Y$ between metric spaces X and Y has a graph that is closed in $X \times Y$ as long as the product is endowed with a suitable metric (Q 8.13). We observed in 3.3.5 that the graph of the continuous function $\sin(1/x)$, although it does not include the whole of its boundary in $\mathbb{R}^2$ and is therefore not closed in $\mathbb{R}^2$, has nonetheless closed graph in $\mathbb{R}\backslash\{0\} \times \mathbb{R}$. This prompts a question that we shall answer when we have discussed continuity. What condition on a subset of $\mathbb{R}$ makes every continuous real function defined on that subset have a graph that is closed in $\mathbb{R}^2$ (Q 8.8)?

Example 4.1.13

Let X be an arbitrary metric space. The empty set has no boundary points, so it has the unusual property that it contains all of its boundary points while containing none of them. In short, it is both open and closed in X. The same is true of X itself. It should, of course, be clear to the reader that these cases of subsets that are both open and closed, though not the only ones we shall encounter, are exceptional—indeed, how else but by having an empty boundary could a subset both include its boundary and contain no point of it?

Is the closure of a set closed? It would be foolish not to check that it is. One naturally expects it to be; one might expect further that the closure of a subset S of a metric space X would be the smallest superset of S that is a closed subset of X. We have already voiced this expectation in slightly different terms in Chapter 3. In the same place, we expressed our hope that the interior of S might be the largest open subset of X that is included in S. We now realize these expectations in 4.1.14.

Theorem 4.1.14

Suppose X is a metric space and $S \subseteq X$. Then

(i) S° is the largest subset of S that is open in X; and

(ii) $\overline{S}$ is the smallest superset of S that is closed in X.

Proof

By 3.7.2, $\overline{\overline{S}} = \overline{S}$ and $(S^\circ)^\circ = S^\circ$. So $\overline{S}$ is closed in X and S° is open in X, by 4.1.3. If U is an open subset of X and $U \subseteq S$, then, using 3.7.1, $U = U^\circ \subseteq S^\circ$; and if F is a closed subset of X with $S \subseteq F$, then, also by 3.7.1, $\overline{S} \subseteq \overline{F} = F$. $\quad\square$

4.2 Dense Subsets

The rational numbers are densely packed along the real line. This was presented as a fact about order in B.6.11. Subsequently, it was translated into the language of distance by saying, in 2.2.3, that every point of $\mathbb{R}$ is zero distance from $\mathbb{Q}$. The concept of closure enabled us, in 3.6.3, to encapsulate the latter statement succinctly in the *density* formula $\overline{\mathbb{Q}} = \mathbb{R}$. This idea of density we now extend to an arbitrary metric space, with or without ordering. In 4.2.1, we show also that it can be formulated as well in terms of open sets as in terms of closure.

Theorem 4.2.1 (Criteria for Being Dense)

Suppose X is a metric space and S is a subset of X. The following statements are equivalent:

(i)   For every $x \in X$, $\mathrm{dist}(x, S) = 0$.

(ii) $\overline{S} = X$.

(iii) S has non-empty intersection with every non-empty open subset of X.

Proof

The statements are all true if $X = \varnothing$, so we suppose otherwise. That (i) and (ii) are equivalent follows immediately from 3.6.10. To show that (ii) implies (iii), we suppose there exists a non-empty open subset U of X such that $S \cap U = \varnothing$. Then $S \subseteq U^c$ and 3.7.1 gives $\overline{S} \subseteq \overline{U^c}$. But U^c is closed by 4.1.4, so that $\overline{U^c} = U^c$ (4.1.2). Therefore $\overline{S} \subseteq U^c$, whence $\overline{S} \cap U = \varnothing$ and therefore $\overline{S} \neq X$. So (ii) implies (iii). Last, we show that (iii) implies (ii). Suppose (iii). Since $(\overline{S})^c$ is clearly disjoint from S and is open in X by 4.1.2 and 4.1.4, we have $(\overline{S})^c = \varnothing$ and therefore $\overline{S} = X$. So (iii) implies (ii). $\quad\square$

Definition 4.2.2

Suppose X is a metric space and S is a subset of X. Then S is said to be *dense* in X if, and only if, $\overline{S} = X$.

Following this definition and 4.2.1, we can use any one of the criteria of 4.2.1 as a criterion for density.

Example 4.2.3

If X is a metric space, then, of course, X is dense in itself. But the prime non-trivial example of a dense subset is $\mathbb{Q}$ in $\mathbb{R}$. The set of irrational numbers is also dense in $\mathbb{R}$ because $\overline{\mathbb{R} \backslash \mathbb{Q}} = \mathbb{R}$ (3.6.3).

Example 4.2.4

Let $n \in \mathbb{N}$. Then $\mathbb{Q}^n$ is dense in $\mathbb{R}^n$ with the Euclidean metric. To see this, suppose $\epsilon \in \mathbb{R}^+$ and $x \in \mathbb{R}^n$. Select $q \in \mathbb{Q}^n$ as follows. For each $i \in \mathbb{N}_n$, pick $q_i \in \mathbb{Q}$ such that $|x_i - q_i| < \epsilon/\sqrt{n}$, which is possible because $\overline{\mathbb{Q}} = \mathbb{R}$. It follows that $\mu_2(x, q) < \epsilon$ (1.6.1) and then that $\mathrm{dist}_{\mathbb{R}^n}(x, \mathbb{Q}^n) < \epsilon$. Since ϵ is arbitrary in $\mathbb{R}^+$, we get $\mathrm{dist}_{\mathbb{R}^n}(x, \mathbb{Q}^n) = 0$ and, by 3.6.10, $x \in \overline{\mathbb{Q}^n}$. Since x is arbitrary in $\mathbb{R}^n$, $\overline{\mathbb{Q}^n} = \mathbb{R}^n$ and $\mathbb{Q}^n$ is dense in $\mathbb{R}^n$.

Example 4.2.5

Dense subsets may be thought to be large in a metric sense, but they are not necessarily large in other ways. $\mathbb{R}$ is an uncountable set and many of its dense subsets are, like $\mathbb{Q}$, countable (B.17). Actually, $\mathbb{R}$ has many dense subsets that are much smaller, in the sense of inclusion, than $\mathbb{Q}$: the set $\{m/2^n \mid m \in \mathbb{Z}, n \in \mathbb{N}\}$ of *dyadic rational numbers* is dense in $\mathbb{R}$, and the reader can, no doubt, find even smaller ones.

4.3 Topologies

The collection of open subsets of a metric space is so important that it is given a special name: it is called the *topology* of the space. The Greek word τόπος means simply *place*. The derivative English word *topology* has two meanings in mathematics. It is, as we have said, the collection of open subsets of a particular space. It is also the name of a branch of mathematics, sometimes referred to as *rubber sheet geometry*, that involves the study of those mathematical concepts the analysis of which depends directly on open sets rather than on any metric that might have produced them. The most fundamental fact about topologies—that they are algebraically closed (B.20.5) under all unions and under finite intersections—is proved in 4.3.2 together with a complementary statement about the collection of closed subsets.

Definition 4.3.1

Suppose (X, d) is a metric space. The collection of open subsets of X is called the *topology* determined by the metric d.

Theorem 4.3.2

Suppose X is a metric space and $\mathcal{C}$ is a non-empty collection of subsets of X.

(i)   If each member of $\mathcal{C}$ is closed in X, then $\bigcap \mathcal{C}$ is closed in X.

(ii) If $\mathcal{C}$ is finite and each member of $\mathcal{C}$ is closed in X, then $\bigcup \mathcal{C}$ is closed in X.

(iii) If each member of $\mathcal{C}$ is open in X, then $\bigcup \mathcal{C}$ is open in X.

(iv) If $\mathcal{C}$ is finite and each member of $\mathcal{C}$ is open in X, then $\bigcap \mathcal{C}$ is open in X.

Proof

Suppose first that every member of $\mathcal{C}$ is closed in X. For each $A \in \mathcal{C}$, we have $A = \overline{A}$ by 4.1.2, so the first part of 3.8.1 can be written $\overline{\bigcap \mathcal{C}} \subseteq \bigcap \mathcal{C}$. Since the reverse inclusion is certainly true, $\bigcap \mathcal{C}$ is closed in X, by 4.1.2, proving (i). Similarly, the second part of 3.8.1 tells us that if $\mathcal{C}$ is finite, then $\bigcup \mathcal{C} = \overline{\bigcup \mathcal{C}}$, so that, again by 4.1.2, $\bigcup \mathcal{C}$ is closed in X, proving (ii).

Now suppose that every member of $\mathcal{C}$ is open in X. By 4.1.1, $A = A^\circ$ for each A in $\mathcal{C}$, so the third part of 3.8.1 can be written $\bigcup \mathcal{C} \subseteq (\bigcup \mathcal{C})^\circ$ and, since the reverse inclusion is certainly true, $\bigcup \mathcal{C}$ is open in X by 4.1.1. The fourth part of 3.8.1 says that if $\mathcal{C}$ is finite, then $(\bigcap \mathcal{C})^\circ = \bigcap \mathcal{C}$, and it follows, again by 4.1.1, that $\bigcap \mathcal{C}$ is open in X. □

Question 4.3.3

Is finiteness of the collection necessary in the second and fourth parts of 4.3.2? There are, of course, infinite collections of closed sets with closed unions and infinite collections of open sets with open intersections, but such occurrence is not universal. Consider the collection of closed intervals $\{[r, 1] \mid r \in (0, 1)\}$ of $\mathbb{R}$. It is easy to check that its union is $(0, 1]$, which is not closed in $\mathbb{R}$. Similarly, we might consider the collection $\{(-r, r) \mid r \in \mathbb{R}^+\}$ of open intervals; its intersection, $\{0\}$, is closed, and not open, in $\mathbb{R}$.

Question 4.3.4

Every interval of the type (a, b) for $a, b \in \mathbb{R}$ with $a < b$ is an open subset of $\mathbb{R}$ (4.1.9). So, by 4.3.2, every union of such intervals is open in $\mathbb{R}$. Are there any other open subsets of $\mathbb{R}$? In other words, can every open subset of $\mathbb{R}$ be expressed as a union of open intervals? We shall answer this question in 5.2.4.

Question 4.3.5

Can a metric be recovered from the topology it produces? Specifically, suppose X is a set endowed with a metric that generates the topology $\mathcal{U}$. Knowing $\mathcal{U}$, can we decide what metric generated it? If X has more than one member, the answer is always *no*. If d is one metric on X that produces the topology $\mathcal{U}$, then $(a, b) \mapsto d(a, b)/2$ is another; $(a, b) \mapsto d(a, b)/(1 + d(a, b))$ is yet another (Q 4.11). Metrics that produce the same topology are said to be *topologically equivalent*. We shall discuss this concept in some detail in Chapter 13.

Example 4.3.6

When a non-empty set X is endowed with the discrete metric, each singleton set, being of distance 1 from every point in its complement, has empty boundary and so is open in X. Since every subset of X is the union of its singleton subsets, 4.3.2 implies that every subset of X is open in X. Then 4.1.4 implies that every subset of X is also closed in X. If one were of a humorous disposition, one might say that there is no metric less discreet (more open) than the discrete metric. But that would be uncomplimentary to discrete metrics; the complementary thing to say is that every subset of a space with the discrete metric is closed.

There are metric spaces with metrics that may differ from the discrete metric yet generate the same topology, namely the power set. Such spaces are collectively called *discrete metric spaces*. An example is given in 4.3.8.

Definition 4.3.7

A metric space (X, d) is called a *discrete metric space* if, and only if, all its subsets are open (and therefore also closed) in X.

Example 4.3.8

Every finite metric space is a discrete space. $\mathbb{N}$ with its usual metric inherited from $\mathbb{R}$ is a discrete metric space. $\mathbb{N}$ with the metric $(m, n) \mapsto |m^{-1} - n^{-1}|$ is a discrete metric space. But not all countable metric spaces are discrete (Q 4.15).

Example 4.3.9

When $\tilde{\mathbb{N}} = \mathbb{N} \cup \{\infty\}$ (see B.7.1) is endowed with the inverse metric of 1.1.12, its subspace $\mathbb{N}$ is the last discrete metric space of 4.3.8. Suppose U is an open subset of $\tilde{\mathbb{N}}$ with this metric and $\infty \in U$. Let $r = \mathrm{dist}(\infty, \tilde{\mathbb{N}} \backslash U)$. Then $r > 0$ and, for each $n \in \mathbb{N}$ with $1/n < r$, we have $n \in U$. So $\tilde{\mathbb{N}} \backslash U$ is finite. It follows that the topology of $\tilde{\mathbb{N}}$ consists of all subsets of $\mathbb{N}$ together with all subsets of $\tilde{\mathbb{N}}$ that have finite complement.

4.4 Topologies on Subspaces and Superspaces

Let X be a metric space and Z be a metric subspace of X. The metric on Z is a restriction of the metric on X. However, the topology on Z is not usually a subset of the topology on X. And when it is—that is, when the open subsets of Z are all open in X—it is hardly ever true that the closed subsets of Z are all closed in X (Q 4.7). The relationship between the topology on X and that on Z is, however, very simply stated (4.4.1).

Theorem 4.4.1

Suppose X is a metric space and Z is a metric subspace of X. The topology of Z is $\{U \cap Z \mid U$ open in $X\}$.

Proof

Suppose first that U is an open subset of X. We want to show that $U \cap Z$ is open in Z. This is true if $U \cap Z$ is empty, so we suppose otherwise. Suppose $x \in U \cap Z$. Then $x \in U$ and, since U is open in X, we have $\mathrm{dist}(x, X \backslash U) > 0$ by 3.6.10; but $Z \backslash (U \cap Z) \subseteq X \backslash U$, so that, by 2.3.1, $\mathrm{dist}(x, Z \backslash (U \cap Z)) > 0$ also, and, using 3.6.10 again, we get $x \in \mathrm{Int}_Z(U \cap Z)$. Since x is arbitrary in $U \cap Z$, it follows by 4.1.1 that $U \cap Z$ is open in Z.

For the converse, suppose that V is a subset of Z and that V is open in Z. Let $U = \mathrm{Int}_X((X \backslash Z) \cup V)$. Then U is open in X by 4.1.14. We claim that $V = U \cap Z$. Certainly $U \subseteq (X \backslash Z) \cup V$, so that $U \cap Z \subseteq V$. Since $V \subseteq Z$, it is sufficient therefore to verify that $V \subseteq U$. Suppose $x \in V$. Then, because V is open in Z, we have $x \in \mathrm{Int}_Z(V)$ by 4.1.1, and

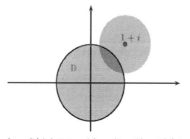

$\{z \in \mathbb{C} \mid |z| \le 1 \text{ and } |z - (1 + i)| < 3/4\}$ is open in the closed unit disc $\mathbb{D}$.

$\mathrm{dist}(x, Z\backslash V) > 0$ by 3.6.10. But $Z\backslash V = Z \cap (X\backslash V) = X\backslash((X\backslash Z) \cup V)$ by B.11.2. So $\mathrm{dist}(x, X\backslash((X\backslash Z) \cup V)) > 0$ and $x \in \mathrm{Int}_X((X\backslash Z) \cup V) = U$ by 3.6.10. Since x is arbitrary in V, this establishes that $V \subseteq U$, as required. $\square$

Corollary 4.4.2

Suppose Z is a metric space and X is a metric superspace of Z. Then the topology of X includes that of Z if, and only if, Z is open in X.

Proof

If the topology of X includes that of Z, then, since Z is open in Z, we have Z open in X as well. For the converse, if Z is open in X, then, for each subset U of X that is open in X, we have $Z \cap U$ open in X by 4.3.2. So the topology of Z is included in that of X by 4.4.1. $\square$

Example 4.4.3

Every open subset of $\mathbb{R}$ is the intersection with $\mathbb{R}$ of an open subset of $\mathbb{C}$. Moreover, such intersections always produce open subsets of $\mathbb{R}$. But no non-empty open subset of $\mathbb{R}$ is open in $\mathbb{C}$.

Note 4.4.4

Readers who noticed that we used expressions such as $\mathrm{dist}(x, A)$ in the proof of 4.4.1 without being explicit about whether these distances were being calculated in X or in Z will probably know why we did so. For others, we offer the explanation that, since the metric on Z is merely a restriction of that on X, distances calculated in Z are precisely the same as those calculated in X.

4.5 Topologies on Product Spaces

There are many ways in which a finite product of metric spaces can be endowed with a metric. We have confined our attention largely to what we have called conserving metrics (1.6.2) because they have a close relationship to the metrics on the individual spaces. We see now that all conserving metrics produce exactly the same topology and that this topology can be expressed very neatly in terms of the topologies on the individual spaces that make up the product.

Theorem 4.5.1

Suppose $n \in \mathbb{N}$ and, for each $i \in \mathbb{N}_n$, (X_i, τ_i) is a metric space. Endow the product $P = \prod_{i=1}^{n} X_i$ with a conserving metric d. The topology on P is the collection of all unions of members of the set $\{\prod_{i=1}^{n} U_i \mid U_i \text{ open in } X_i\}$.

Proof

Suppose that, for each $i \in \mathbb{N}_n$, U_i is open in X_i. Suppose $x \in \prod_{i=1}^{n} U_i$. Then, for each $i \in \mathbb{N}_n$, $x_i \in U_i$ and, since U_i is open in X_i, we have $x_i \notin \partial U_i$, whence $s_i = \text{dist}(x_i, X_i \backslash U_i) > 0$. The subset $\{s_i \mid i \in \mathbb{N}_n\}$ of $\mathbb{R}^+$ is finite and so has a minimum element t and $t > 0$. But d is a conserving metric. So, for each $y \in P$ with $d(x, y) < t$, we have, for all $i \in \mathbb{N}_n$, $\tau_i(x_i, y_i) \leq d(x, y) < t \leq s_i$, which gives $y_i \in U_i$ and therefore $y \in \prod_{i=1}^{n} U_i$. So $\text{dist}(x, P \backslash (\prod_{i=1}^{n} U_i)) \geq t$ and $x \notin \partial(\prod_{i=1}^{n} U_i)$. Since x is arbitrary in the product $\prod_{i=1}^{n} U_i$, this product is open in P and, by 4.3.2, so are all unions of such products.

For the converse, suppose W is an open subset of (P, d). The empty set and P itself are certainly unions of the prescribed type, so we suppose $W \neq \varnothing$ and $W \neq P$. For each $x \in W$, we have $x \notin \partial W$, so that $r = \text{dist}(x, P \backslash W) > 0$. For each $i \in \mathbb{N}_n$, let $V_{x,i} = \{a \in X_i \mid \tau_i(a, x_i) < r/n\}$. Then $V_{x,i}$ is open in X_i by 4.1.6. Also, because d is a conserving metric, for each $v \in \prod_{i=1}^{n} V_{x,i}$, we have $d(x, v) \leq \sum_{i=1}^{n} \tau_i(x_i, v_i) < r$, whence $v \in W$. Since v is arbitrary in $\prod_{i=1}^{n} V_{x,i}$, we get $\prod_{i=1}^{n} V_{x,i} \subseteq W$. Since x is arbitrary in W, it follows that $\bigcup\{\prod_{i=1}^{n} V_{x,i} \mid x \in W\} \subseteq W$. Since the reverse inclusion clearly holds, we then have $\bigcup\{\prod_{i=1}^{n} V_{x,i} \mid x \in W\} = W$, as required. $\qquad \square$

Definition 4.5.2

Suppose $n \in \mathbb{N}$ and, for each $i \in \mathbb{N}_n$, (X_i, τ_i) is a metric space. The collection of all unions of members of $\{\prod_{i=1}^{n} U_i \mid U_i \text{ open in } X_i\}$ will be called the *product topology* on P. Any metric on P that generates the product topology will be called a *product metric* on $\prod_{i=1}^{n} X_i$.

Note 4.5.3

Subsets of products are not usually products; the subset $\{(0, 1), (1, 0)\}$ of $\mathbb{R}^2$, for example, is not a product. Members of the product topology can all be expressed as unions of products; this in itself makes them special. But most members of the product topology are not products. For example, the open disc $\{x \in \mathbb{R}^2 \mid x_1^2 + x_2^2 < 1\}$ is a member of the product topology on $\mathbb{R}^2$, but it is not a product of open intervals of $\mathbb{R}$. In fact, to express it as a union of products of open intervals, we need to use an infinite collection of such products (Q 4.4).

Example 4.5.4

Every conserving metric is a product metric by 4.5.1. But there are many product metrics that are not conserving. This is true even in relatively simple cases. Although the definition of a product metric is designed for non-trivial products, it applies also when there is only one space (1.6). In such cases, there is only one conserving metric but there may be many product metrics. So, a *product* metric on $\mathbb{R}$ (inappropriate though the name may be in this case) is one that generates the same open sets as the usual metric. There are many such metrics. Consider, for example, the function d defined on $\mathbb{R} \times \mathbb{R}$ by

$$d(b,a) = d(a,b) = \begin{cases} 2b - 2a, & \text{if } 0 \le a \le b; \\ 2b - a, & \text{if } a < 0 \le b; \\ b - a, & \text{if } a \le b < 0. \end{cases}$$

This function stretches the positive part of the real line and is easily shown to be a metric on $\mathbb{R}$ that produces the same open sets as the Euclidean metric (Q 4.12). It is different from the original metric and so is not conserving.

4.6 Universal Openness and Universal Closure

Are there any non-empty metric spaces that are universally open or closed, in the sense that they are open or closed in every possible superspace? The first question we shall answer in the negative in 4.6.1; the second we shall answer in the affirmative by giving a very important example in 4.6.2. A metric space that is closed in every metric superspace is said to be *complete*. We shall characterize complete metric spaces in a number of other ways and make a study of them in Chapter 10.

Example 4.6.1

Suppose (X, d) is a non-empty metric space. Suppose $X \cap \mathbb{R}^+ = \varnothing$.[1] Pick $z \in X$. Let $Y = X \cup \mathbb{R}^+$. Extend d to $Y \times Y$, setting $d(a,b) = |a - b|$ for $a,b \in \mathbb{R}^+$ and $d(c,b) = d(b,c) = d(c,z) + b$ for each $c \in X$ and $b \in \mathbb{R}^+$. It is easy to check that d, thus extended, is a metric on Y. Moreover, $\mathrm{dist}_Y(z, \mathbb{R}^+) = 0$ because $\inf\{b \mid b \in \mathbb{R}^+\} = 0$, so that $z \in X \cap \partial_Y X$, ensuring that X is not open in Y.

For $c \in X$ and $b \in \mathbb{R}^+$, $d(c,b) = d(c,z) + b$.

---

[1] If this is not so, the argument can be modified using, instead of $\mathbb{R}^+$, some order-isomorphic copy of $\mathbb{R}^+$ that has empty intersection with X (B.21.1). It can be shown within set theory that there is such a copy.

Theorem 4.6.2

Suppose (X, d) is a metric superspace of $\mathbb{R}$ with its usual metric. Then $\mathbb{R}$ is a closed subset of X.

Proof

Suppose $z \in \text{Cl}_X(\mathbb{R})$. Then $\text{dist}(z, \mathbb{R}) = 0$, so, by 2.8.4, there exists $w \in \mathbb{R}$ such that $d(z, w) = 0$. Therefore $z = w \in \mathbb{R}$. Since z is arbitrary in $\text{Cl}_X(\mathbb{R})$, it follows that $\mathbb{R}$ is closed in X. □

Definition 4.6.3

Suppose X is a metric space. Then X is called a *complete metric space* if, and only if, X is closed in every metric superspace of X.

What we have shown in 4.6.2 is that $\mathbb{R}$ is a complete metric space. The name is appropriate and reflects the fact that the order completeness of $\mathbb{R}$ (B.6.7) was crucial in the proof of 2.8.4, of which 4.6.2 is a mere corollary. We shall see, however, that completeness as a property of metric spaces is quite independent of any ordering the spaces may have; indeed, most of the interesting complete metric spaces are not equipped with any standard ordering.

4.7   Nests of Closed Subsets

A nest (B.2.3) of non-empty subsets of a metric space may have empty intersection; $\bigcap\{(0, r) \mid r \in \mathbb{R}^+\}$, for example, is empty. We might expect non-empty intersection of the closures of the sets, but this need not be the case either, even in $\mathbb{R}$ (4.7.3). There is, however, a simple condition on such a nest—namely that there are sets in the nest of arbitrarily small diameter—that ensures non-empty intersection of their closures in a suitably extended metric superspace of the given space. It follows that non-empty intersection of a nest of non-empty closed subsets is assured in a complete metric space provided only that the small-diameter condition is satisfied (4.7.2).

Theorem 4.7.1 (Cantor's Intersection Theorem)

Suppose (X, d) is a metric space and $\mathcal{F}$ is a nest of non-empty subsets of X for which $\inf\{\text{diam}(A) \mid A \in \mathcal{F}\} = 0$. Suppose $\bigcap\{\overline{A} \mid A \in \mathcal{F}\} = \varnothing$. Then, given $z \notin X$, d can be extended to be a metric on $X' = X \cup \{z\}$ in such a way that $\text{Cl}_{X'}(A) = \text{Cl}_X(A) \cup \{z\}$ for all $A \in \mathcal{F}$. Thus $\bigcap\{\text{Cl}_{X'}(A) \mid A \in \mathcal{F}\} = \{z\}$.

Proof

Extend d to $X' \times X'$ by setting $d(z,z) = 0$ and, for each $x \in X$, $d(x,z)$ and $d(z,x)$ to have the value $\sup\{\operatorname{dist}(x,A) \mid A \in \mathcal{F}\}$. Note that $d(z,x) > 0$ because there is at least one $A \in \mathcal{F}$ for which $x \notin \overline{A}$, and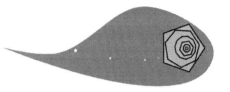
then $\operatorname{dist}(x,A) > 0$ by 3.6.10. Note also that $d(z,x) < \infty$: pick $S \in \mathcal{F}$ with $\operatorname{diam}(S) < \infty$; then, by 2.3.1, $\operatorname{dist}(x,A) \leq \operatorname{dist}(x,S) + \operatorname{diam}(S)$ for all $A \in \mathcal{F}$ because either $A \subseteq S$ or $S \subseteq A$, so that, by its definition, $d(x,z) \leq \operatorname{dist}(x,S) + \operatorname{diam}(S) < \infty$. So d is a real symmetric function with the positive property of 1.1.1.

We want to show that d is a metric on X'. Let $\epsilon \in \mathbb{R}^+$ and let $C \in \mathcal{F}$ with $\operatorname{diam}(C) < \epsilon$. Suppose $a,b \in X$ are arbitrary. Using 2.3.2, we have $d(a,b) \leq \operatorname{dist}(a,C) + \operatorname{diam}(C) + \operatorname{dist}(b,C) \leq d(a,z) + \epsilon + d(z,b)$, which, since ϵ is arbitrary in $\mathbb{R}^+$, yields $d(a,b) \leq d(a,z) + d(z,b)$. Also, using 2.3.2 again, $\operatorname{dist}(a,A) \leq d(a,b) + \operatorname{dist}(b,A) \leq d(a,b) + d(b,z)$ for each $A \in \mathcal{F}$, so that, by its definition, $d(a,z) \leq d(a,b) + d(b,z)$. These two calculations give us the triangle inequality for the extended function d. Therefore d is a metric on X'.

Let $A \in \mathcal{F}$ be arbitrary. Since $\mathcal{F}$ is a nest, $A \cap C$ is either A or C and is thus non-empty. Let $x \in A \cap C$. Then, for each $B \in \mathcal{F}$, $B \cap C$ is the smaller of B and C, so that, since $x \in C$, $\operatorname{dist}(x,B) \leq \operatorname{dist}(x,B \cap C) \leq \operatorname{diam}(C) < \epsilon$. Because B is arbitrary in $\mathcal{F}$, it follows, by definition, that $d(z,x) \leq \epsilon$. This in turn, because $x \in A$, implies $\operatorname{dist}(z,A) \leq \epsilon$. But ϵ is arbitrary in $\mathbb{R}^+$, so that $\operatorname{dist}(z,A) = 0$. It follows from 3.6.10 that $\operatorname{Cl}_{X'}(A) = \operatorname{Cl}_X(A) \cup \{z\}$ and, since A is arbitrary in $\mathcal{F}$, this proves the theorem. $\qquad\square$

The condition imposed on the diameters of the members of the nest $\mathcal{F}$ of 4.7.1 ensures that $\bigcap\{\overline{A} \mid A \in \mathcal{F}\}$ has no more than one element (Q 4.14). It is also an immediate consequence of 4.7.1 that if $\bigcap\{\overline{A} \mid A \in \mathcal{F}\} = \varnothing$, then X is not complete (4.6.3) because 4.7.1 then yields $z \in \operatorname{Cl}_{X'}(X)$. Is the converse also true? We deal with this question now, getting a necessary and sufficient condition for completeness to begin the list we shall make in Chapter 10.

Theorem 4.7.2

Suppose X is a metric space. The following statements are equivalent:

(i) X is complete.

(ii) Every nest $\mathcal{F}$ of non-empty closed subsets of X that has the property that $\inf\{\operatorname{diam}(A) \mid A \in \mathcal{F}\} = 0$ has singleton intersection.

Proof

That completeness of X implies non-empty intersection of all the nests is, as noted above, immediate from 4.7.1. That the intersection is a singleton is an easy consequence (Q 4.14). For the converse, we suppose that Y is a metric superspace of X in which X is not closed and let $z \in \mathrm{Cl}_Y(X) \backslash X$. For each $r \in \mathbb{R}^+$, let $B_r = \mathrm{Cl}_X(\{x \in X \mid d(x, z) < r\})$. It is easily checked that $\{B_r \mid r \in \mathbb{R}^+\}$ is a nest of non-empty closed subsets of X, that $\mathrm{diam}(B_r) \leq 2r$ for each $r \in \mathbb{R}^+$ and that $\bigcap\{B_r \mid r \in \mathbb{R}^+\} = \varnothing$. $\qquad\square$

Example 4.7.3

It may appear strange that the diameter condition of 4.7.2 is necessary not simply to prove that the intersection is a singleton set but also to ensure its non-emptiness. This is so even in $\mathbb{R}$. If the sets of the nest are too big, the intersection can be empty. An example is the nest $\{[n, \infty) \mid n \in \mathbb{N}\}$ of closed subsets of $\mathbb{R}$, each of infinite diameter.

It is evident from 4.7.3 that $\mathbb{R}$, despite its completeness, does not have the property that every nest of non-empty closed subsets has non-empty intersection. The spaces that do have this nice property are called *compact* metric spaces. They will be examined in detail in Chapter 12.

Summary

Open and closed subsets of a metric space are at the heart of this chapter. The concept of density has also been introduced. We have explained the notion of a metric topology. We have shown in detail how the topology of a metric subspace relates to that of an enveloping superspace and have determined the topology for conserving metrics on a finite product. We have demonstrated that $\mathbb{R}$ with its usual metric is complete—in other words, universally closed—and have begun a related discussion about nests of closed subsets.

EXERCISES

Q 4.1 Let A be a subset of a metric space X. Show that $\partial A = \varnothing$ if, and only if, A is both open and closed in X.

†Q 4.2 Suppose X is a metric space and $S \subseteq X$. Show that S has empty interior if, and only if, S has dense complement.

Q 4.3   Find two disjoint closed subsets of $\mathbb{R}$ that are zero distance apart.

Q 4.4   Express the disc $U = \left\{ x \in \mathbb{R}^2 \mid x_1^2 + x_2^2 < 1 \right\}$ as a union of products of pairs of open intervals of $\mathbb{R}$ (see 4.5.3).

†Q 4.5   Suppose X is a metric space and Z is a metric subspace of X. Show that the collection of closed subsets of Z is $\{ F \cap Z \mid F \text{ closed in } X \}$.

†Q 4.6   Suppose X is a metric space and Z is a metric subspace of X. Show that the collection of closed subsets of Z is included in the collection of closed subsets of X if, and only if, Z is closed in X.

†Q 4.7   Suppose X is a metric space and Z is a metric subspace of X. Show that the open subsets of Z are all open in X and the closed subsets of Z are all closed in X if, and only if, $\partial_X Z = \varnothing$.

Q 4.8   Show that $\{ x \in \mathbb{Q} \mid x \in [0,1] \}$ is a closed subset of $\mathbb{Q}$ with its usual metric.

Q 4.9   Show that, as in 2.5.2, the points of a discrete metric space are all isolated.

†Q 4.10   Give an example of a metric subspace of $\mathbb{R}$ in which all open subsets are open in $\mathbb{R}$ but not all closed subsets are closed in $\mathbb{R}$.

Q 4.11   Suppose (X, d) is a metric space. Show that the metric e of Q 1.11, namely $(a, b) \mapsto d(a, b)/(1 + d(a, b))$, generates the same topology on X as d.

†Q 4.12   Show that the function d defined on $\mathbb{R} \times \mathbb{R}$ in 4.5.4 is a metric on $\mathbb{R}$ and that it generates the same topology as the Euclidean metric.

†Q 4.13   Suppose X is a metric space and Z is a metric subspace of X. Suppose $S \subseteq Z$. Show that the topology on S as a subspace of X is the same as the topology on S as a subspace of Z.

†Q 4.14   Suppose that (X, d) is a metric space and that $\mathcal{F}$ is a nest of non-empty subsets of X for which $\inf\{ \operatorname{diam}(A) \mid A \in \mathcal{F} \} = 0$. Show that either $\bigcap \mathcal{F} = \varnothing$ or $\bigcap \mathcal{F}$ is a singleton set.

†Q 4.15   Find a countable metric space that is not a discrete metric space.

Q 4.16   A subset A of a metric space X is called a PERFECT set if $A = \operatorname{acc}(A)$. Show that perfect sets are precisely those that are closed and have no isolated points.

Q 4.17   Suppose A is a closed subset of a metric space X. Show that A can be expressed as the disjoint union of its set of isolated points and its set of accumulation points.

Q 4.18 Suppose X is a metric space. Let $\mathcal{C}$ denote the collection of all dense subsets of X. Show that $\bigcap \mathcal{C} = \mathrm{iso}(X)$.

†Q 4.19 Suppose X is a metric space and S is a subset of X. We say that S is *nowhere dense* in X if, and only if, the closure of S in X has empty interior. Show that every nowhere dense subset of X has dense complement and that every closed dense subset of X has nowhere dense complement.

Q 4.20 Show that $\mathbb{N}$ is nowhere dense in $\mathbb{R}$ (Q 4.19).

Q 4.21 Show that although $\mathbb{Q}$ is dense in $\mathbb{R}$, its complement $\mathbb{R} \backslash \mathbb{Q}$ is not nowhere dense in $\mathbb{R}$ (Q 4.19).

†Q 4.22 Show that no non-empty metric space has a dense subset that is also nowhere dense (Q 4.19).

†Q 4.23 Suppose $n \in \mathbb{N}$ and, for each $i \in \mathbb{N}_n$, (X_i, τ_i) is a non-empty metric space and D_i is a subspace of X_i. Endow $P = \prod_{i=1}^{n} X_i$ with any product metric. Show that $\prod_{i=1}^{n} D_i$ is dense in P if, and only if, D_i is dense in X_i for all $i \in \mathbb{N}_n$.

Q 4.24 Let $n \in \mathbb{N}$ and, for each $i \in \mathbb{N}_n$, suppose X_i is a metric space. Endow $P = \prod_{i=1}^{n} X_i$ with a product metric. Suppose $S \subseteq P$. Show that if S is dense in P, then $\pi_i(S)$ is dense in X_i for each $i \in \mathbb{N}_n$. Demonstrate that the converse need not hold.

†Q 4.25 Can a subset of a product space with a product metric be dense without including a product of dense subsets of the coordinate spaces?

†Q 4.26 A metric space is said to be *separable* if, and only if, it has a countable (B.17.3) dense subset. Let $n \in \mathbb{N}$. Show that $\mathbb{Q}^n$ is dense in $\mathbb{R}^n$ when $\mathbb{R}^n$ is endowed with any product metric. Deduce that $\mathbb{R}^n$ is separable when endowed with any product metric.

†Q 4.27 Suppose X is a metric space and $\mathcal{C}$ is a countable collection of separable (Q 4.26) subspaces of X. Show that $\bigcup \mathcal{C}$ is also separable.

Q 4.28 Let X be the set of sequences in $[-1, 1]$. Define d on $X \times X$ to be $(a, b) \mapsto \sup\{|a_n - b_n| \mid n \in \mathbb{N}\}$. Then d is a metric on X; the proof is similar to that of 1.1.17. Show that (X, d) is not separable (Q 4.26).

Q 4.29 Show that a finite product of non-empty metric spaces, when endowed with a product metric, is separable (Q 4.26) if, and only if, each of the coordinate spaces is separable.

Q 4.30 Show that every metric subspace of a separable metric space is separable (Q 4.26).

5
Balls

A ball is determined by a centre and a radius, the prototypes being the discs of the complex plane. In a metric space, open balls may be regarded as the fundamental open sets because every open set is a union of open balls (5.2.2). They are useful because many properties that depend on the topology can be tested using only open balls rather than all open sets.

5.1 Open and Closed Balls

An open ball of radius r centred at a in a metric space X is the set of all points of X of distance less than r from a. Geometrically, this idea is quite intuitive. We shall see, however, that balls do not always have the shape we expect and that centres and radii may not always be well defined.

The open ball $\flat[x\,;r)$.

Definition 5.1.1

Suppose (X,d) is a metric space and $x \in X$. For each $r \in \mathbb{R}^+$, we define

- the *open ball in X centred at the point x and with radius r* to be the set
  $\flat[x\,;r) = \{y \in X \mid d(x,y) < r\}$; and

- the *closed ball in X centred at the point x and with radius r* to be the set
  $\flat[x\,;r] = \{y \in X \mid d(x,y) \leq r\}$.

Where it is considered necessary in order to avoid ambiguity, we may augment our notation with a subscript, as in $\flat_X[x\,;r)$ or $\flat_d[x\,;r)$.

Example 5.1.2

The open balls of $\mathbb{R}$ with the usual metric are the open intervals of the type (a,b), where $a, b \in \mathbb{R}$ with $a < b$; in fact, $(a,b) = \flat[(a+b)/2\,;(b-a)/2]$. The closed balls of $\mathbb{R}$ are the closed intervals of the type $[a,b]$, where $a, b \in \mathbb{R}$ with $a < b$. This interval is $\flat[(a+b)/2\,;(b-a)/2]$. No other interval is a ball.

Example 5.1.3

The open balls of $\mathbb{C}$ with the usual metric are the open discs of the complex plane. They are all circular in shape.

Example 5.1.4

Square balls may seem a little odd. Consider the metric $(a,b) \mapsto |b_1 - a_1| + |b_2 - a_2|$ on $\mathbb{R}^2$, which we mentioned in 1.1.13. The open balls of this space are all squares with sides that are at 45 degrees to the axes. The picture shows the open ball of radius $3/4$ centred at the point $(1, 1)$.

Example 5.1.5

Consider $\mathbb{R}^2$ endowed with the metric of 1.1.15. There are balls of three different shapes. First, we have the single-point balls, $\flat[(0,0)\,;r) = \{(0,0)\}$, for all $r \in (0,1]$; second, we have those balls that are precisely the same as the balls of the Euclidean metric, such as the ball $\flat[(0,0)\,;2) = \{x \in \mathbb{R}^2 \mid x_1^2 + x_2^2 < 4\}$; and third, we have balls that are punctured at the origin, such as the pictured example, $\flat[(1/2,1/2)\,;1) = \{x \in \mathbb{R}^2\backslash\{(0,0)\} \mid (x_1 - 1/2)^2 + (x_2 - 1/2)^2 < 1\}$.

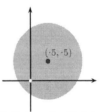

$(0,0)$ is not in this ball.

Example 5.1.6

Consider the metric space $\mathcal{F}$ of functions from $[0,1]$ to $[0,1]$ with the supremum metric $(f,g) \mapsto \sup\{|f(x) - g(x)| \mid x \in [0,1]\}$ discussed in 1.1.17. For each $f \in \mathcal{F}$ and $r \in \mathbb{R}^+$, the open ball of radius r centred at f is the set of functions $\{g \in \mathcal{F} \mid |f(x) - g(x)| < r \text{ for all } x \in [0,1]\}$. The diagram shows the graph of a

function f; the ball $\flat[f\,;0\cdot125]$ is the set of all functions defined on $[0,1]$ with graphs that lie in the grey area around the graph of f. The reader who thinks the diagram is wrong may care to check that the height of the grey region is constant throughout its length—except where the head of the region is cut off.

Are closed balls closed? We know that open balls of a metric space are open subsets of that space (4.1.6), but we have not checked that closed balls are closed. We do so now and also restate formally the openness of open balls.

Theorem 5.1.7

Suppose (X, d) is a metric space, $a \in X$ and $r \in \mathbb{R}^+$. Then

(i) $\partial(\flat[a\,;r)) \subseteq \{x \in X \mid d(x, a) = r\}$;

(ii) $\partial(\flat[a\,;r]) \subseteq \{x \in X \mid d(x, a) = r\}$;

(iii) $\flat[a\,;r)$ is open in X; and

(iv) $\flat[a\,;r]$ is closed in X.

Proof

Let B be either ball. Suppose $z \in \partial B$. Then $\mathrm{dist}(z, B) = 0$ and $\mathrm{dist}(z, B^c) = 0$. Set $s = d(z, a)$. For each $w \in B$, we have $d(a, w) \le r$. From 1.1.2, we then have $d(z, w) \ge d(z, a) - d(a, w) \ge s - r$. Therefore

$$0 = \mathrm{dist}(z, B) = \inf\{d(z, w) \mid w \in B\} \ge s - r.$$

So $s \le r$. Similarly, for $v \in B^c$, we have $d(v, a) \ge r$, which, again by 1.1.2, implies that $d(z, v) \ge d(v, a) - d(z, a) \ge r - s$, yielding

$$0 = \mathrm{dist}(z, B^c) = \inf\{d(z, v) \mid v \in B^c\} \ge r - s.$$

So $r \le s$. The two inequalities give $r = s$ and, since z is arbitrary in ∂B, we have proved (i) and (ii). The two other parts follow by definition because $\flat[a\,;r)$ contains none of these boundary points and $\flat[a\,;r]$ contains all of them.   □

Question 5.1.8

Why did we not simplify matters in 5.1.7 by saying that the boundary of $\flat[a\,;r)$ is the set $\{x \in X \mid d(x, a) = r\}$ or that $\flat[a\,;r]$ is the closure of $\flat[a\,;r)$? The short answer is that neither of these statements need be true. Consider the metric space $X = [0, 1] \cup [7, 9]$ with the usual metric inherited from $\mathbb{R}$. The closed ball $\flat_X[8\,;1] = [7, 9]$ has no boundary point in X despite the fact that $\{x \in X \mid d(8, x) = 1\} = \{7, 9\}$. Similarly, the open ball $\flat_X[8\,;7) = [7, 9]$ has no boundary point in X, despite the fact that $\{x \in X \mid d(8, x) = 7\} = \{1\}$. Notice also that $\overline{\flat_X[8\,;7)} = [7, 9]$, whereas $\flat_X[8\,;7] = [7, 9] \cup \{1\}$. These unpleasant situations can occur and we need to be aware of them. Not all is lost, however, for we salvage something in the inclusions of 5.1.9.

Theorem 5.1.9

Suppose X is a metric space, $a \in X$ and $r \in \mathbb{R}^+$. Then

(i) $\overline{\flat[a\,;r)} \subseteq \flat[a\,;r]$; and

(ii) $\flat[a\,;r) \subseteq \operatorname{Int}(\flat[a\,;r])$.

Proof

Certainly it is true by definition that $\flat[a\,;r) \subseteq \flat[a\,;r]$. Since $\flat[a\,;r)$ is open in X and $\flat[a\,;r]$ is closed in X (5.1.7), both assertions are immediate consequences of 4.1.14.   □

Is the diameter of a ball equal to twice its radius? Not necessarily. Indeed, the diameter can be smaller than the radius; it can even be zero. With reference to the example of 5.1.8, notice that $\flat_X[8\,;7)$ has diameter 2 but specified radius 7. We do, however, have the relationships of 5.1.10.

Theorem 5.1.10

Suppose (X,d) is a metric space, $a, z \in X$ and $r, s \in \mathbb{R}^+$. Then

(i) $\operatorname{diam}(\flat[a\,;r)) \leq 2r$;

(ii) $\operatorname{diam}(\flat[a\,;r]) \leq 2r$;

(iii) if $z \in \flat[a\,;r)$, then $\flat[a\,;s) \subseteq \flat[z\,;r+s)$; and

(iv) if $z \in \flat[a\,;r]$, then $\flat[a\,;s] \subseteq \flat[z\,;r+s]$.

Proof

 Suppose $x, y \in X$ and $d(a,x) \leq r$ and $d(a,y) \leq r$. Then the triangle inequality ensures that $d(x,y) \leq 2r$ and establishes (i) and (ii). For (iii), suppose $z \in \flat[a\,;r)$. Then $d(z,a) < r$, so that $d(z,x) \leq d(z,a) + d(x,a) < r+s$ for all $x \in \flat[a\,;s)$, whence also $x \in \flat[z\,;r+s)$. Since x is arbitrary in $\flat[a\,;s)$, this proves (iii). Part (iv) is proved similarly.   □

Question 5.1.11

Given a point x in a metric space X and a positive real number r, the ball $\flat[x\,;r)$ is well defined as the set of points of X of distance less than r from x. But, given an arbitrary ball of X, are its centre and its radius well defined? In other words, is it possible for $\flat[x\,;r)$ to be equal to $\flat[y\,;s)$ when $x \neq y$ and $r \neq s$? Unfortunately, it is. Consider, for example, a non-empty set X with the discrete metric. For each $x \in X$, the balls $\flat[x\,;r)$ for $r \in (0\,,1]$ are all equal to the singleton set $\{x\}$. On the other hand, the balls $\flat[x\,;r)$ for $r \in (1\,,\infty)$ are all equal to X for all $x \in X$. Such examples do not occur in nice spaces like $\mathbb{R}^2$, but they can occur in subspaces of $\mathbb{R}^2$ so we must be careful. To put the matter in a nutshell, we cannot define *the centre* or *the radius* of an arbitrary ball because they may not be unique. However, when a ball has been specified in the usual way as a ball of radius r centred at x—as $\flat[x\,;r)$ or $\flat[x\,;r]$—we shall happily refer to *the centre* and *the radius*, meaning the specified centre x and the specified radius r.

Question 5.1.12

We have seen in 4.3.5 that, except in trivial cases, it is not possible to recover a metric from the topology it generates. Can it be recovered from the collection of open balls it produces? Specifically, suppose X is a metric space and $\mathcal{B}$ is the collection of open balls of X. Knowing $\mathcal{B}$, can we decide what metric is on X? Of course, if radii for the balls are not stated, then there is no way of distinguishing a metric d from any of its scalar multiples λd for $\lambda \in \mathbb{R}^+$, and if centres are not stated for the balls, we may have difficulty distinguishing a metric d from a metric $(a,b) \mapsto d(f(a), f(b))$ for some injective function. (See Q 1.15 and Q 5.12.) Let us therefore modify the question as follows: given $\flat[x\,;r)$ for all $x \in X$ and $r \in \mathbb{R}^+$, can we discover d? This we are able to do, for $d(a,b) = \inf\{r \in \mathbb{R}^+ \mid b \in \flat[a\,;r)\}$ for all $a, b \in X$.

5.2 Using Balls

In the theory of metric spaces, properties that can be tested using open sets can generally be tested equally well using open balls. Open sets themselves are, in fact, always expressible as unions of open balls (5.2.2).

Theorem 5.2.1

Suppose X is a metric space, $x \in X$ and S is a subset of X. Then

(i) $x \in \overline{S}$ if, and only if, every open ball of X centred at x has non-empty intersection with S;

(ii) $x \in \partial S$ if, and only if, every open ball of X centred at x has non-empty intersection with both S and S^c; and

(iii) $x \in S^\circ$ if, and only if, S includes an open ball of X centred at x.

Proof

If $\operatorname{dist}(x, S) = 0$, then, for each $r \in \mathbb{R}^+$, there exists $z \in S$ such that $d(x, z) < r$. Then $z \in \flat[x\,;r)$ and $S \cap \flat[x\,;r) \neq \varnothing$. Conversely, if $\operatorname{dist}(x, S) \neq 0$, then either $S = \varnothing$ or $\flat[x\,;\operatorname{dist}(x, S)) \cap S = \varnothing$. So $\operatorname{dist}(x, S) = 0$ if, and only if, every open ball centred at x has non-empty intersection with S. Then (i) follows because $x \in \overline{S}$ if, and only if, $\operatorname{dist}(x, S) = 0$ by 3.6.10. And (ii) follows by applying the argument to both S and S^c because $x \in \partial S$ if, and only if, $\operatorname{dist}(x, S) = 0$ and $\operatorname{dist}(x, S^c) = 0$. For (iii), $x \in S^\circ$ if, and only if, $x \in S$ and $x \notin \partial S$ (3.6.1); by (ii), $x \notin \partial S$ if, and only if, there is an open ball of X centred at x that is included in either S or S^c; and since such a ball contains x, it cannot be included in S^c. So (iii) holds. $\square$

Theorem 5.2.2

Suppose X is a metric space and S is a non-empty subset of X. The following statements are equivalent:

(i) S is open in X.

(ii) For each $x \in S$, S includes an open ball of X centred at x.

(iii) S is a union of open balls of X.

Proof

Suppose first that S is open in X. Then $S = S^\circ$ by 4.1.1 and (ii) is obtained from 5.2.1. So (i) implies (ii).

Now suppose (ii) is true. Let $\mathcal{U}$ be the collection of all open balls of X that are included in S. Then certainly $\bigcup \mathcal{U} \subseteq S$. Also, by hypothesis, for each $x \in S$, there is some member of $\mathcal{U}$ that contains x, so that $x \in \bigcup \mathcal{U}$. Since x is arbitrary in S, this gives $S \subseteq \bigcup \mathcal{U}$. The two inclusions yield $S = \bigcup \mathcal{U}$. So (ii) implies (iii).

That (iii) implies (i) is an immediate consequence of 5.1.7 and 4.3.2. $\square$

Example 5.2.3

Suppose (X, d) is a metric space and $a \in X$. Then, for each $x \in X$, we have $x \in \flat[a\,;d(a, x) + 1)$, so that $X = \bigcup\{\flat[a\,;r) \mid r \in \mathbb{R}^+\}$, reflecting the fact that X is open in itself.

Example 5.2.4

Every open subset of $\mathbb{R}$ can be expressed as a union of open intervals of $\mathbb{R}$ of the type (a, b) with $a < b$. Specifically, the empty set is the empty union; $\mathbb{R}$ itself is the union of all intervals (a, b) for $a, b \in \mathbb{R}$ with $a < b$; and, for each non-empty proper open subset U of $\mathbb{R}$, for each $x \in U$, we have $\alpha_x = \text{dist}(x, U^c) > 0$, and it follows easily that $U = \bigcup \{(x - \alpha_x, x + \alpha_x) \mid x \in U\}$.

Example 5.2.5

The open upper half-plane $U = \{z \in \mathbb{C} \mid \Im z > 0\}$ can be expressed as the union $\bigcup \{\flat[z; \Im z] \mid z \in U\}$.

Example 5.2.6

Suppose $n \in \mathbb{N}$ and, for each $i \in \mathbb{N}_n$, (X_i, τ_i) is a metric space. A product metric on $\prod_{i=1}^{n} X_i$ is a metric that generates the topology made up of all unions of members of $\{\prod_{i=1}^{n} U_i \mid U_i \text{ open in } X_i\}$ (4.5.2). It is an easy exercise (Q 5.14) to show, using this and 5.2.2, that every open subset of $\prod_{i=1}^{n} X_i$ with a product metric can be expressed as a union of members of $\{\prod_{i=1}^{n} B_i \mid B_i \text{ is an open ball of } X_i\}$. Suppose, for example, that S is an open subset of $\mathbb{R}^2$ with its usual Euclidean metric. For each $x \in S$, define $\beta_x = \text{dist}(x, \mathbb{R}^2 \backslash S) / \sqrt{2}$. Then $S = \bigcup \{\flat_{\mathbb{R}}[x_1; \beta_x] \times \flat_{\mathbb{R}}[x_2; \beta_x] \mid x \in S\}$. (The balls of the last expression are more familiarly expressed as the intervals $(x_1 - \beta_x, x_1 + \beta_x)$ and $(x_2 - \beta_x, x_2 + \beta_x)$.)

5.3 Balls in Subspaces and in Products

The balls of a subspace Z of a metric space X are merely the intersections with Z of balls of X centred in Z. Balls of product spaces are much more elusive.

Theorem 5.3.1

Suppose X is a metric space and Z is a metric subspace of X. Suppose $x \in X$ and $r \in \mathbb{R}^+$. If $x \in Z$, then

(i) $Z \cap \flat_X[x; r)$ is the open ball $\flat_Z[x; r)$ of Z;

(ii) $Z \cap \flat_X[x; r]$ is the closed ball $\flat_Z[x; r]$ of Z;

and all the open and closed balls of Z are of this form.

Proof

For each $x \in Z$ and $r \in \mathbb{R}^+$, it follows immediately from the definition of an open ball that $\flat_Z[x; r)$ is $Z \cap \flat_X[x; r)$. Conversely, every ball of Z has a centre in Z and so is of the prescribed form. The same is true for the closed balls. □

Example 5.3.2

Endow the set $\mathbb{R}^+$ with the inverse metric $e : (a, b) \mapsto \left| a^{-1} - b^{-1} \right|$ of 1.1.12, and let d denote the usual metric on $\mathbb{R}^+$. The open balls of $(\mathbb{R}^+, d)$ are the open intervals (a, b) with $a, b \in \mathbb{R}^+$ and $a < b$ (5.1.2 and 5.3.1). Each of these balls is an open ball of $(\mathbb{R}^+, e)$ as well. Specifically, $(a, b) = \flat_e [2ab/(a+b) \, ; (b-a)/2ab]$. There are, however, open balls of $(\mathbb{R}^+, e)$ that, though they are open in $(\mathbb{R}^+, d)$, are not balls in that space. Specifically, for $x, r \in \mathbb{R}^+$, the ball $\flat_e[x \, ; r]$ is the interval $(x/(1+rx), x/(1-rx))$ if $rx < 1$ and is $(x/(1+rx), \infty)$ otherwise. Note that, since all the open balls of $(\mathbb{R}^+, d)$ are open in $(\mathbb{R}^+, e)$ and all the open balls of $(\mathbb{R}^+, e)$ are open subsets of $(\mathbb{R}^+, d)$, 5.2.2 and 4.3.2 ensure that these two spaces have the same topology. It is nonetheless true that the balls $\flat_e[x \, ; r]$ do not even have finite diameter in (X, d) if $rx \geq 1$.

Example 5.3.3

The balls of a product of two or more metric spaces may differ considerably when the product is endowed with different metrics. This is so even for conserving metrics. Consider, for example, the open balls of $\mathbb{R}^2$ with the Euclidean metric. None of them is a product of open intervals of $\mathbb{R}$ but merely an infinite union of such products. The metrics μ_∞ (1.6.1) are the only product metrics that have a conserving property for balls (5.3.4).

Theorem 5.3.4

Suppose $n \in \mathbb{N}$ and, for each $i \in \mathbb{N}_n$, (X_i, τ_i) is a metric space. Endow the product $P = \prod_{i=1}^{n} X_i$ with the metric $\mu_\infty : (a, b) \mapsto \sup\{\tau_i(a_i, b_i) \mid i \in \mathbb{N}_n\}$. Then, for each $x \in P$ and $r \in \mathbb{R}^+$, we have $\flat_P[x \, ; r] = \prod_{i=1}^{n} \flat_{X_i}[x_i \, ; r]$.

Proof

Suppose $x \in P$ and $r \in \mathbb{R}^+$. Then $a \in \flat[x \, ; r] \Leftrightarrow \mu_\infty(x, a) < r$, which occurs if, and only if, $\tau_i(x_i, a_i) < r$ for all $i \in \mathbb{N}_n$ or, equivalently, if, and only if, $a_i \in \flat[x_i \, ; r]$ for all $i \in \mathbb{N}_n$; that is, if, and only if, $a \in \prod_{i=1}^{n} \flat_{X_i}[x_i \, ; r]$. $\qquad \square$

5.4 Balls in Normed Linear Spaces

The really useful facts about open balls in a given normed linear space are that they are *convex* and have identical shape, each being a *translation* of every other open ball of the same radius (5.4.3). The same is true of closed balls.

Definition 5.4.1

Suppose V is a normed linear space and $C \subseteq V$. Then C is said to be *convex* if, and only if, for each $a, b \in C$, the *line segment* $\{(1-t)a + tb \mid t \in [0,1]\}$ joining a and b is included in C.

The set S is not convex. The line segment joining u and v has points outside S itself.

Definition 5.4.2

Suppose V is a normed linear space.

- The ball $\flat[0\,;1)$ is called the *open unit ball* of V.
- The ball $\flat[0\,;1]$ is called the *closed unit ball* of V.

Theorem 5.4.3

Suppose V is a normed linear space, a is a non-zero vector in V and $r \in \mathbb{R}^+$. Then, in the notation of B.20.3,

(i) $\flat[a\,;r) = a + r\flat[0\,;1)$;

(ii) $\flat[a\,;r] = a + r\flat[0\,;1]$; and

(iii) $\flat[a\,;r)$ and $\flat[a\,;r]$ are convex.

Proof

$x \in a + r\flat[0\,;1) \Leftrightarrow (x - a)/r \in \flat[0\,;1) \Leftrightarrow \|x - a\| < r \Leftrightarrow x \in \flat[a\,;r)$. This proves (i); (ii) is proved similarly. For convexity, suppose $x, y \in \flat[a\,;r)$. Then $\|x - a\| < r$ and $\|y - a\| < r$, and, for each $t \in [0,1]$, we have $\|(tx + (1-t)y) - a\| \leq t\|x - a\| + (1-t)\|y - a\| < r$, so $tx + (1-t)y \in \flat[a\,;r)$. Therefore $\flat[a\,;r)$ is convex. By a similar argument, $\flat[a\,;r]$ is also convex. $\square$

Example 5.4.4

If $\flat[a\,;r)$ and $\flat[b\,;r)$ are open balls of radius r in a normed linear space, then the second is the image of the first under the *translation* $x \mapsto x + (b - a)$. Such translations depend on the algebraic structure of the linear space, so we do not expect any general result of the same type to hold in arbitrary metric spaces. To get nice counterexamples, however, we do not need to wander into arbitrary metric spaces; we need look no further than $\mathbb{R}^2$ with a metric very nearly the same as the Euclidean metric. Example 5.1.5 shatters any hope we might have had for translation invariance, for shape similarity or indeed for convexity of a ball in a linear space endowed with a metric that is not determined by a norm since a ball with a hole in it is not convex.

Example 5.4.5

Example 5.4.4 involves a metric on $\mathbb{R}^2$ that is not determined by a norm. But even in metric subspaces of normed linear spaces—where the metric is determined by a norm—the shape of a ball is not entirely predictable. Suppose V is a normed linear space and S is a non-empty subset of V endowed with the restriction of the metric determined by the norm on V. Suppose $a \in S$ and $r \in \mathbb{R}^+$. Then it is true that $\flat_S[a\,;r] = S \cap (a + r\flat[0\,;1))$, but this tells us very little, generally speaking, about $\flat_S[a\,;r]$. If, for example, $V = \mathbb{R}^2$ and $S = \left\{ x \in \mathbb{R}^2 \mid x_2 \geq 0 \right\} \cup \{(0,-1)\}$, then $\flat[(0,-1)\,;1)$ is the singleton set $\{(0,-1)\}$ and $\flat[(0,-1)\,;1]$ is the set $\{(0,-1),(0,0)\}$.

Example 5.4.6

The reader should not be fooled into thinking that the ball that is the subject of the illustration in 5.1.6 is not convex. The shaded region on the page is not convex, of course, but the set of functions that it illustrates is. Specifically, if g and h are two functions from $[0,1]$ into $[0,1]$ with graphs that lie in that shaded region, then $\sup\{|f(x) - g(x)| \mid x \in [0,1]\} < 0{\cdot}125$ and $\sup\{|f(x) - h(x)| \mid x \in [0,1]\} < 0{\cdot}125$, from which it follows that, for all $t \in [0,1]$, $\sup\{|f(x) - (tg(x) + (1-t)h(x))| \mid x \in [0,1]\} < 0{\cdot}125$.

Question 5.4.7

What general restrictions are there on the shape of open balls of a real normed linear space? Since open balls with respect to a given norm all have the same shape (5.4.3), we can simplify the question: which subsets of a given real linear space V are open unit balls with respect to some norm on the space? The restrictions are, in fact, not as stringent as we might expect. We can list the properties that we know such an open unit ball must have:

- It must be convex (5.4.3).
- It must be *balanced* in the sense that, for each $a \in U$, we have $-a \in U$ because we want $-a$ to have the same norm as a.
- For each $x \in V \backslash \{0\}$, the set $\{t \in \mathbb{R}^+ \mid tx \in U\}$ must be non-empty to ensure that x has some real norm, its supremum s must be real to ensure that x does not have zero norm, and, to ensure that U has a chance of being open when V is endowed with an appropriate norm, we must have $sx \notin U$.

Actually, these properties are also sufficient to ensure that U is an open unit ball with respect to some norm, and it is not difficult to show that the norm is unique. The interested reader can check these things.

Example 5.4.8

Norms are abundant even on $\mathbb{R}^2$. To get one, all we need to do is draw a convex set with the properties of 5.4.7. We can then use it to measure the norm of each vector. Here are some examples. (Note that the larger circular ball goes with the factor 0·75 and the smaller one with the factor 2.)

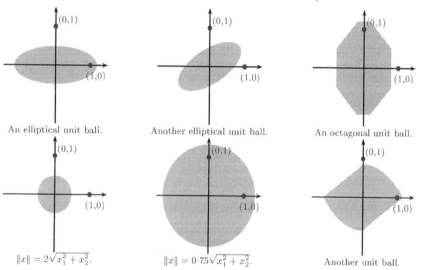

An elliptical unit ball.   Another elliptical unit ball.   An octagonal unit ball.

$\|x\| = 2\sqrt{x_1^2 + x_2^2}.$ $\|x\| = 0{\cdot}75\sqrt{x_1^2 + x_2^2}.$ Another unit ball.

Summary

After introducing open and closed balls, we showed that all open sets are unions of open balls and that boundary, closure and interior can be identified using open balls. We showed that balls in normed linear spaces are all convex and balanced and that, in any given space, they all have the same shape.

EXERCISES

Q 5.1   Show that the set $S = \big\{ a \in \mathbb{R}^3 \mid a_1 + a_3^2 \sin(a_1 + a_2) \geq a_3 \big\}$ is closed in $\mathbb{R}^3$ with the Euclidean metric.

†Q 5.2   Suppose X is a metric space, U is a non-empty open subset of X, $u \in U$ and $r \in \mathbb{R}^+$. Show that there exists $s \in (0, r)$ such that $b[u; s] \subseteq U$.

Q 5.3   Suppose X is a metric space, $a \in X$ and $r \in \mathbb{R}^+$. Show that $b[a; r)$ and $b[a; r]$ can be expressed, respectively, in the following ways:
$\bigcup\{b[a; s) \mid s \in (0, r)\} = \bigcup\{b[a; s] \mid s \in (0, r)\} = \bigcup\{b[a; s) \mid s \in (0, r]\}$; and
$\bigcap\{b[a; s) \mid s \in (r, \infty)\} = \bigcap\{b[a; s] \mid s \in [r, \infty)\} = \bigcap\{b[a; s] \mid s \in (r, \infty)\}$.

†Q 5.4 Suppose (X, d) is a metric space and $z \in X$. Show that $z \in \mathrm{iso}(X)$ if, and only if, there exists $r \in \mathbb{R}^+$ such that $\flat[z\,;r] = \{z\}$.

Q 5.5 Suppose X is a metric space. Show that every open ball of X is included in a closed ball of X and that every closed ball of X is included in an open ball of X.

Q 5.6 Show that the square $\{a \in \mathbb{R}^2 \mid a_1, a_2 \in (-1\,,1)\}$ is an open subset of $\mathbb{R}^2$ with the Euclidean metric.

Q 5.7 Show that the open subsets of the subspace $\mathbb{R} \times \{0\}$ of $\mathbb{R}^2$ with the usual metric are precisely those sets $U \times \{0\}$ where U is open in $\mathbb{R}$. Show also that none of these sets, except the empty set, is open in $\mathbb{R}^2$.

Q 5.8 Endow $\mathbb{R}^2$ with the metric $(a, b) \mapsto \max\{|a_1 - b_1|\,, |a_2 - b_2|\}$ and show that $\{a \in \mathbb{R}^2 \mid a_1^2 + a_2^2 < 1\}$ is open in $\mathbb{R}^2$ with this metric.

†Q 5.9 Suppose X is a metric space and E and F are disjoint closed subsets of X. Show that there exist disjoint open subsets U and V of X such that $E \subseteq U$ and $F \subseteq V$. Does this imply that $\mathrm{dist}(E\,,F) > 0$?

†Q 5.10 Define $e(x, y) = \max\left\{|x_1 - y_1|\,, \sqrt{(x_2 - y_2)^2 + (x_3 - y_3)^2}\right\}$ for each $x, y \in \mathbb{R}^3$. Show that e is a metric on $\mathbb{R}^3$ and describe the shape of the ball $\flat[0\,;1]$ it produces, where 0 denotes the origin $(0, 0, 0) \in \mathbb{R}^3$.

Q 5.11 Suppose X is a non-empty set and d is an ultrametric on X (see Q 1.9). Suppose B is an open ball of (X, d). Show that every point of B is a centre for B.

Q 5.12 Show that the stretching metric d on $\mathbb{R}$ described in Q 1.15 produces exactly the same set of open balls of $\mathbb{R}$ as the Euclidean metric does.

†Q 5.13 Suppose X is a metric space and $S \subseteq X$. Show that S is dense in X if, and only if, S has non-empty intersection with every open ball of X.

†Q 5.14 Suppose $n \in \mathbb{N}$ and, for each $i \in \mathbb{N}_n$, (X_i, τ_i) is a metric space. Show that every member of the product topology on $P = \prod_{i=1}^{n} X_i$ is a union of members of $\{\prod_{i=1}^{n} \flat[x_i\,;r] \mid r \in \mathbb{R}^+,\ x \in \prod_{i=1}^{n} X_i\}$.

Q 5.15 Describe the shape of the open unit ball of $\mathbb{R}^3$ when it is endowed with each of the norms $\|\cdot\|_1$, $\|\cdot\|_2$ and $\|\cdot\|_\infty$ (see 1.7.4, 1.7.6).

†Q 5.16 Suppose X is a normed linear space and $S \subseteq X$. For each subset A of X, write $-A = \{-a \mid a \in A\}$. Suppose $S \subseteq X$. Show that $\partial(-S) = -\partial S$, $\mathrm{Int}(-S) = -\mathrm{Int}(S)$ and $\overline{-S} = -\overline{S}$.

Q 5.17 Suppose X is a normed linear space and C is a convex subset of X. Show that C° and $\overline{C}$ are both convex and that ∂C need not be convex.

6

Convergence

As for everything else, so for a mathematical theory:
beauty can be perceived but not explained.   *Arthur Cayley, 1821–1895*

Sequences play an important role in the theory of metric spaces just as they do in the analysis of the real line. Here we present an introduction to convergence of sequences in arbitrary metric spaces. We consider what property of a sequence ensures its convergence—if not in the metric space under consideration, then in some metric superspace. We also characterize in several ways precisely those points that occur as limits of subsequences (B.18.2) of a given sequence in a metric space.

6.1 Definition of Convergence for Sequences

Below in 6.1.2 are some equivalent statements, any one of which might be used in a definition of convergence in metric spaces. In 6.1.4 we add another, which relates the theory of convergence of sequences in arbitrary metric spaces to the theory of *null sequences*, sequences that converge to 0, in $\mathbb{R}^{\oplus}$. The formal definition we actually adopt (6.1.3) is one that will admit the greatest generalization in more advanced courses in mathematics.

Definition 6.1.1

Suppose X is a non-empty set and $x = (x_n)$ is a sequence in X. For each $m \in \mathbb{N}$, the set $\{x_n \mid n \in \mathbb{N},\ n \geq m\}$ is called the mth *tail* of the sequence (x_n). It will be denoted by $\mathrm{tail}_m(x)$.

Theorem 6.1.2 (Criteria for Convergence)

Suppose X is a metric space, $z \in X$ and (x_n) is a sequence in X. The following statements are equivalent:

(i)   (CLOSURE CRITERION I) $\bigcap \left\{ \overline{\{x_n \mid n \in S\}} \;\middle|\; S \subseteq \mathbb{N},\, S \text{ infinite} \right\} = \{z\}$.

(ii) (CLOSURE CRITERION II) $z \in \bigcap \left\{ \overline{\{x_n \mid n \in S\}} \;\middle|\; S \subseteq \mathbb{N},\, S \text{ infinite} \right\}$.

(iii) (DISTANCE CRITERION) $\operatorname{dist}(z, \{x_n \mid n \in S\}) = 0$ for every infinite subset S of $\mathbb{N}$.

(iv) (BALL CRITERION) Every open ball centred at z includes a tail of (x_n).

(v)   (OPEN SET CRITERION) Every open subset of X that contains z includes a tail of (x_n).

Proof

That (i) implies (ii) is clear. That (ii) implies (iii) follows from 3.6.10. Suppose that (iii) holds. Then, given any $r \in \mathbb{R}^+$, the set $\{n \in \mathbb{N} \mid d(x_n, z) \geq r\}$ is finite, so that $\flat[z\,;r]$ includes a tail of (x_n). So (iii) implies (iv). Since every open subset of X that contains z includes a ball centred at z (5.2.2), (iv) implies (v).

All except a finite number of terms are covered by the ball.

To show that (v) implies (i), we suppose (v) holds and proceed as follows. Suppose S is an infinite subset of $\mathbb{N}$. Then $X \backslash \overline{\{x_n \mid n \in S\}}$ is an open subset of X (4.1.14 and 4.1.4) that certainly does not include a tail of (x_n) and does not therefore contain z, by hypothesis. So $z \in \overline{\{x_n \mid n \in S\}}$ and, since S is an arbitrary infinite subset of $\mathbb{N}$, $z \in \bigcap \left\{ \overline{\{x_n \mid n \in S\}} \;\middle|\; S \subseteq \mathbb{N},\, S \text{ infinite} \right\}$. Moreover, z is unique in this intersection, for, if $w \in X \backslash \{z\}$, then (v) implies that $\flat[z\,; d(z,w)/2]$ includes a tail of (x_n), and w, being at least $d(z,w)/2$ distant from this ball, is not in the closure of that tail. $\square$

Definition 6.1.3

Suppose X is a metric space, $z \in X$ and (x_n) is a sequence in X. We say that (x_n) *converges* to z in X, written $x_n \to z$, if, and only if, every open subset of X that contains z includes a tail of (x_n).

Now 6.1.2 allows us to use any of the criteria listed there as a criterion for convergence. It is also worth noting at this point that the closure criteria need every infinite subset of $\mathbb{N}$; the intersection of the closures of the tails of a sequence may be a singleton set without the sequence converging (Q 6.1).

Theorem 6.1.4

Suppose (X, d) is a metric space, $z \in X$ and (x_n) is a sequence in X. Then (x_n) converges to z in X if, and only if, the real sequence $(d(x_n, z))_{n \in \mathbb{N}}$ converges to 0 in $\mathbb{R}$.

Proof

Suppose $r \in \mathbb{R}^+$. Then the ball $\flat_X[z\,;r]$ includes a tail of (x_n) if, and only if, the interval $[0, r)$ includes the corresponding tail of $(d(x_n, z))$, which in turn occurs if, and only if, that tail is included in $(-r, r)$, which is $\flat_{\mathbb{R}}[0\,;r)$ because d has no negative values. But (x_n) converges to z in X if, and only if, the first of these equivalent conditions holds for all $r \in \mathbb{R}^+$; and the sequence $(d(x_n, z))$ converges to 0 in $\mathbb{R}$ if, and only if, the last of them holds for all $r \in \mathbb{R}^+$. □

Example 6.1.5

A singleton set in a metric space is, of course, included in every ball centred at its only point. A sequence that has a singleton set for a tail is said to be *eventually constant*; such sequences must converge in any metric space to which they belong. A *constant sequence*, in which all terms are the same, is a special case of an eventually constant sequence and converges to its single value in any metric space to which it belongs.

Example 6.1.6

In a discrete metric space X, it is very difficult for a sequence to converge. Each singleton set $\{a\}$ is open and the only way that $\{a\}$ can include a tail of a sequence (x_n) is if the sequence is eventually constant with $x_n = a$ for all sufficiently large $n \in \mathbb{N}$.

Example 6.1.7

The sequence $(1/n)$ of inverses of the natural numbers converges to 0 because for each $r \in \mathbb{R}^+$ there exists $k \in \mathbb{N}$ such that $1/k < r$ and then $\flat[0\,;r)$ includes the kth tail of $(1/n)$.

6.2 Limits

The reader will recall that convergent sequences of real numbers can converge to at most one point. Is this so for sequences in arbitrary metric spaces? The question is not an idle one, and the reader who delves far enough into abstract mathematics will encounter situations in which sequences may converge to more than one point. In metric spaces, that never happens, as 6.2.1 below shows. The reason for this, as the reader can check by looking at the proof of the closure criterion in 6.1.2, is that distinct points are separated by a positive distance.

Theorem 6.2.1

Suppose (X, d) is a metric space and (x_n) is a sequence in X that converges in X. Then (x_n) converges to exactly one point in X.

Proof

Suppose that $z, w \in X$ and that (x_n) converges to z and to w. Then, by the first closure criterion for convergence of 6.1.2, $\{z\} = \{w\}$, whence $z = w$. $\square$

Definition 6.2.2

Suppose X is a metric space, $z \in X$ and (x_n) is a sequence in X that converges to z in X. Then z is called the *limit* of (x_n) and is denoted by $\lim x_n$.

When different metrics are placed on a given set, the sequences that converge may differ and limits may differ too. For example, when $\mathbb{R}$ is endowed with the discrete metric, the sequence $(1/n)$ does not converge. However, the formal definition of convergence (unlike the distance criterion and the ball criterion of 6.1.2) does not rely on a fixed metric but only on the open sets produced by the metric. It follows that, on any given set, metrics that produce the same topology produce also the same convergent sequences with the same limits. We state this formally in 6.2.3.

Theorem 6.2.3

Suppose X is a set, $z \in X$ and (x_n) is a sequence in X. Suppose d and e are metrics on X that produce the same topology. Then (x_n) converges to z in (X, d) if, and only if, (x_n) converges to z in (X, e).

6.3 Superior and Inferior Limits of Real Sequences

Every real sequence, whether or not it converges, has a *limit inferior* and a *limit superior* in the set $\tilde{\mathbb{R}}$ of extended real numbers (B.7), and the former does not exceed the latter.

Definition 6.3.1

Suppose $x = (x_n)$ is a sequence in $\mathbb{R}$. We define

- the *limit superior* of (x_n), $\limsup x_n$, to be $\inf\{\sup \operatorname{tail}_n(x) \mid n \in \mathbb{N}\}$; and
- the *limit inferior* of (x_n), $\liminf x_n$, to be $\sup\{\inf \operatorname{tail}_n(x) \mid n \in \mathbb{N}\}$.

Example 6.3.2

Define a sequence (x_n) by $x_{2n-1} = 1$ and $x_{2n} = 1/n$ for all $n \in \mathbb{N}$. Then $\operatorname{tail}_n(x)$ is $\{1/m \mid m \in \mathbb{N}, 2m \geq n\} \cup \{1\}$; its supremum is 1 and its infimum is 0. This is true for all $n \in \mathbb{N}$, so that $\inf\{\sup \operatorname{tail}_n(x) \mid n \in \mathbb{N}\} = 1$ and $\sup\{\inf \operatorname{tail}_n(x) \mid n \in \mathbb{N}\} = 0$. Thus $\limsup x_n = 1$ and $\liminf x_n = 0$.

Notation 6.3.3

Suppose (x_n) is a sequence in $\mathbb{R}$. We extend the notation used for convergence as follows. We write $x_n \to \infty$ if, and only if, for every $s \in \mathbb{R}$, the interval (s, ∞) includes a tail of (x_n). Similarly, we write $x_n \to -\infty$ if, and only if, for every $s \in \mathbb{R}$, the interval $(-\infty, s)$ includes a tail of (x_n).

Theorem 6.3.4

Suppose (x_n) is a sequence of real numbers and $z \in \tilde{\mathbb{R}}$. Then $x_n \to z$ if, and only if, $\liminf x_n = z = \limsup x_n$.

Proof

We leave the cases when $z = \infty$ and $z = -\infty$ to the reader (Q 6.6) and suppose that $z \in \mathbb{R}$. If (x_n) converges to $z \in \mathbb{R}$ and $\epsilon \in \mathbb{R}^+$, then there exists $k \in \mathbb{N}$ such that $\operatorname{tail}_k(x) \subseteq (z - \epsilon, z + \epsilon)$, and it follows that $z - \epsilon \leq \inf \operatorname{tail}_k(x)$ and $\sup \operatorname{tail}_k(x) \leq z + \epsilon$, whence $z - \epsilon \leq \liminf x_n \leq \limsup x_n \leq z + \epsilon$. Since ϵ is arbitrary in $\mathbb{R}^+$, we then get $\liminf x_n = z = \limsup x_n$, as required.

For the converse, we suppose that $\liminf x_n = \limsup x_n = w \in \mathbb{R}$ and let $\epsilon \in \mathbb{R}^+$. Then there exist $k, l \in \mathbb{N}$ such that $w - \epsilon \leq \inf \operatorname{tail}_k(x)$ and $\sup \operatorname{tail}_l(x) \leq w + \epsilon$. Let $m = \max\{k, l\}$. Then $w - \epsilon \leq \inf \operatorname{tail}_k(x) \leq \inf \operatorname{tail}_m(x)$ and $\sup \operatorname{tail}_m(x) \leq \sup \operatorname{tail}_l(x) \leq w + \epsilon$, whence $\operatorname{tail}_m(x) \subseteq \flat[w\,;\epsilon]$. Since ϵ is arbitrary in $\mathbb{R}^+$, this means that (x_n) converges to w.   □

Example 6.3.5

Consider the real sequence (x_n) given by $x_n = -n$ for each $n \in \mathbb{N}$. For each $m \in \mathbb{N}$, we have $\operatorname{tail}_m(x) = \{-m \mid m \leq n\}$, so that $\sup \operatorname{tail}_m(x) = -m$ and $\inf \operatorname{tail}_m(x) = -\infty$. So $\inf\{\sup \operatorname{tail}_m(x) \mid m \in \mathbb{N}\} = \inf\{-m \mid m \in \mathbb{N}\} = -\infty$

and $\sup\{\inf \operatorname{tail}_m(x) \mid m \in \mathbb{N}\} = \sup\{-\infty\} = -\infty$. Therefore this sequence has $-\infty$ for both its limit inferior and limit superior. In short, $x_n \to -\infty$.

6.4 Convergence in Subspaces and Superspaces

A sequence may converge in a metric space yet fail to converge in a subspace to which every term of it belongs simply because the limit in the larger space does not belong to the subspace. The sequence $(1/n)$, for example, converges to 0 in $\mathbb{R}$ with its usual metric but does not converge in its subspace $(0, 1]$.

Theorem 6.4.1

Suppose X is a metric space, $w \in X$ and Z is a metric subspace of X. Suppose (x_n) is a sequence in Z that converges to w in X. Then (x_n) converges in Z if, and only if, $w \in Z$, and, in that case, its limit in Z is w.

Proof

Of course, (x_n) cannot converge in Z to a point outside Z, by definition. However, if (x_n) converges to w in X, then, for each $r \in \mathbb{R}^+$, the ball $\flat_X[w\,;r]$ includes a tail of (x_n), and, since every term of (x_n) is in Z, $\flat_X[w\,;r] \cap Z$ includes the same tail. If also $w \in Z$, then $\flat_Z[w\,;r] = \flat_X[w\,;r] \cap Z$ (5.3.1) and, since r is arbitrary in $\mathbb{R}^+$, we deduce that (x_n) converges to w in Z also. $\square$

Theorem 6.4.2

Suppose X is a metric space, $w \in X$ and (x_n) is a sequence in X that converges to w. Suppose Y is a metric superspace of X. Then (x_n) converges to w in Y.

Proof

For each $r \in \mathbb{R}^+$, the ball $\flat_Y[w\,;r]$ includes the ball $\flat_X[w\,;r]$, which in turn includes a tail of (x_n). $\square$

6.5 Convergence in Product Spaces

Let us turn now to products. Is there a relationship between convergence of a sequence in a finite product of metric spaces and convergence of the coordinate sequences in the coordinate spaces? The question is imprecise because there are many metrics on a product that have little relationship with the metrics on the coordinate spaces. However, if the metric on the product is a product metric (4.5.2), then the nicest possible relationship exists (6.5.1).

Theorem 6.5.1

Suppose $n \in \mathbb{N}$ and, for each $i \in \mathbb{N}_n$, (X_i, τ_i) is a metric space. Denote the product $\prod_{i=1}^{n} X_i$ by P. Endow P with any product metric (4.5.2). Suppose $(x_m)_{m \in \mathbb{N}}$ is a sequence in P. Then (x_m) converges in P if, and only if, $(\pi_i(x_m))_{m \in \mathbb{N}}$ converges in X_i for every $i \in \mathbb{N}_n$, where π_i denotes the natural projection of P onto X_i. Moreover, if this occurs, then $\pi_i(\lim x_m) = \lim \pi_i(x_m)$ for all $i \in \mathbb{N}_n$.

Proof

Suppose first that (x_m) converges in the product, and let z be its limit. For each $r \in \mathbb{R}^+$, the product $\prod_{i=1}^{n} \flat[z_i\,;r)$ is open in P by definition of the product topology (4.5.1), so it includes a tail of (x_m), whence each ball $\flat[z_i\,;r)$ includes the corresponding tail of $(\pi_i(x_m))$. So $(\pi_i(x_m))$ converges to z_i in X_i for each $i \in \mathbb{N}_n$.

For the converse, we suppose that, for each $i \in \mathbb{N}_n$, $(\pi_i(x_m))$ converges in X_i; we denote the limit by w_i and set $w = (w_1, \ldots, w_n) \in P$. Let $s \in \mathbb{R}^+$. Then, by Q 5.14 and 5.2.2, $\flat_P[w\,;s)$ includes a product $\prod_{i=1}^{n} \flat[w_i\,;r_i)$, where, for each $i \in \mathbb{N}_n$, $r_i \in \mathbb{R}^+$. Each of the balls $\flat[w_i\,;r_i)$ includes a tail, say the k_ith tail of $(\pi_i(x_m))$, by 6.1.2. The set $\{k_i \mid i \in \mathbb{N}_n\}$ is a finite set of natural numbers and so has a maximum element, say t (B.6.4). Then $\prod_{i=1}^{n} \flat[w_i\,;r_i)$, and hence also $\flat_P[w\,;s)$, includes the tth tail of (x_m). Since s is arbitrary in $\mathbb{R}^+$, we conclude that (x_m) converges in P to w. □

Question 6.5.2

We have no wish to define arbitrary infinite products in this book, but there is at least one that is easy enough to get used to. Consider the collection P of all sequences (x_n) with terms in $[0, 1]$. This set is the product of a countably infinite number of copies of $[0, 1]$ and may be written $\prod_{i=1}^{\infty} [0, 1]$. Does the nice situation of 6.5.1 extend to this infinite product? Alas, no. Mimicking 1.1.17, for each $x, y \in P$, we set $s(x, y) = \sup\{|x_i - y_i| \mid i \in \mathbb{N}\}$, and this is easily shown to be a metric on P. Whatever similarity there may be between s and conserving metrics on finite products, the analogue of 6.5.1 does not hold in P, as we now see. For each $m \in \mathbb{N}$, let $e(m) \in P$ be the sequence in which the mth term is 1 and the other terms are all zero. Then, for each $i \in \mathbb{N}$, the sequence $(\pi_i(e(m)))_{m \in \mathbb{N}}$ (here, $\pi_i(e(m))$ is the ith term of the sequence $e(m)$) has all except one term equal to 0 and so certainly converges to 0. However, the sequence $(e(m))_{m \in \mathbb{N}}$ of sequences $e(m)$ does not converge to the zero sequence in P because the distance from $e(m)$ to the zero sequence is 1 for every $m \in \mathbb{N}$. What about the converse? If a sequence in P converges in P, do its coordinate sequences converge in the coordinate spaces (Q 6.14)?

6.6 Convergence Criteria for Interior and Closure

A standard test for closure of a subset S of a metric space is to check that every convergent sequence that lies in S has its limit in S (6.6.3).

Theorem 6.6.1

Suppose X is a metric space, $z \in X$ and S is a non-empty subset of X. The following statements are equivalent:

(i) $z \in S^\circ$.

(ii) S includes a tail of every sequence in X that converges to z in X.

(iii) No sequence in S^c converges to z in X.

Proof

Suppose first that $z \in S^\circ$. Then every sequence that converges to z in X has a tail in S° by definition (6.1.3) because S° is open in X (4.1.14). So (i) implies (ii). That (ii) implies (iii) is immediate. To complete the proof, we suppose that (i) does not hold, so that $z \notin S^\circ$. Then, by 5.2.1, for each $n \in \mathbb{N}$, $A_n = S^c \cap b[z\,;1/n]$ is not empty. Choose a sequence (x_n) with $x_n \in A_n$ for each $n \in \mathbb{N}$ (B.19.1). Then (x_n) is a sequence in S^c and, since $d(x_n, z) < 1/n$ for all $n \in \mathbb{N}$, (x_n) converges to z in X by 6.1.4; so (iii) does not hold. Therefore (iii) implies (i).   □

Corollary 6.6.2

Suppose X is a metric space, $z \in X$ and S is a non-empty subset of X. The following statements are equivalent:

(i) $z \in \overline{S}$.

(ii) There exists a sequence in S that converges in X to z.

Proof

$z \in \overline{S} \Leftrightarrow z \notin (S^c)^\circ$ by 3.6.9. But $z \notin (S^c)^\circ$ if, and only if, there is a sequence in S that converges to z in X, by 6.6.1.   □

Corollary 6.6.3

Suppose X is a metric space and S is a subset of X. The following statements are equivalent:

(i) S is closed in X.

(ii) Every sequence in S that converges in X has its limit in S.

Proof

Suppose first that S is closed in X and (x_n) is a sequence in S that converges to $z \in X$. Then $z \in \overline{S} = S$ by 6.6.2 and 4.1.2. So (i) implies (ii). Conversely, if S is not closed in X, then there exists $z \in \overline{S} \backslash S$, and by 6.6.2, there exists a sequence in S that converges to z. Therefore (ii) implies (i).   □

6.7   Convergence of Subsequences

Every subsequence of a convergent sequence converges to the same limit as the parent sequence. A sequence that does not converge, however, may have many convergent subsequences with various limits. We discover in this section what those limits are.

Theorem 6.7.1

Suppose X is a metric space, $z \in X$ and (x_n) is a sequence in X that converges to z. Suppose (x_{m_n}) is a subsequence of (x_n). Then the sequence (x_{m_n}) also converges to z.

Proof

Since every tail of (x_n) includes a tail of (x_{m_n}), this is implicit in the definition of convergence.   □

A sequence in a metric space can have at most one limit. It is clear, however, from looking at sequences such as $((-1)^n)$, that it may have more than one convergent subsequence and that such convergent subsequences may have different limits. Is there any general condition that determines which points of a metric space occur as limits of subsequences of a given sequence? Theorem 6.7.2 provides an answer.

Theorem 6.7.2

Suppose X is a metric space, $z \in X$ and (x_n) is a sequence in X. The following statements are equivalent:

(i)   There exists a subsequence of (x_n) that converges to z.

(ii) Every ball centred at z contains an infinite number of terms[1] of (x_n).

(iii) z is in the closure of every tail of (x_n).

(iv) Either $\{n \in \mathbb{N} \mid x_n = z\}$ is infinite or z is an accumulation point of every tail of (x_n).

(v)   Either $\{n \in \mathbb{N} \mid x_n = z\}$ is infinite or $z \in \mathrm{acc}(\{x_n \mid n \in \mathbb{N}\})$.

(vi) Either $\{n \in \mathbb{N} \mid x_n = z\}$ is infinite or z is an accumulation point of some subset of $\{x_n \mid n \in \mathbb{N}\}$.

Proof

Suppose first that there exists a subsequence of (x_n) that converges to z. Then every ball centred at z includes a tail of such a subsequence and therefore contains an infinite number of terms of (x_n). So (i) implies (ii).

Suppose (ii) holds. Suppose S is any tail of (x_n). For each $r \in \mathbb{R}^+$, the ball $\flat[z\,;r]$ contains a member of S so that $\mathrm{dist}(z\,,S) < r$ and therefore, since r is arbitrary in $\mathbb{R}^+$, $\mathrm{dist}(z\,,S) = 0$ and $z \in \overline{S}$ by 3.6.10. So (ii) implies (iii).

Suppose (iii) holds. If z is in every tail of (x_n), then $\{n \in \mathbb{N} \mid x_n = z\}$ is infinite. Otherwise, for each tail S of (x_n) with $z \notin S$ we have, by hypothesis, $z \in \overline{S}$, so that $z \in \mathrm{acc}(S)$ (3.6.8). All other tails include such tails S, whence z is an accumulation point of all tails of (x_n) (2.6.5). So (iii) implies (iv). It is obvious that (iv) implies (v) and that (v) implies (vi).

Suppose now that (vi) holds. If z occurs an infinite number of times as a term of (x_n), then (x_n) has a constant subsequence with all terms equal to z. This subsequence converges to z (6.1.5), so (i) is satisfied. If, on the other hand, z is an accumulation point of some subset of $\{x_n \mid n \in \mathbb{N}\}$, then $z \in \mathrm{acc}(\{x_n \mid n \in \mathbb{N}\})$ also, so that $\mathrm{dist}(z\,, \{x_n \mid n \in \mathbb{N}\} \setminus \{z\}) = 0$. We recursively define (B.19) a subsequence (x_{m_n}) of (x_n) as follows. Let m_1 be the smallest natural number such that $x_{m_1} \neq z$ and, for each $n \in \mathbb{N} \setminus \{1\}$, let m_n be the smallest natural number such that $0 < d(x_{m_n}, z) < \min\{1/n, d(x_{m_{n-1}}, z)\}$. Clearly $m_n > m_{n-1}$ for all $n \in \mathbb{N} \setminus \{1\}$, so that (x_{m_n}) is a subsequence of (x_n). By construction, the sequence $(d(x_{m_n}, z))$ converges to 0, so that (x_{m_n}) converges to z by 6.1.4. So (i) is satisfied in this case, too. We have thus shown that (vi) implies (i). $\square$

---

[1] The reader should note that when we say that a set contains *an infinite number of terms of a sequence*, it is the number of different indices that concerns us; in other words, each value is counted as many times as it occurs as a term of the sequence.

Corollary 6.7.3

Suppose X is a metric space and (x_n) is a sequence in X that has no sub-sequence that converges in X. Then every subset S of $\{x_n \mid n \in \mathbb{N}\}$ consists entirely of isolated points of S and is closed in X.

Proof

By 6.7.2, no subset of $\{x_n \mid n \in \mathbb{N}\}$ has any accumulation point in X. Then 3.6.8 and 2.6.4 give the result. $\square$

Example 6.7.4

Suppose (x_n) is a sequence of real numbers and $z = \limsup x_n$ is real (6.3.1). Let $r \in \mathbb{R}^+$. Then, by the definition of limit superior (6.3.1), there is some tail of (x_n) that has no point in the interval $[z + r, \infty)$, and that interval therefore contains only a finite number of terms of (x_n). In consequence, if the ball $\flat[z \, ; r)$ had in it only a finite number of terms of (x_n), we should have only a finite number of terms of (x_n) in the interval $(z - r, \infty)$, contradicting the fact that $z = \limsup x_n$. We deduce that $\flat[z \, ; r)$ contains an infinite number of terms of (x_n) and, since r is arbitrary in $\mathbb{R}^+$, it follows from 6.7.2 that (x_n) has a subsequence that converges to z. Similarly, if $\liminf x_n \in \mathbb{R}$, then (x_n) has a subsequence that converges to $\liminf x_n$.

The reader can check that (x_n) has no subsequence that converges to any number less than $\liminf x_n$ or greater than $\limsup x_n$ (Q 6.12). Of course, (x_n) converges if, and only if, $\limsup x_n$ and $\liminf x_n$ are equal and real (6.3.4).

Example 6.7.5

Consider the real sequence (x_n) given by $x_{2n-1} = (-1)^n/n$ and $x_{2n} = 1$ for each $n \in \mathbb{N}$. This sequence does not converge but has many convergent subsequences. Evidently, $\limsup x_n = 1$ and $\liminf x_n = 0$. The number 1 is not an accumulation point of $\{x_n \mid n \in \mathbb{N}\}$, but it occurs an infinite number of times as a term of (x_n). In fact, (x_{2n}) and all its subsequences, amongst others, converge to 1. Also, 0 is an accumulation point of $\{x_n \mid n \in \mathbb{N}\}$ and the subsequence (x_{2n+1}) of (x_n) converges to 0—note that whether or not 0 occurs as a term of the sequence is irrelevant here. The fact that there are no real numbers other than 0 and 1 that either occur an infinite number of times as terms of (x_n) or are accumulation points of $\{x_n \mid n \in \mathbb{N}\}$ means that 0 and 1 are the only numbers that occur as limits of subsequences of (x_n).

6.8 Cauchy Sequences

The definition of a convergent sequence uses the limit, so that, in order to use it to test a sequence for convergence, we need first to guess the limit. This situation is not wholly satisfactory. We should like an alternative formulation that does not mention the limit. We could, of course, use the second closure criterion of 6.1.2, saying simply that a sequence (x_n) converges if, and only if, $\bigcap \left\{ \overline{\{x_n \mid n \in S\}} \;\middle|\; S \subseteq \mathbb{N},\, S \text{ infinite} \right\}$ is non-empty, but this is a little unwieldy and we might prefer to stay closer to the actual definition. An attempt to do this could easily lead us to the definition given below of a Cauchy sequence. It turns out, however, that a Cauchy sequence need not converge (6.8.5). It is of great importance, therefore, to know which are the metric spaces in which we can be sure that Cauchy sequences converge; we shall see presently (6.11.3) that these are exactly the same metric spaces as those that are universally closed, the complete metric spaces (4.6.3). In a precise sense that we shall clarify in 6.11, a Cauchy sequence is one that ought to converge and fails to do so only because there is a hole in the space where its limit should be.

Definition 6.8.1

Suppose X is a metric space and (x_n) is a sequence in X. We say that (x_n) is a *Cauchy sequence* in X if, and only if, for every $r \in \mathbb{R}^+$, there is a ball of X of radius r that includes a tail of (x_n).

Theorem 6.8.2

Suppose X is a metric space. Every convergent sequence in X is Cauchy.

Proof

The ball criterion of 6.1.2 clearly implies the condition of 6.8.1.   □

A convergent sequence can have no subsequence converging to any limit but its own (6.7.1). Can a Cauchy sequence have convergent subsequences without itself converging to the same limit? We see in 6.8.3 that it cannot.

Theorem 6.8.3

Suppose X is a metric space and (x_n) is a Cauchy sequence in X. The following statements are equivalent:

(i) (x_n) converges in X.

(ii) (x_n) has a subsequence that converges in X.

Proof

Certainly, if (x_n) converges, then (x_n) has a convergent subsequence, namely itself. For the converse, we suppose that (x_n) has a convergent subsequence with limit z. Let $r \in \mathbb{R}^+$. Since (x_n) is Cauchy, there is a ball B of radius $r/2$ that includes a tail of (x_n). The distance from z to this tail is 0, by 6.7.2, so that $\operatorname{dist}(z, B) = 0$ also, by 2.3.1. From 5.1.10, we get $B \subseteq \flat[z\,;r]$. So $\flat[z\,;r]$ includes a tail of (x_n). Since r is arbitrary in $\mathbb{R}^+$, it follows from 6.1.2 that (x_n) converges to z. $\qquad\square$

Corollary 6.8.4

Suppose X is a metric space and (x_n) is a Cauchy sequence that does not converge in X. Then every subset of $\{x_n \mid n \in \mathbb{N}\}$ is closed in X.

Proof

By 6.8.3, (x_n) has no subsequence that converges in X, so 6.7.3 gives the result. $\qquad\square$

Example 6.8.5

We have already noted that in $\mathbb{R}$ with the usual metric, the sequence $(1/n)$ converges to 0, and that this sequence does not converge in the metric subspace $(0, 1]$ of $\mathbb{R}$. It is, however, Cauchy in $(0, 1]$ because, for each $r \in (0, 1]$, the ball $\flat_{(0,1]}[r\,;r)$ includes a tail of $(1/n)$.

Example 6.8.6

Here is a more elaborate example of a non-convergent Cauchy sequence than the one given in 6.8.5. Let X be the collection of continuous real functions defined on $[0, 1]$ with the metric d discussed in Q 1.13: for each $f, g \in X$, $d(f, g) = \int_0^1 |f(t) - g(t)|\, dt$. For each $n \in \mathbb{N}$ and $t \in [0, 1]$, let $f_n(t) = 1/\sqrt{t}$ if $1/n^2 \le t \le 1$ and $f_n(t) = tn^3$ otherwise. Then $f_n \in X$ for each $n \in \mathbb{N}$. Let $r \in \mathbb{R}^+$ and let $k \in \mathbb{N}$ be such that $2kr > 3$. For all $m \in \mathbb{N}$ with $m > k$, we have

$$f_m(t) - f_k(t) = \begin{cases} 0, & \text{when } 1/k^2 \le t \le 1 \\ t^{-1/2} - tk^3, & \text{when } 1/m^2 < t < 1/k^2 \\ t(m^3 - k^3), & \text{when } 0 \le t \le 1/m^2. \end{cases}$$

The graph of f_2.

Then $f_m(t) \geq f_k(t)$ for all $t \in [0, 1]$. A simple calculation yields

$$d(f_m, f_k) = \int_0^1 |f_m(t) - f_k(t)| \, dt = \int_0^1 f_m(t) - f_k(t) \, dt = \frac{3}{2k} - \frac{3}{2m} < r,$$

and it follows that the kth tail of (f_n) is included in $\flat[f_k \, ; r)$. Since r is arbitrary in $\mathbb{R}^+$, this means that (f_n) is Cauchy.

Does (f_n) converge in X? Let us suppose that it does and that $g \in X$ is its limit. Let $p \in \mathbb{N}$. Then, $f_n(t) = 1/\sqrt{t}$ for all $n \geq p$ and all $t \in [1/p^2, 1]$. So

$$d(g, f_n) = \int_0^1 |g(t) - f_n(t)| \, dt \geq \int_{1/p^2}^1 \left| g(t) - \frac{1}{\sqrt{t}} \right| \, dt \geq 0.$$

Since $d(g, f_n) \to 0$, this yields

$$\int_{1/p^2}^1 \left| g(t) - \frac{1}{\sqrt{t}} \right| \, dt = 0.$$

Because g is continuous, we get $g(t) - 1/\sqrt{t} = 0$ for all $t \in [1/p^2, 1]$ (see Q 1.13). This is true for all $p \in \mathbb{N}$, so we have $g(t) = 1/\sqrt{t}$ for all $t \in (0, 1]$. Therefore $g(t) \to \infty$ as $t \to 0$, so that, whatever real value g has at 0, g is not continuous at 0. This contradicts the assumption that $g \in X$.

Question 6.8.7

Metrics that produce the same topology produce also the same convergent sequences with the same limits (6.2.3). Do they produce the same Cauchy sequences? Let us examine this question. The reason that convergent sequences are preserved under the change of metric is that convergence depends only on the open subsets of the space. Cauchy sequences depend on a little more, namely the size of open balls. We have already observed (5.3.2) that even when two metrics on a single set produce exactly the same topology, balls produced by one of the metrics may have infinite diameter when measured by the other. Indeed, 5.3.2 furnishes us with a nice counterexample to our question. When $\mathbb{R}^+$ is endowed with the inverse metric $(a, b) \mapsto |a^{-1} - b^{-1}|$, the sequence (n) of natural numbers is Cauchy and the sequence $(1/n)$ of their inverses is not. This contrasts sharply with the situation in which $\mathbb{R}^+$ is given its usual Euclidean metric, despite the fact (5.3.2) that the two metrics produce the same topology and therefore the same convergent sequences.

6.9 Cauchy Sequences in Subspaces

We have observed that the sequence $(1/n)$, although it fails to converge in the subspace $(0,1]$ of $\mathbb{R}$, remains a Cauchy sequence in the subspace. This situation is typical, as we see in 6.9.1.

Theorem 6.9.1

Suppose X is a metric space and (x_n) is a Cauchy sequence in X. Then

(i) (x_n) is Cauchy in every superspace of X; and

(ii) (x_n) is Cauchy in every subspace of X that contains all terms of (x_n).

Proof

For each $w \in X$ and $r \in \mathbb{R}^+$, we have $\flat_X[w\,;r] \subseteq \flat_Y[w\,;r]$ for every superspace Y of X and (i) follows immediately.

Suppose Z is a subspace of X and $x_n \in Z$ for every $n \in \mathbb{N}$. Let $r \in \mathbb{R}^+$. Since (x_n) is Cauchy in X, there exists $a \in X$ such that $\flat_X[a\,;r/2]$ includes a tail of (x_n). Let $k \in \mathbb{N}$ be such that $x_k \in \flat_X[a\,;r/2]$. Then, by 5.1.10,

$$Z \cap \flat_X[a\,;r/2] \subseteq Z \cap \flat_X[x_k\,;r] = \flat_Z[x_k\,;r]\,,$$

so that the ball $\flat_Z[x_k\,;r]$ of Z includes a tail of (x_n). Since r is arbitrary in $\mathbb{R}^+$, it follows that (x_n) is Cauchy in Z. □

6.10 Cauchy Sequences in Product Spaces

In 6.5.1, we unveiled the very close relationship between convergence in a finite product space with a product metric (4.5.1) and convergence in the coordinate spaces. Cauchy sequences in product spaces do not behave quite so well. To get an analogous result, we restrict our attention in 6.10.1 to conserving metrics.

Theorem 6.10.1

Suppose $n \in \mathbb{N}$ and, for each $i \in \mathbb{N}_n$, (X_i, τ_i) is a metric space. Denote the product $\prod_{i=1}^{n} X_i$ by P. Endow P with any conserving metric e. Suppose $(x_m)_{m \in \mathbb{N}}$ is a sequence in P. Then (x_m) is Cauchy in P if, and only if, $(\pi_i(x_m))_{m \in \mathbb{N}}$ is Cauchy in X_i for every $i \in \mathbb{N}_n$, where π_i denotes the natural projection of P onto X_i.

Proof

Suppose first that (x_m) is Cauchy in the product. For each $r \in \mathbb{R}^+$, there exists a ball $\flat_P[z\,;r)$ that includes a tail of (x_m). For each $x_k \in \flat_P[z\,;r)$, we have, since e is conserving,

$$\tau_i(z_i, \pi_i(x_k)) \leq \sup\{\tau_j(z_j, \pi_j(x_k)) \mid j \in \mathbb{N}_n\} \leq e(z, x_k) < r$$

for all $i \in \mathbb{N}_n$, yielding $\pi_i(x_k) \in \flat_{X_i}[z_i\,;r)$, so that $\flat_{X_i}[z_i\,;r)$ includes a tail of $(\pi_i(x_m))_{m \in \mathbb{N}}$. But r is arbitrary in $\mathbb{R}^+$, so $(\pi_i(x_m))_{m \in \mathbb{N}}$ is Cauchy in X_i for each $i \in \mathbb{N}_n$.

Towards the converse, we suppose that $(\pi_i(x_m))_{m \in \mathbb{N}}$ is Cauchy in X_i for each $i \in \mathbb{N}_n$. So, for each $r \in \mathbb{R}^+$ and each $i \in \mathbb{N}_n$, there exist $w_i \in X_i$ and $k_i \in \mathbb{N}$ such that $\flat_{X_i}[w_i\,;r/n)$ includes the k_ith tail of $(\pi_i(x_m))$. Set $w = (w_1, \ldots, w_n)$. The set $\{k_i \mid i \in \mathbb{N}\}$ is finite and so has a maximum member (B.6.4); call it t. Now, because e is conserving, we have, for each $p \in \mathbb{N}$ with $p \geq t$,

$$e(x_p, w) \leq \sum_{i=1}^{n} \tau_i(\pi_i(x_p), w_i) < r,$$

so that $x_p \in \flat_P[w\,;r)$. Then $\flat_P[w\,;r)$ includes the tth tail of (x_m). But r is arbitrary in $\mathbb{R}^+$, so (x_m) is Cauchy in P.   □

Example 6.10.2

The problem with trying to get a version of 6.10.1 for non-conserving product metrics is that we may have no way of guaranteeing that the radius of a ball in the product is related to the radii of balls in the coordinate spaces. This lack of relationship has nothing particular to do with products; indeed, we have already observed (5.3.2) that even when two metrics on a single set produce exactly the same topology, balls produced by one of the metrics may have infinite diameter in relation to the other. This fact was crucial in our example (6.8.7) of metrics that give the same topology but not the same Cauchy sequences. Drawing on 6.8.7, we can easily find a product with a product metric that does not have the property of 6.10.1. We proceed as follows.

Begin with the space $\mathbb{R}^+$ with its usual metric, form the trivial product $\mathbb{R}^+$ itself (there is just one space in this product), and endow this product with the inverse metric $(a, b) \mapsto |a^{-1} - b^{-1}|$. This is not a conserving metric but it is a product metric because, as we saw in 5.3.2, it produces the same topology as the usual metric. However, as we observed in 6.8.7, its Cauchy sequences differ from those of the original space, namely $\mathbb{R}^+$ with the usual metric.

6.11 Forcing Convergence of Cauchy Sequences

We have said that the only reason that a Cauchy sequence might not converge is that what ought to be its limit is in some sense missing from the space (6.8). We make this statement precise and confirm it now in 6.11.1.

Theorem 6.11.1

Suppose (X, d) is a metric space and (a_n) is a Cauchy sequence in X that does not converge in X. Suppose $z \notin X$ and let $X' = X \cup \{z\}$. Then X' can be made into a metric superspace of X in which (a_n) converges to z.

Proof

The tails of (a_n) are closed in X (6.8.4); moreover, they form a nest (B.2.3) and, since (a_n) is Cauchy, the infimum of their diameters is 0. Since (a_n) does not converge, it certainly has no constant subsequence (6.8.3), so that the intersection of the tails of (a_n) is empty. Therefore, by 4.7.1, X' can be made into a metric superspace of X in which $z \in \mathrm{Cl}_{X'}\{a_n \mid n \in \mathbb{N}\}$. Since $z \notin X$, 3.6.8 implies that $z \in \mathrm{acc}_{X'}(\{a_n \mid n \in \mathbb{N}\})$, and 6.7.2 then implies that some subsequence of (a_n) converges to z in X'. But (a_n) is Cauchy in X' (6.9.1), so, by 6.8.3, (a_n) itself converges to z in X'. $\qquad\square$

Example 6.11.2

With reference to 6.8.6, define $h: [0, 1] \to \mathbb{R}$ by $h(t) = 1/\sqrt{t}$ for $t \in (0, 1]$ and $h(0) = \alpha$ for arbitrary $\alpha \in \mathbb{R}$. h is not a continuous function, so $h \notin X$. h is, however, integrable on $[0, 1]$. Let $X' = X \cup \{h\}$, and extend the metric of X to X' by using exactly the same formula as before: $d(f, g) = \int_0^1 |f(t) - g(t)| \, dt$ for all $f, g \in X'$. The sequence (f_n) of 6.8.6 then converges in X' to h. In this example, instead of using an arbitrary z and extending the metric as we did in the general theorem 6.11.1, we have constructed a specific function h and made the extension of d conform to the formula for d itself; h is, however, not unique, in that α may be taken to be any real number at all. Why does this not involve a contradiction to the uniqueness of limits in a metric space?

By 6.11.1, every non-convergent Cauchy sequence in a metric space can be made to converge by the addition of a single extra point to the space and by a judicious extension of the metric. We now ask whether or not it is possible to extend a metric space in such a way that all non-convergent Cauchy sequences are simultaneously made to converge. We should like our extension to be minimal in the sense that it does not use any more extra points than are essential, and we should like also to be certain that no new non-convergent Cauchy sequences are produced by our extension. All of this is possible, as we shall show by and by (10.12.2). We content ourselves now by demonstrating that a necessary and sufficient condition for completeness (4.6.3) of a metric space X is that every Cauchy sequence in X converge in X. This, in fact, is the standard definition of completeness.

Theorem 6.11.3

Suppose X is a metric space. Then X is complete if, and only if, every Cauchy sequence in X converges in X.

Proof

Suppose that X is closed in every metric superspace and that (x_n) is a Cauchy sequence in X. If (x_n) failed to converge in X, it would converge in a metric superspace of X, by 6.11.1, and, since X is closed in that superspace, the limit would be in X by 6.6.3, yielding a contradiction. So (x_n) converges in X.

For the converse, suppose that every Cauchy sequence in X converges in X and that Y is an arbitrary metric superspace of X. Suppose $z \in \mathrm{Cl}_Y(X)$. By 6.6.2, there exists a sequence (x_n) in X that converges to z in Y. Then (x_n) is Cauchy in Y by 6.8.2 and therefore also in X by 6.9.1, and so converges in X by hypothesis. So $z \in X$ because (x_n) cannot have two limits in Y (6.2.1). $\square$

Summary

In this chapter, we have introduced convergent and Cauchy sequences in an arbitrary metric space and have developed criteria for convergence of sequences and subsequences. We have identified the points that are limits of subsequences of any given sequence. We have shown how Cauchy sequences can be made to converge. We have shown the relationship between convergence and closure. We have also demonstrated the equivalence of two criteria for the very important concept of *completeness*, a concept that we are going to discuss in detail in Chapter 10.

EXERCISES

†Q 6.1   Find a non-convergent sequence (x_n) of real numbers that satisfies
$$\bigcap\{\mathrm{tail}_k(x) \mid k \in \mathbb{N}\} = \{0\}.$$

Q 6.2 Define a real sequence recursively by the following equations: $x_1 = 0$, $x_{2n} = x_{2n-1}/2$ and $x_{2n+1} = x_{2n} + 1/2$ for each $n \in \mathbb{N}$. Find $\limsup x_n$ and $\liminf x_n$.

Q 6.3 Suppose (x_n) is a sequence in $\mathbb{R}$ and $k \in \mathbb{N}$. Show that

(i) $\limsup x_n = \inf\{\sup \text{tail}_n(x) \mid n \in \mathbb{N},\ n \geq k\}$;

(ii) $\liminf x_n = \sup\{\inf \text{tail}_n(x) \mid n \in \mathbb{N},\ n \geq k\}$; and

(iii) $\liminf x_n \leq \limsup x_n$.

†Q 6.4 Suppose that (x_n) is a sequence of positive real numbers. Show that
$$\liminf \frac{x_{n+1}}{x_n} \leq \liminf x_n^{\frac{1}{n}} \leq \limsup x_n^{\frac{1}{n}} \leq \limsup \frac{x_{n+1}}{x_n}.$$

Q 6.5 For each $n \in \mathbb{N}$, define x_n to be $1/n^2$ if n is odd and to be $2/n^2$ if n is even. Show that $\liminf \dfrac{x_{n+1}}{x_n} < \limsup x_n^{\frac{1}{n}} < \limsup \dfrac{x_{n+1}}{x_n}$.

†Q 6.6 Suppose (x_n) is a sequence of real numbers. Show that $x_n \to \infty$ if, and only if, $\liminf x_n = \infty$ and that, in that case, $\limsup x_n = \infty$ also. Show also that $x_n \to -\infty$ if, and only if, $\limsup x_n = -\infty$ and that, in that case, $\liminf x_n = -\infty$ also.

†Q 6.7 (ROOT TEST) Suppose $\sum_{n \in \mathbb{N}} a_n$ is a series with terms that are all in $\mathbb{R}^+$. Show that $\sum_{n \in \mathbb{N}} a_n$ converges if $\limsup a_n^{1/n} < 1$ and does not converge if $\limsup a_n^{1/n} > 1$.

†Q 6.8 (RATIO TEST) Suppose $\sum_{n \in \mathbb{N}} a_n$ is a series with terms that are all in $\mathbb{R}^+$. Show that $\sum_{n \in \mathbb{N}} a_n$ converges if $\limsup(a_{n+1}/a_n) < 1$ and does not converge if $\liminf(a_{n+1}/a_n) > 1$.

Q 6.9 For what values of $z \in \mathbb{C}$ does the sequence (z^n) converge in $\mathbb{C}$? For those values of z, what is $\lim z^n$?

Q 6.10 Find a metric space (X, d) and a sequence (x_n) in X that has no convergent subsequence in X but for which the infimum of the set $\{d(x_m, x_n) \mid m, n \in \mathbb{N},\ m \neq n\}$ is zero.

Q 6.11 Suppose X is a metric space, $z \in X$ and (x_n) is a sequence in X. Show that if X has a subsequence that converges to z, then $\text{dist}(z, \{x_n \mid n \in \mathbb{N}\}) = 0$, and show also that the converse need not be true.

Q 6.12 Suppose (x_n) is a sequence in $\mathbb{R}$. Show that no subsequence of (x_n) converges to any number less than $\liminf x_n$ or greater than $\limsup x_n$.

†Q 6.13 Suppose X is a metric space and $S \subseteq X$. Show that S is dense in X if, and only if, for each $x \in X$, there is a sequence in S that converges to x in X.

†Q6.14 As in 6.5.2, let $P = \prod_{i=1}^{\infty} X_i$, where $X_i = [0,1]$ for each $i \in \mathbb{N}$, and endow P with the supremum metric $(x,y) \mapsto \sup\{|x_i - y_i| \mid i \in \mathbb{N}\}$. For each $m \in \mathbb{N}$, let $a_m \in P$ and suppose the sequence of sequences $(a_m)_{m \in \mathbb{N}}$ converges in P. Does the sequence $(\pi_i(a_m))_{m \in \mathbb{N}}$ converge in $[0,1]$ for each $i \in \mathbb{N}$, where π_i denotes the natural projection of P onto X_i?

Q6.15 Construct a metric on $\mathbb{R}$ in which the sequence $(1/n)$ of inverses of natural numbers converges to a limit other than 0.

7
Bounds

The terms *unbounded* and *infinite* tend to be used in common parlance as if
they meant the same thing. Mathematically, they are quite different. A non-
empty subset of a metric space, whether it is finite or infinite, is *bounded* if,
and only if, its diameter is finite. Some bounded sets have an even stronger
property called *total boundedness*. A function is bounded if, and only if, its
range is bounded, and sets of bounded functions can be endowed with a very
natural metric.

7.1 Bounded Sets

We present in 7.1.1 three equivalent conditions, any of which can be used in a
definition of boundedness. Here, as so often happens in this abstract theory, we
must take some care because the meaning of the term *bounded* is dependent on
the metric. A set that is bounded with respect to one metric may be unbounded
with respect to another.

Theorem 7.1.1 (Criteria for Boundedness)

Suppose X is a non-empty metric space, $z \in X$ and $S \subseteq X$. The following
statements are equivalent:

(i) $\operatorname{diam}(S) < \infty$.

(ii)   There is a ball of X centred at z that includes S.

(iii)   There is a ball of X that includes S.

Proof

The empty set has diameter $-\infty$ and is included in every ball of X. Suppose therefore that $S \neq \varnothing$. If $\operatorname{diam}(S)$ is finite, let $a \in S$. Then, for each $x \in S$, we have $d(x, z) \leq d(x, a) + d(a, z) \leq \operatorname{diam}(S) + d(a, z)$, so that $x \in \flat[z\,; \operatorname{diam}(S) + d(a, z)]$. So (i) implies (ii). Clearly, (ii) implies (iii). Finally, if S is included in a ball $\flat[w\,; r]$ of X, then $\operatorname{diam}(S) \leq 2r$ (5.1.10), so that (iii) implies (i). $\qquad\square$

Definition 7.1.2

A subset S of a metric space X is called a *bounded subset* of X if, and only if, $S = X = \varnothing$ or S is included in some ball of X. A metric space X is said to be *bounded* if, and only if, it is a bounded subset of itself.

Now any one of the criteria of 7.1.1 may be used as a necessary and sufficient condition for boundedness.

Example 7.1.3

By 2.1.5, the bounded intervals of $\mathbb{R}$ are the degenerate ones together with those of type $(a\,, b)$, $[a\,, b)$, $(a\,, b]$ and $[a\,, b]$, where a and b are real numbers and $a < b$. A subset of $\mathbb{R}$ is bounded if, and only if, it is included in a bounded interval (5.1.2).

Example 7.1.4

Even quite simply stated metrics can turn unbounded sets into bounded ones and vice versa. When $\mathbb{R}^+$ is endowed with the metric $(a, b) \mapsto |a^{-1} - b^{-1}|$, the set $\mathbb{N}$ of natural numbers has diameter 1 and the set $\{1/n \mid n \in \mathbb{N}\}$ of inverses of the natural numbers is unbounded (Q 7.4).

Example 7.1.5

Every ball is bounded. Every subset of a bounded set is bounded; in particular, an intersection of sets is bounded if one of the sets is bounded. The closure of every bounded set is bounded (Q 7.3). Every finite set is bounded and every finite union of bounded sets is bounded (Q 7.1). These are simple facts about any metric space and it is left to the reader to check them.

7.2 Finite Products of Bounded Sets

We show next that finite products of bounded sets are bounded, provided the metric on the product space is conserving (7.2.1).

Theorem 7.2.1

Suppose $n \in \mathbb{N}$ and, for each $i \in \mathbb{N}_n$, (X_i, τ_i) is a metric space. Denote the product $\prod_{i=1}^{n} X_i$ by P and suppose S is a subset of P. Endow P with any conserving metric e. Then S is bounded in P if, and only if, $\pi_i(S)$ is bounded in X_i for all $i \in \mathbb{N}_n$, where π_i denotes the natural projection of P onto the coordinate space X_i (B.13.3).

Proof

Suppose first that S is bounded in P. Then, since e is conserving, we have $\tau_i(a_i, b_i) \leq e(a, b) \leq \operatorname{diam}_P(S)$ for all $i \in \mathbb{N}_n$ and all $a, b \in S$. It follows that $\operatorname{diam}_{X_i}(\pi_i(S)) \leq \operatorname{diam}_P(S) < \infty$ for all $i \in \mathbb{N}_n$. For the converse, we suppose that $\pi_i(S)$ is bounded for each $i \in \mathbb{N}_n$. Then, for each $a, b \in S$, we have $e(a, b) \leq \sum_{i=1}^{n} \tau_i(a_i, b_i) \leq \sum_{i=1}^{n} \operatorname{diam}_{X_i}(\pi_i(S)) < \infty$, which then yields $\operatorname{diam}_P(S) < \infty$. $\qquad\square$

Example 7.2.2

A subset S of $\mathbb{R}^3$ with a conserving metric is bounded if, and only if, S can be enclosed in a rectangular box the sides of which are of finite length.

Example 7.2.3

Product metrics that are not conserving need not preserve boundedness. The inverse metric $(a, b) \mapsto \left| a^{-1} - b^{-1} \right|$ is a product metric on the trivial product $\mathbb{R}^+$ but does not produce the same bounded sets as the Euclidean metric (7.1.4).

7.3 The Hausdorff Metric

In 1.1.18 and 2.7.4, we toyed with the idea of defining a metric on a collection of subsets of a metric space in terms of the original metric. In 1.1.18, we succeeded in doing this for the closed bounded intervals of $\mathbb{R}$. At the same time we recognized that the specified metric could not be extended to unbounded intervals or to intervals that are not closed. The question we ask now is whether or not this idea can be extended to all non-empty closed bounded subsets of an arbitrary metric space X. In short, is there a metric on the collection of all non-empty closed bounded subsets of a metric space that mimics, on the singleton sets, the original metric? The answer is that there is.

Theorem 7.3.1

Suppose X is a non-empty set and d is a metric on X. Let $\mathcal{S}(X)$ denote the collection of all non-empty closed bounded subsets of X. For each A and B in $\mathcal{S}(X)$, define $h(A, B)$ to be $\max\{\sup\{\operatorname{dist}(b, A) \mid b \in B\}, \sup\{\operatorname{dist}(a, B) \mid a \in A\}\}$. Then h is a metric on $\mathcal{S}(X)$. It is called the *Hausdorff metric*.

Proof

h is non-negative, and its symmetry is built into its definition. If $A, B \in \mathcal{S}(X)$ and $h(A, B) = 0$, then $\sup\{\operatorname{dist}(b, A) \mid b \in B\} = 0$, so that $\operatorname{dist}(b, A) = 0$ for all $b \in B$, which, in turn, yields $B \subseteq \overline{A}$. Also, $\sup\{\operatorname{dist}(a, B) \mid a \in A\} = 0$, which similarly yields $A \subseteq \overline{B}$. Since A and B are closed, we then have $A = B$.

Towards the triangle inequality for h, suppose $A, B, C \in \mathcal{S}(X)$. Let $r \in \mathbb{R}^+$ and $u \in A$. There exists $v \in B$ such that $d(u, v) \leq \operatorname{dist}(u, B) + r/2$. Then also there exists $w \in C$ such that $d(v, w) \leq \operatorname{dist}(v, C) + r/2$. So

$$
\begin{aligned}
\operatorname{dist}(u, C) &\leq d(u, w) \\
&\leq d(u, v) + d(v, w) \\
&\leq \operatorname{dist}(u, B) + \operatorname{dist}(v, C) + r \\
&\leq h(A, B) + h(B, C) + r.
\end{aligned}
$$

Since u is arbitrary in A, $\sup\{\operatorname{dist}(a, C) \mid a \in A\} \leq h(A, B) + h(B, C) + r$. Because r is arbitrary in $\mathbb{R}^+$ and $\inf \mathbb{R}^+ = 0$, it follows from this that $\sup\{\operatorname{dist}(a, C) \mid a \in A\} \leq h(A, B) + h(B, C)$. Using a similar argument, we get $\sup\{\operatorname{dist}(c, A) \mid c \in C\} \leq h(A, B) + h(B, C)$, and the two inequalities together give $h(A, C) \leq h(A, B) + h(B, C)$, as required. $\qquad\square$

Example 7.3.2

Consider the discs $A = \{z \in \mathbb{C} \mid |z| \leq 2\}$ and $B = \{z \in \mathbb{C} \mid |z| \leq 1\}$. These are closed bounded subsets of $\mathbb{C}$. To calculate the Hausdorff distance between them, we must first of all look at $\operatorname{dist}(a, B)$ and $\operatorname{dist}(b, A)$ for all $a \in A$ and $b \in B$. Since $B \subseteq A$, $\operatorname{dist}(b, A) = 0$ for all $b \in B$. For each $a \in A \cap B$, we also have $\operatorname{dist}(a, B) = 0$. But for $a \in A \backslash B$ we get $\operatorname{dist}(a, B) = |a| - 1$. The supremum of all these values is 1. So $h(A, B) = 1$.

7.4   Spaces of Bounded Functions

A function into a metric space is said to be *bounded* if its range is bounded; real and complex bounded functions are of special interest. Given any set X, with or without a metric, and a metric space Y, the set of all bounded functions from X into Y is normally endowed with the supremum metric as its standard metric.

Definition 7.4.1

Suppose X is a set, Y is a metric space and $f\colon X \to Y$. Then f is called a *bounded function* from X to Y if, and only if, $f(X)$ is a bounded subset of Y. The set of bounded functions from X to Y will be denoted by $B(X, Y)$.

Theorem 7.4.2

Suppose X is a non-empty set and (Y, e) is a non-empty metric space. The function $s\colon B(X, Y) \times B(X, Y) \to \mathbb{R}$ given by the formula $s(f, g) = \sup\{e(f(x), g(x)) \mid x \in X\}$ for all $f, g \in B(X, Y)$ is a metric on $B(X, Y)$. (This *supremum metric* is the standard metric on $B(X, Y)$ and we shall assume that $B(X, Y)$ is endowed with it unless we state otherwise.)

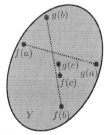

$e(f(x), g(x))$ for $x = a, b, c \in X$

Proof

We must check first that s is a real function; in other words, that it does not take the value ∞. Suppose $f, g \in B(X, Y)$. Let $w \in Y$ and let $\alpha, \beta \in \mathbb{R}^+$ be such that $f(X) \subseteq \flat_Y[w\,;\alpha]$ and $g(X) \subseteq \flat_Y[w\,;\beta]$ (7.1.1). Then, for each $x \in X$, $e(f(x), g(x)) \leq e(f(x), w) + e(w, g(x)) \leq \alpha + \beta$, so that $s(f, g) \leq \alpha + \beta \in \mathbb{R}$. So s is a real function. Certainly s is symmetric, and s is non-negative because e is non-negative. If $s(f, g) = 0$, then $e(f(x), g(x)) = 0$ for all $x \in X$, so that $f(x) = g(x)$ for all $x \in X$ and therefore $f = g$. Towards the triangle inequality, suppose $f, g, h \in B(X, Y)$. Then, for all $x \in X$, we have

$$e(f(x), g(x)) \leq e(f(x), h(x)) + e(h(x), g(x)) \leq s(f, h) + s(h, g).$$

Since this holds for all $x \in X$, we then have $s(f, g) \leq s(f, h) + s(h, g)$, as required. $\square$

Example 7.4.3

If X is any non-empty set, we can form the metric space $B(X, \mathbb{R})$. The fact that $\mathbb{R}$ is equipped with a well-defined metric is sufficient to establish that $B(X, \mathbb{R})$ is a metric space. The metric has nothing whatever to do with any metric that may or may not be applied to X. This situation is analogous to that mentioned in B.20.7; indeed, the observation made in B.20.7 is applicable here because $\mathbb{R}$ has algebraic structure as well as metric structure. Just as the metric structure of $\mathbb{R}$ bestows on $B(X, \mathbb{R})$ a standard metric, so the algebraic structure of $\mathbb{R}$ bestows on $B(X, \mathbb{R})$ algebraic structure: functions $f, g \in B(X, \mathbb{R})$ are added and multiplied using the formulae $(f + g)(x) = f(x) + g(x)$ and $(fg)(x) = f(x)g(x)$ and are multiplied by scalars λ using the formula $(\lambda f)(x) = \lambda f(x)$, all the resulting functions $f + g$, fg and λf being members of $B(X, \mathbb{R})$. Simply put, $B(X, \mathbb{R})$, as well as being a metric space, is a subalgebra of the algebra (B.20.4) of all real functions with domain X. An analogous statement can be made for $B(X, \mathbb{C})$.

Example 7.4.4

Sequences are functions (B.18.1); bounded sequences are bounded functions. The metric space of real bounded sequences $B(\mathbb{N}, \mathbb{R})$, with the linear structure bestowed on it by $\mathbb{R}$, is denoted by $\ell_\infty(\mathbb{R})$ or simply by ℓ_∞. The supremum metric of 7.4.2, namely $(x, y) \mapsto \sup\{|x_n - y_n| \mid n \in \mathbb{N}\}$, is the usual metric on ℓ_∞. It is determined (1.7.2) by the norm $x \mapsto \sup\{|x_n| \mid n \in \mathbb{N}\}$, which we usually denote, as in the finite case, by $\|\cdot\|_\infty$.

Example 7.4.5

There are a number of distinguished subsets of ℓ_∞. The set of real sequences $x = (x_n)$ for which the series $\sum_{n=1}^{\infty} |x_n|$ converges is usually denoted by $\ell_1(\mathbb{R})$, or simply by ℓ_1, and is endowed with a norm quite different from that of its superset ℓ_∞: for each $x \in \ell_1$, we define $\|x\|_1 = \sum_{n=1}^{\infty} |x_n|$. Similarly, the set of real sequences $x = (x_n)$ for which the series $\sum_{n=1}^{\infty} |x_n|^2$ converges is usually denoted by $\ell_2(\mathbb{R})$, or simply by ℓ_2, and is endowed with yet another norm: for each $x \in \ell_2$, we define $\|x\|_2 = \sqrt{\sum_{n=1}^{\infty} |x_n|^2}$. That these functions and many similar ones are norms is a consequence of a general theorem that we shall present in Chapter 12 (12.11.3, Q 12.25).

7.5 Attainment of Bounds

Some bounded real functions attain both a maximum and a minimum value. Amongst these are functions that have closed range in $\mathbb{R}$.

Definition 7.5.1

Suppose X is a set and $f\colon X \to \mathbb{R}$. We say that f *attains its bounds* if, and only if, f is bounded and there exist $a, b \in X$ such that $f(a) = \inf f(X)$ and $f(b) = \sup f(X)$.

Evidently, a real function attains its bounds if, and only if, it has a maximum and a minimum value in $\mathbb{R}$. This happens for every real bounded function that has closed range (7.5.2) but happens also for some functions that do not have closed range (Q 7.5).

Theorem 7.5.2

Suppose X is a non-empty set and $f\colon X \to \mathbb{R}$ is bounded and has closed range in $\mathbb{R}$. Then f attains its bounds.

Proof

By 2.2.5, both $\inf \operatorname{ran}(f)$ and $\sup \operatorname{ran}(f)$ are zero distance from $\operatorname{ran}(f)$. Because $\operatorname{ran}(f)$ is closed in $\mathbb{R}$, they are therefore members of $\operatorname{ran}(f)$ (3.6.10).   □

7.6   Convergence and Boundedness

All Cauchy sequences are bounded, but the converse is not true.

Theorem 7.6.1

Suppose (X, d) is a metric space and (x_n) is a sequence in X.

(i)   If (x_n) is Cauchy, then (x_n) is bounded in X.

(ii)   If (x_n) converges in X, then (x_n) is bounded in X.

Proof

If (x_n) is Cauchy, then there is a ball $\flat[a\,;1]$ of radius 1 that includes a tail of (x_n). So the set $\{d(a, x_n) \mid n \in \mathbb{N},\ x_n \notin \flat[a\,;1]\}$ is finite. If it is empty, we let $m = 1$; otherwise, it has a maximum real value (B.6.4) and we let m be this value. Then $x_n \in \flat[a\,;m]$ for all $n \in \mathbb{N}$. This proves (i); (ii) follows immediately because convergent sequences are Cauchy (6.8.2).   □

Example 7.6.2

Bounded sequences need not be Cauchy; an example is the sequence $((-1)^n)$.
So the collection of convergent real sequences is a proper subset of the collection
of bounded real sequences. It is often denoted by $c(\mathbb{R})$, or simply by c, and its
usual metric is the supremum metric $(a, b) \mapsto \sup\{|a_n - b_n| \mid n \in \mathbb{N}\}$, making
it a metric subspace of $\ell_\infty(\mathbb{R})$ (7.4.4). A smaller metric subspace of $\ell_\infty(\mathbb{R})$ is
the space $c_0(\mathbb{R})$, or c_0, of real sequences that converge to 0—the *null sequences*.
These spaces inherit algebraic structure as well as metric structure from $\ell_\infty(\mathbb{R})$,
and there are corresponding metric subspaces $c(\mathbb{C})$ and $c_0(\mathbb{C})$ of $\ell_\infty(\mathbb{C})$ with
similar algebraic structure.

7.7   Uniform and Pointwise Convergence

The metric spaces $B(X, Y)$ of bounded functions are spaces in which the points
are functions, and those functions have values that belong to the metric space
Y. This prompts us to ask whether or not there is a relationship between
convergence of a sequence of functions in $B(X, Y)$ and convergence in Y of the
sequences of values of those functions at specified points in their domain. The
fundamental result in this area is 7.7.1.

Theorem 7.7.1

Suppose X is a non-empty set and (Y, e) is a non-empty metric space. Endow
the collection $B(X, Y)$ of bounded functions from X to Y with its usual supremum metric (7.4.2), which we label s. Suppose (f_n) is a sequence in $B(X, Y)$
that converges to g in $B(X, Y)$. Then, for each $z \in X$, the sequence $(f_n(z))$
converges to $g(z)$ in Y.

Proof

For each $z \in X$, $0 \le e(f_n(z), g(z)) \le \sup\{e(f_n(x), g(x)) \mid x \in X\} = s(f_n, g)$.
Since $s(f_n, g)$ converges to 0 by 6.1.4, it follows that $e(f_n(z), g(z))$ also converges to 0. Then, again by 6.1.4, $f_n(z) \to g(z)$, as required. $\square$

The converse of 7.7.1 is not in general true. It is certainly not true if the
sequence (f_n) is unbounded in $B(X, Y)$ (7.7.2), but it may fail even when (f_n)
is a bounded sequence (7.7.3). The best we can say in this regard is that if
a bounded sequence in $B(X, Y)$ has the property that each of the sequences
$(f_n(z))$ converges in Y, then the function $z \mapsto \lim f_n(z)$ must be a member of
$B(X, Y)$ (7.7.4).

Example 7.7.2

The clearest reason for the failure of the converse of 7.7.1 is that a sequence (f_n) in $B(X,Y)$ may be unbounded and yet have the property that $(f_n(x))$ converges in Y for each $x \in X$; this happens, for example, when, for each $n \in \mathbb{N}$, $f_n \colon \mathbb{R}^+ \to \mathbb{R}$ is given by $f_n(x) = n$ if $x \in (0, 1/n)$ and $f_n(x) = 1/x$ if $x \in [1/n, \infty)$. This sequence (f_n), being unbounded, cannot possibly converge in $B(X,Y)$ (7.6.1), but it is nonetheless true that $f_n(x) \to 1/x$ for all $x \in \mathbb{R}^+$.

Example 7.7.3

Even a bounded sequence (f_n) in some $B(X,Y)$ may have the property that there exists $g \in B(X,Y)$ such that $f_n(x) \to g(x)$ for all $x \in X$ without this entailing the convergence of (f_n) to g. For each $n \in \mathbb{N}$, define $f_n \colon [0,1] \to \mathbb{R}$ by $f_n(x) = x^n$ and let $g \in B([0,1], \mathbb{R})$ be given by $g(1) = 1$ and $g(x) = 0$ for all $x \in [0,1)$. Then $(f_n(x))$ converges to $g(x)$ for all $x \in [0,1]$. Note that g is bounded, a fact ensured by 7.7.4 below. However, $\sup\{|f_n(x) - g(x)| \mid x \in [0,1]\} = 1$ for

The graphs of f_2, f_3, f_4 and f_8.

all $n \in \mathbb{N}$, so that (f_n) does not converge to g in $B([0,1], \mathbb{R})$. Because limits in $\mathbb{R}$ are unique, 7.7.1 ensures that (f_n) cannot converge to anything other than g in $B([0,1], \mathbb{R})$. So (f_n) does not converge in $B([0,1], \mathbb{R})$.

Theorem 7.7.4

Suppose X is a non-empty set and (Y, e) is a metric space. Suppose (f_n) is a bounded sequence in $B(X,Y)$, with its usual supremum metric s, such that, for each $x \in X$, the sequence $(f_n(x))$ converges in Y. Then the function $g \colon X \to Y$ given by $g(x) = \lim f_n(x)$ for each $x \in X$ is bounded.

Proof

Let $r \in \mathbb{R}^+$ be such that $\operatorname{diam}_{B(X,Y)}(\{f_n \mid n \in \mathbb{N}\}) < r$. Then, for each $x \in X$, $\operatorname{diam}_Y(\{f_n(x) \mid n \in \mathbb{N}\}) < r$ and, since $g(x) \in \overline{\{f_n(x) \mid n \in \mathbb{N}\}}$ (6.1.2), we have $e(g(x), f_1(x)) < r$ (3.6.11) and therefore $\operatorname{dist}_Y(g(x), f_1(X)) < r$. Since this is true for all $x \in X$ and $f_1(X)$ is bounded, it follows that $g(X)$ is bounded in Y and therefore that $g \in B(X,Y)$, as required.   □

Example 7.7.5

The metric used on $B(X,Y)$ in 7.7.1 is significant. Let us consider the subset of $B([0,1], \mathbb{R})$ consisting of all continuous bounded real functions defined on $[0,1]$. We shall label this set $\mathcal{C}([0,1])$ and mark such sets for discussion later in the book (8.9.1). Because the members of $\mathcal{C}([0,1])$ are continuous, they can all be integrated. This permits us to endow $\mathcal{C}([0,1])$ with a metric different from the standard supremum metric inherited from $B([0,1],\mathbb{R})$, namely the integral metric $(h,k) \mapsto \int_0^1 |h(x) - k(x)|\, dx$ (Q 1.13). If a sequence (f_n) in $\mathcal{C}([0,1])$ converges to a function g in $\mathcal{C}([0,1])$ with respect to the supremum metric, then it follows, by 7.7.1, that $(f_n(z))$ converges to $g(z)$ in $\mathbb{R}$ for all $z \in [0,1]$ because $\mathcal{C}([0,1])$ with the supremum metric is a metric subspace of $B([0,1],\mathbb{R})$. If, however, (f_n) converges to g in $\mathcal{C}([0,1])$ with respect to the integral metric, then there is no guarantee that $f_n(z) \to g(z)$ in $\mathbb{R}$ for all $z \in [0,1]$. Here is a counterexample. Consider the functions f_n defined, for each $n \in \mathbb{N}$, by

$$f_n(t) = \begin{cases} 1 - nt, & \text{if } 0 \le t < 1/n \\ 0, & \text{if } 1/n \le t \le 1. \end{cases}$$

Each of these functions is in $\mathcal{C}([0,1])$ and, for all $n \in \mathbb{N}$,

$$\int_0^1 f_n(t)\, dt = \int_0^{1/n} 1 - nt\, dt = \frac{1}{2n},$$

(0,1)

(1/7, 0)   (1, 0)

The graph of f_7.

so that, with respect to the integral metric, (f_n) converges to the zero function. Does $(f_n(z))$ converge to 0 for all $z \in [0,1]$? Not only does it not do so, but there is, in fact, a different bounded function g, not continuous, such that $f_n(z) \to g(z)$ for all $z \in [0,1]$; g is the function that has value 1 at 0 and value 0 on $(0,1]$.

Convergence of the sequence of values of a sequence of functions at every point of their common domain is generally called *pointwise convergence* to distinguish it from convergence of the sequence of functions, which is then styled *uniform convergence*. These terms allow us to extend the ideas we have studied in this section slightly to incorporate some sequences of unbounded functions (7.7.6).

Definition 7.7.6

Suppose X is a non-empty set and Y is a non-empty metric space. Suppose (f_n) is a sequence of functions from X to Y and $g \colon X \to Y$. We say that

- (f_n) *converges pointwise* to g if, and only if, $(f_n(z))$ converges to $g(z)$ in Y for all $z \in X$; and
- (f_n) *converges uniformly* to g if, and only if, $\sup\{e(f_n(x),g(x)) \mid x \in X\}$ is real for each $n \in \mathbb{N}$ and the sequence $(\sup\{e(f_n(x),g(x)) \mid x \in X\})_{n \in \mathbb{N}}$ converges to zero in $\mathbb{R}$.

Note 7.7.7

Definition 7.7.6 encompasses sequences of bounded functions and sequences of unbounded functions as well. A requirement of the definition is that the functions be bounded *relative to* the limit function; in particular, it is not possible for a sequence of unbounded functions to converge uniformly to a bounded function (Q 7.9). Of course, if the functions are bounded, then uniform convergence is exactly the same thing as convergence in the space of bounded functions; indeed, it is easy to check that the uniform limit of a sequence of bounded functions is necessarily bounded (Q 7.6). But the definition allows us to consider sequences of functions that are not bounded when we need to do so. The proof of 7.7.8 below is exactly the same as the proof of 7.7.1.

Theorem 7.7.8

Suppose X is a non-empty set and Y is a non-empty metric space. Suppose (f_n) is a sequence of functions from X to Y and $g\colon X \to Y$. If (f_n) converges uniformly to g, then (f_n) converges pointwise to g.

Example 7.7.9

Let $r \in \mathbb{R}^{\oplus}$. For each $n \in \mathbb{N}$, define f_n on (r, ∞) by $f_n(x) = 1/nx$. Then (f_n) converges pointwise to the zero function whatever the value of r and so cannot converge uniformly to any function other than the zero function. In fact, if $r > 0$, the functions are all bounded and (f_n) does converge uniformly to 0. If, however, $r = 0$, the functions f_n are all unbounded and so the sequence (f_n) cannot converge uniformly to the bounded function 0 (Q 7.9).

7.8   Totally Bounded Sets

We sometimes make use of a concept stronger than boundedness called *total boundedness*. Whereas a set is bounded if it is included in a single ball, it is totally bounded if it can be covered (see B.11.1) by a finite set of balls of arbitrarily small radius. We shall see in 7.8.2 that the concept is closely related to that of Cauchy sequences. The claim that total boundedness is a stronger concept than mere boundedness must, of course, be verified (7.8.1).

Theorem 7.8.1

Suppose (X, d) is a metric space and S is a subset of X that is covered by a finite collection of balls of X. Then S is bounded in X.

Proof

If S is empty, then certainly it is bounded. Suppose $S \neq \varnothing$. Suppose that $n \in \mathbb{N}$ and $\{\flat[a_i\,; r_i] \mid i \in \mathbb{N}_n\}$ covers S. Let $s = \max\{d(a_1, a_i) + r_i \mid i \in \mathbb{N}_n\}$. Then, for each $z \in S$, there exists $j \in \mathbb{N}_n$ such that $z \in \flat[a_j\,; r_j]$, so that $d(a_1, z) \leq d(a_1, a_j) + d(a_j, z) \leq d(a_1, a_j) + r_j \leq s$. Therefore $z \in \flat[a_1\,; s]$ and, because z is arbitrary in S, $S \subseteq \flat[a_1\,; s]$.   □

Theorem 7.8.2 (Criteria for Total Boundedness)

Suppose (X, d) is a metric space and S is a subset of X. The following statements are equivalent:

(i)   (INTERNAL CRITERION) For each $r \in \mathbb{R}^+$, there is a finite collection of balls of S of radius r that covers S.

(ii) (GLOBAL CRITERION) For each $r \in \mathbb{R}^+$, there is a finite collection of balls of X of radius r that covers S.

(iii) (CAUCHY CRITERION) Every sequence in S has a Cauchy subsequence.

Proof

It is clear that (i) implies (ii). Suppose that S satisfies (ii). Then S is bounded by 7.8.1. Suppose (x_n) is an arbitrary sequence in S. For each infinite subset J of $\mathbb{N}$, we shall, for convenience, denote the set $\{x_n \mid n \in J\}$ by $\Xi(J)$. $\Xi(J)$, being a subset of S, is bounded and can, by hypothesis, be covered by a finite number of balls of radius $\operatorname{diam}(\Xi(J))/4$. Since J is infinite and the number of balls is finite, at least one of the balls contains x_n for an infinite number of values of $n \in J$; in other words, such a ball includes $\Xi(K)$ for some infinite subset K of J. It follows that the domain of the relation

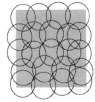

A set S covered by a finite number of balls of equal radius. Note that any sequence in S must have an infinite number of terms in at least one of the balls.

$$\{(J, K) \mid K \subseteq J \subseteq \mathbb{N},\ J, K\ \text{infinite}, \operatorname{diam}(\Xi(K)) \leq \operatorname{diam}(\Xi(J))/2\}$$

consists of all infinite subsets of $\mathbb{N}$ and therefore includes its range. So, by the Axiom of Dependent Choice (B.19.2), there exists a sequence (I_n) of infinite subsets of $\mathbb{N}$ such that $I_{n+1} \subseteq I_n$ and $\operatorname{diam}(\Xi(I_{n+1})) \leq \operatorname{diam}(\Xi(I_n))/2$ for each $n \in \mathbb{N}$. It then follows by induction that $\operatorname{diam}(\Xi(I_n)) \leq \operatorname{diam}(S)/2^{n-1}$ for all $n \in \mathbb{N}$. Now let $m_1 = \min I_1$ and, for each $n \in \mathbb{N}$ with $n > 1$, let $m_n = \min(I_n \backslash \{m_k \mid k \in \mathbb{N}_{n-1}\})$, thus inductively producing an increasing subsequence (m_n) of $\mathbb{N}$. Then (x_{m_n}) is a subsequence of (x_n). We claim that (x_{m_n}) is Cauchy.

Suppose $r \in \mathbb{R}^+$ is arbitrary. Let $p \in \mathbb{N}$ be such that $\text{diam}(S) < 2^{p-1}r$. The pth tail of (x_{m_n}) is, by the way it was formed, included in $\Xi(I_p)$, and, since $\text{diam}(\Xi(I_p)) \leq \text{diam}(S)/2^{p-1} < r$, this tail is included in a ball of radius r. Since r is arbitrary in $\mathbb{R}^+$, this means that (x_{m_n}) is a Cauchy subsequence of (x_n), as claimed. Therefore (ii) implies (iii).

We now show that (iii) implies (i). Suppose that S does not satisfy (i), and let $r \in \mathbb{R}^+$ be such that no finite collection of balls of S of radius r covers S. Then the domain of the relation $\{(F, F \cup \{z\}) \mid F \subset S,\, F \text{ finite}, \text{dist}(z, F) > r\}$ consists of all finite subsets of S, and so certainly includes the range. By the Axiom of Dependent Choice (B.19.2), there exists a sequence (F_n) of finite subsets of X such that, for each $n \in \mathbb{N}$, $F_{n+1} \backslash F_n$ is a singleton set and its only element, which we shall label x_n, satisfies $\text{dist}(x_n, F_n) > r$. Then each ball of X of radius $r/2$ contains at most one term of (x_n), and so certainly does not include a tail of any subsequence of (x_n). Therefore (x_n) has no Cauchy subsequence. So S does not satisfy (iii). It follows that (iii) implies (i). □

Definition 7.8.3

A subset S of a metric space X is called a *totally bounded subset* of X if, and only if, for each $r \in \mathbb{R}^+$, there is a finite collection of balls of X of radius r that covers S. A metric space X is said to be *totally bounded* if, and only if, it is a totally bounded subset of itself.[1]

Following this definition, any one of the equivalent conditions of 7.8.2 can now be regarded as a criterion for total boundedness.

Example 7.8.4

It is an easy but useful consequence of 7.8.2 that any sequence with terms that can all be included in a finite number of balls of arbitrarily small radius necessarily has a Cauchy subsequence. It need not, of course, have convergent subsequences; the sequence $(1/n)$ in $(0, 1]$ is a counterexample.

There are special spaces in which the concepts of boundedness and total boundedness coincide. This happens on the real line, for example (7.8.5). We shall see presently (7.11.1) that the concepts coincide in many spaces other than $\mathbb{R}$, in particular in all spaces that have the nearest-point property that we discussed in 2.8.

---

[1] Note that the empty space is totally bounded because it is covered by an empty collection of balls.

Theorem 7.8.5

Every bounded subset of $\mathbb{R}$ is totally bounded.

Proof

Suppose S is a bounded subset of $\mathbb{R}$, and let $a, b \in \mathbb{R}$ be such that $a < b$

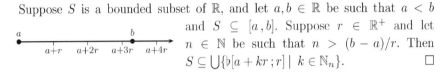

and $S \subseteq [a, b]$. Suppose $r \in \mathbb{R}^+$ and let $n \in \mathbb{N}$ be such that $n > (b - a)/r$. Then $S \subseteq \bigcup \{ b[a + kr\,;r] \mid k \in \mathbb{N}_n \}$.   □

Example 7.8.6

Bounded subsets of metric spaces need not be totally bounded. Every infinite subset of a space with the discrete metric is bounded but not totally bounded.

7.9   Total Boundedness in Subspaces and Superspaces

Total boundedness carries over to subspaces and to finite unions.

Theorem 7.9.1

Suppose X is a totally bounded metric space and $A \subseteq X$. Then

(i) A is totally bounded in every metric superspace of A; and

(ii) for each metric subspace Z of X, $A \cap Z$ is totally bounded in Z.

Proof

Every sequence in A or in $A \cap Z$ is a sequence in X. Then 6.9.1 and the Cauchy criterion for total boundedness give the result.   □

Theorem 7.9.2

Suppose X is a metric space and $\mathcal{C}$ is a finite collection of totally bounded subsets of X. Then $\bigcup \mathcal{C}$ is also totally bounded.

Proof

Suppose $r \in \mathbb{R}^+$. For each $A \in \mathcal{C}$, there is a finite collection $\mathcal{B}_A$ of balls of X of radius r such that $A \subseteq \bigcup \mathcal{B}_A$. Then $\{B \in \mathcal{B}_A \mid A \in \mathcal{C}\}$ is finite and covers $\bigcup \mathcal{C}$. Since r is arbitrary in $\mathbb{R}^+$, $\bigcup \mathcal{C}$ is totally bounded.   □

7.10 Total Boundedness in Product Spaces

Product metrics need not preserve total boundedness, but conserving metrics always do.

Example 7.10.1

Consider the infinite product space P discussed in 6.5.2. The set $\{e(m) \mid m \in \mathbb{N}\}$ given there is bounded in the space P because each $e(n)$ is of distance 1 from the zero sequence. The set $\{e(m) \mid m \in \mathbb{N}\}$, however, is not totally bounded because the distance between distinct elements of it is 1. In 7.10.2, we see that finite products with conserving metrics are much better behaved, but, as in 6.10.2, it is important that the metric not be merely a product metric.

Theorem 7.10.2

Suppose $n \in \mathbb{N}$ and, for each $i \in \mathbb{N}_n$, (X_i, τ_i) is a metric space. Denote the product $\prod_{i=1}^n X_i$ by P and suppose S is a subset of P. Endow P with any conserving metric e. Then S is totally bounded in P if, and only if, $\pi_i(S)$ is totally bounded in X_i for each $i \in \mathbb{N}_n$, where each π_i is the natural projection of P onto the coordinate space X_i (B.13.3).

Proof

If any of the coordinate spaces is empty, then $P = S = \varnothing$ and the theorem holds, so we suppose otherwise.

Suppose that $\pi_i(S)$ is totally bounded for each $i \in \mathbb{N}_n$. Let $r \in \mathbb{R}^+$. For each $i \in \mathbb{N}_n$, let $\mathcal{B}_i$ be a finite collection of balls of X_i of radius $r/2n$ that covers $\pi_i(S)$. Consider the collection $\{\prod_{i=1}^n A_i \mid A_i \in \mathcal{B}_i\}$ of subsets of P. It is finite because each $\mathcal{B}_i$ is finite. It covers $\prod_{i=1}^n \pi_i(S)$ and therefore also S because $S \subseteq \prod_{i=1}^n \pi_i(S)$. Each member of $\{\prod_{i=1}^n A_i \mid A_i \in \mathcal{B}_i\}$ has diameter at most r because each A_i has diameter at most r/n (5.1.10) and the metric is conserving; specifically, for $u, v \in \prod_{i=1}^n A_i$, where $A_i \in \mathcal{B}_i$ for each $i \in \mathbb{N}_n$, we have $e(u, v) \leq \sum_{i=1}^n \tau_i(u_i, v_i) \leq \sum_{i=1}^n \operatorname{diam}(A_i) \leq r$. The members of $\{\prod_{i=1}^n A_i \mid A_i \in \mathcal{B}_i\}$ may not be balls of P, but, since each has diameter at most r, each is included in a ball of radius r. So a finite collection of balls of P of radius r covers S.

For the converse, suppose S is totally bounded in P. Suppose $r \in \mathbb{R}^+$ and let $\mathcal{C}$ be a finite collection of balls of radius $r/2$ that covers S. Then, for each $j \in \mathbb{N}_n$, $\{\pi_j(D) \mid D \in \mathcal{C}\}$ is a finite collection of subsets of X_j each of which has diameter at most r because the metric is conserving; specifically, if $D \in \mathcal{C}$, then, for each $u, v \in D$ and $j \in \mathbb{N}_n$, we have $\tau_j(u_j, v_j) \leq \sup\{\tau_i(u_i, v_i) \mid i \in \mathbb{N}_n\} \leq e(u, v) \leq r$.

So each member of $\{\pi_j(D) \mid D \in \mathcal{C}\}$ is included in a ball of X_j of radius r. But $\{\pi_j(D) \mid D \in \mathcal{C}\}$ covers $\pi_j(S)$, and therefore so does any such collection of balls. $\qquad\square$

Corollary 7.10.3

Suppose $n \in \mathbb{N}$ and, for each $i \in \mathbb{N}_n$, (X_i, τ_i) is a non-empty metric space. Denote the product $\prod_{i=1}^{n} X_i$ by P. Endow P with any conserving metric. Then every bounded subset of P is totally bounded if, and only if, for all $i \in \mathbb{N}_n$, every bounded subset of X_i is totally bounded.

Proof

Because the metric is conserving, this equivalence follows easily using 7.10.2 and 7.2.1. $\qquad\square$

Example 7.10.4

Suppose $n \in \mathbb{N}$ and endow $\mathbb{R}^n$ with a conserving metric. Then every bounded subset of $\mathbb{R}^n$ is totally bounded by 7.8.5 and 7.10.3. The same is true for bounded subsets of $\mathbb{C}^n$ for the same reason.

7.11   Solution to the Nearest-Point Problem

We present now a solution to the problem that arose in 2.8.2 of discovering the circumstances in which nearest points have to exist, incorporating the headway that was already made in 2.8.3. We prove that the non-empty metric spaces that always admit nearest points are precisely those in which all bounded sequences have convergent subsequences. We give also a number of other conditions that are equivalent to this (7.11.1), but our list is not exhaustive, and our exploration of the idea is not complete—we shall return to the matter in 12.6.1.

Theorem 7.11.1 (Criteria for the Nearest-Point Property)

Suppose (X, d) is a metric space. The following statements are equivalent:

(i)   (NEAREST-POINT CRITERION) $X = \varnothing$ or X admits a nearest point to each point in every metric superspace of X.

(ii)   (POINTLIKE CRITERION) Every pointlike function on X attains its minimum value on X.

(iii) (BW CRITERION) Every infinite bounded subset of X has an accumulation point in X.

(iv) (CONVERGENCE CRITERION) Every bounded sequence of X has a subsequence that converges in X.

(v) (CAUCHY CRITERION) X is complete and every bounded subset of X is totally bounded.

Proof

That (i) and (ii) are equivalent was shown in 2.8.3.

Suppose (ii) holds. Suppose S is an infinite bounded subset of X, and choose a sequence (x_n) of distinct terms in S (B.19.3). Then (x_n) is a bounded sequence because S is bounded; let $\alpha \in \mathbb{R}^+$ be such that $\operatorname{diam}(\{x_n \mid n \in \mathbb{N}\}) < 2\alpha$. Define $u: X \to \mathbb{R}^{\oplus}$ by $u(w) = \alpha + \inf\{d(x_n, w) + 1/n \mid n \in \mathbb{N}\}$ for all $w \in X$. We show that u is a pointlike function. Suppose $p, q \in X$ and $\epsilon \in \mathbb{R}^+$. Then there exist $i, j \in \mathbb{N}$ such that $u(p) \geq \alpha + d(x_i, p) + 1/i - \epsilon/2$ and $u(q) \geq \alpha + d(x_j, q) + 1/j - \epsilon/2$, which, because $2\alpha > d(x_i, x_j)$, yield

$$u(p) + u(q) \geq d(x_i, p) + d(x_i, x_j) + d(x_j, q) + 1/i + 1/j - \epsilon > d(p, q) - \epsilon$$

and, because $\alpha + d(x_i, q) + 1/i \geq u(q)$, yield also

$$u(p) + d(p, q) \geq \alpha + d(x_i, p) + 1/i - \epsilon/2 + d(p, q) \geq \alpha + d(x_i, q) + 1/i - \epsilon \geq u(q) - \epsilon.$$

Since ϵ is arbitrary in $\mathbb{R}^+$, these inequalities give $u(p) + u(q) \geq d(p, q)$ and $u(p) + d(p, q) \geq u(q)$, respectively, which together establish that u is a pointlike function.

For each $n \in \mathbb{N}$, we have $\alpha \leq \inf u(X) \leq u(x_n) \leq \alpha + 1/n$, whence $\inf u(X) = \alpha$. By hypothesis, there is a point $w \in X$ such that $u(w) = \alpha$, from which it follows that $\inf\{d(w, x_n) + 1/n \mid n \in \mathbb{N}\} = 0$. Then, for each $r \in \mathbb{R}^+$, $\flat[w\,;r)$ contains some term of (x_n) not equal to w because the x_n are all distinct. Therefore $w \in \operatorname{acc}(S)$. So (ii) implies (iii).

Now suppose that (iii) holds and that (x_n) is a bounded sequence in X. Then either $\{x_n \mid n \in \mathbb{N}\}$ is finite or, by hypothesis, has an accumulation point in X, so that, by 6.7.2, (x_n) has a convergent subsequence. It follows that (iii) implies (iv).

Next, suppose that (iv) holds. Then every Cauchy sequence in X, being bounded (7.6.1), has a convergent subsequence, and so converges by 6.8.3. So X is complete by 6.11.3. Also, if S is a bounded subset of X, then every sequence in S, being bounded, has a convergent subsequence, which is certainly Cauchy (6.8.2), so that S is totally bounded (7.8.2). Therefore (iv) implies (v).

Last, we suppose that (v) holds and prove (ii). Suppose u is a pointlike function on X, and let $s = \inf u(X)$. Then, for each $n \in \mathbb{N}$, the set

$A_n = \{x \in X \mid u(x) < s + 1/n\}$ is non-empty. Choose a sequence (a_n) in X with $a_n \in A_n$ for each $n \in \mathbb{N}$ (B.19.1). Thus $u(a_n) < s + 1/n$ for each $n \in \mathbb{N}$. Then $d(a_n, a_1) \leq u(a_n) + u(a_1) \leq 2s + 2$ for all $n \in \mathbb{N}$, so that $\{a_n \mid n \in \mathbb{N}\}$ is bounded and therefore totally bounded by hypothesis. Since (a_n) is a sequence in $\{a_n \mid n \in \mathbb{N}\}$, there is a Cauchy subsequence (a_{m_n}) of (a_n) (7.8.2). Then (a_{m_n}) converges in X because X is complete (6.11.3). Let $w = \lim a_{m_n}$. Let $\epsilon \in \mathbb{R}^+$. Then there exists $k \in \mathbb{N}$ such that $1/m_k < \epsilon/2$ and $d(a_{m_k}, w) < \epsilon/2$, which together yield $u(w) \leq u(a_{m_k}) + d(a_{m_k}, w) < s + \epsilon$. Since ϵ is arbitrary in $\mathbb{R}^+$ and $s \leq u(w)$, we deduce that $u(w) = s$. So (v) implies (ii) and the proof is finished. □

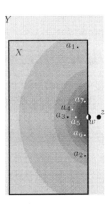

Definition 7.11.2

Suppose X is a metric space. We shall say that X has the *nearest-point property* if, and only if, $X = \varnothing$ or X admits a nearest point to each point in every metric superspace of X.[2]

Because of 7.11.1, we can now say that a metric space has the nearest-point property if, and only if, it satisfies any one of the criteria listed there.

Example 7.11.3

$\mathbb{R}$ has the nearest-point property by 2.8.4 and 7.11.1.

Example 7.11.4

The subset $\mathbb{R}\setminus\{0\}$ of $\mathbb{R}$ has the property that every bounded subset of it, being a bounded subset of $\mathbb{R}$, is totally bounded, but there is no nearest point of $\mathbb{R}\setminus\{0\}$ to $i \in \mathbb{C}$. So the first part of the Cauchy criterion in 7.11.1 is not sufficient to ensure that nearest points always exist. The second part is not sufficient either. For a counterexample, consider any infinite set with the discrete metric. It has no accumulation point and so does not have the nearest-point property, but every Cauchy sequence converges in such a space, quite simply because the only Cauchy sequences are those sequences that are eventually constant.

---

[2] Such spaces are also said to be *boundedly compact*, a term that may be misleading because they need not be compact (9.1.4). We do not use the term in this book.

7.12 Subspaces with the Nearest-Point Property

Every closed subspace of a space with the nearest-point property also has the property, but no other subspace does.

Theorem 7.12.1

Suppose X is a metric space that has the nearest-point property and $S \subseteq X$. Then S has the nearest-point property if, and only if, S is closed in X.

Proof

Suppose S is closed in X and $I \subseteq S$. Then, by 2.6.5, $\mathrm{acc}_X(I) \subseteq \mathrm{acc}_X(S)$ and, since S is closed in X, $\mathrm{acc}_X(S) \subseteq S$ by 4.1.2. So, by 2.6.6, we have $\mathrm{acc}_X(I) = \mathrm{acc}_S(I)$. If I is bounded and infinite, then $\mathrm{acc}_X(I) \neq \varnothing$ because X has the nearest-point property, so $\mathrm{acc}_S(I) \neq \varnothing$, whence S satisfies the BW criterion of 7.11.1 and thus has the nearest-point property.

The converse is true by the Cauchy criterion of 7.11.1 because S must be closed in X in order to be complete. □

Example 7.12.2

Every closed subset of $\mathbb{R}$ has the nearest-point property. Closed intervals have the property, and the Cantor set, being closed in $\mathbb{R}$, also has the property. $\mathbb{N}$ has the nearest-point property. Note that in this case many of the criteria of 7.11.1 are vacuously satisfied. Open intervals of $\mathbb{R}$, other than $\mathbb{R}$ itself, do not have the nearest-point property because they are not closed in $\mathbb{R}$.

Example 7.12.3

There are metric spaces that do not have the nearest-point property but have a similar property in a particular space. The most important examples are the non-empty closed convex subsets of ℓ_2 (5.4.1, 7.4.5). Most of these do not have the nearest-point property, but every such set admits a nearest point to each point of ℓ_2 itself. Moreover, these nearest points are unique (Q 10.16).

7.13 Products with the Nearest-Point Property

Every product of metric spaces with the nearest-point property has the same property if the metric on the product is a conserving one.

Theorem 7.13.1

Suppose $n \in \mathbb{N}$ and, for each $i \in \mathbb{N}_n$, (X_i, τ_i) is a non-empty metric space. Endow the product $P = \prod_{i=1}^{n} X_i$ with a conserving metric. Then P has the nearest-point property if, and only if, X_i has the nearest-point property for all $i \in \mathbb{N}_n$.

Proof

Because the metric is conserving, it is easy to check, using 6.5.1 and 6.10.1, that P is complete (6.11.3) if, and only if, the X_i are all complete (see 10.5.1). Then 7.10.3 and the Cauchy criterion of 7.11.1 clinch the matter. □

Example 7.13.2 (Bolzano–Weierstrass Theorem)

For each $n \in \mathbb{N}$, $\mathbb{R}^n$ with a conserving metric has the property that every bounded infinite subset has an accumulation point: because $\mathbb{R}$ has the nearest-point property (7.11.3), so has $\mathbb{R}^n$, and $\mathbb{R}^n$ therefore satisfies the BW criterion of 7.11.1. Similarly, $\mathbb{C}^n$ with a conserving metric has the nearest-point property. All closed subspaces of these spaces have the nearest-point property by 7.12.1 and therefore also satisfy the BW criterion.

Example 7.13.3

The Bolzano–Weierstrass Theorem for $\mathbb{R}^n$ (7.13.2) is dependent on the fact that $\mathbb{R}^n$ is finite-dimensional (see 12.10.2). The infinite-dimensional linear space $\ell_\infty(\mathbb{R})$ of real bounded sequences with the supremum metric does not have the nearest-point property; indeed, the set S of all real sequences that have exactly one term equal to 1 and all other terms equal to 0 is infinite and has diameter 1 but clearly has no accumulation point in the space.

Example 7.13.4

When the metric on $\mathbb{R}^n$ is merely a product metric, a bounded sequence may fail to have a convergent subsequence and the space thus fails to have the nearest-point property. This failure may occur even when $n = 1$. For example, the sequence of negative integers $(-m)$ is bounded in $\mathbb{R}$ with the metric $(a, b) \mapsto \left| e^b - e^a \right|$ of 1.1.11, but it has no convergent subsequence in $\mathbb{R}$ with that metric. The set of negative integers is, with respect to this metric, both closed and bounded and consists entirely of isolated points.

Example 7.13.5

Every finite-dimensional real normed linear space has the nearest-point property. Since every such space is isomorphic to $\mathbb{R}^n$, where n is its dimension (B.22.5), this will follow immediately (13.3) once we know that all norms on $\mathbb{R}^n$ produce the same topology—such norms are said to be *equivalent* (13.5). It is, however, possible to prove it independently using induction (Q 7.23).

Summary

In this chapter, we have examined the concept of boundedness for subsets of a metric space and the stronger concept of total boundedness. We have introduced bounded functions. We have also developed the Hausdorff metric for measuring distances between non-empty closed bounded subsets of any metric space, and we have presented a solution to the nearest-point problem.

EXERCISES

Q 7.1   Suppose X is a metric space and $\mathcal{C}$ is a finite collection of bounded subsets of X. Show that $\bigcup \mathcal{C}$ is bounded in X.

†Q 7.2   Show that a product of bounded sets need not be bounded when it is endowed with a product metric.

Q 7.3   Suppose X is a metric space and S is a bounded subset of X. Show that $\overline{S}$ is bounded in X.

†Q 7.4   Show that, when $\mathbb{R}^+$ is endowed with the metric $(a, b) \mapsto |a^{-1} - b^{-1}|$ (1.1.12), $\mathbb{N}$ is a bounded subset and $\{1/n \mid n \in \mathbb{N}\}$ is an unbounded subset of $\mathbb{R}^+$.

Q 7.5   Find a bounded function $f \colon \mathbb{R} \to \mathbb{R}$ that attains its bounds but does not have closed range.

†Q 7.6   Suppose X is a non-empty set and (Y, e) is a non-empty metric space. Suppose (f_n) is a sequence in $B(X, Y)$ that converges uniformly to g. Show that $g \in B(X, Y)$.

Q 7.7   Does the sequence (f_n) of 7.7.5 converge in $B([0, 1], \mathbb{R})$?

Q 7.8   Endow the collection $\mathcal{S}$ of closed bounded subsets of $[0, 1]$ with its Hausdorff metric h. Find a sequence (F_n) of finite subsets of $[0, 1]$ that converges to the uncountable set $[0, 1]$ in $(\mathcal{S}, h)$.

†Q 7.9 Show that a sequence of unbounded functions cannot converge uniformly to a bounded function.

Q 7.10 Show by example that a sequence of unbounded functions can converge pointwise to a bounded function.

†Q 7.11 Show by example that the inclusions $c_0(\mathbb{R}) \subset c(\mathbb{R}) \subset \ell_\infty(\mathbb{R})$ are all strict (7.6.2).

Q 7.12 Consider the sequence (f_n) of real functions defined on $[0,1]$ by the equation $f_n(x) = x^n/n$. Show that (f_n) converges uniformly to 0 on $[0,1]$. Show also that the sequence (f'_n) of derivatives does not converge uniformly.

Q 7.13 Let (f_n) be the sequence of real functions defined on $[0,1]$ by setting $f_n(x) = 2nx$ if $0 \le x \le 1/n$ and $f_n(x) = 2n(1-x)/(n-1)$ otherwise. Determine whether or not (f_n) converges uniformly on $[0,1]$.

Q 7.14 For each $n \in \mathbb{N}$, define $g_n \colon [0,\infty) \to \mathbb{R}$ by $g_n(x) = nx/(1+nx)$. Show that (g_n) is uniformly convergent on $[a,\infty)$ for each $a \in \mathbb{R}^+$ but is not uniformly convergent on $[0,\infty)$.

Q 7.15 Suppose (f_n) and (g_n) are uniformly convergent sequences of real-valued functions defined on a metric space X. Prove that (f_n+g_n) is uniformly convergent on X.

†Q 7.16 Find uniformly convergent sequences (f_n) and (g_n) of real functions for which $(f_n g_n)$ does not converge uniformly. For uniform convergence to fail, is it necessary that both (f_n) and (g_n) be unbounded?

Q 7.17 Suppose X is a metric space and S is a totally bounded subset of X. Show that $\overline{S}$ is totally bounded in X.

†Q 7.18 Show that ℓ_∞ is not separable (Q 4.26).

†Q 7.19 Show that every totally bounded metric space is separable.

Q 7.20 Suppose X is a metric space that is covered by a countable collection of totally bounded subsets of X. Show that X is separable.

†Q 7.21 Suppose (X,d) is an unbounded metric space. Show that there exists a sequence in X that has no convergent subsequence.

†Q 7.22 Suppose X is a metric space with the nearest-point property and (x_n) is a bounded sequence in X. Suppose that the limits of all convergent subsequences of (x_n) are equal. Show that (x_n) converges to that limit.

†Q 7.23 Prove using induction that every finite-dimensional normed linear space has the nearest-point property.

8
Continuity

Those properties which, in the theory of ordinary numerical functions, arise only in the handling of the most complicated, most difficult, least practical functions, are the characteristics that are the simplest, most practical, in fact indispensable attributes of functions of generalized variables. Maurice Fréchet, 1878–1973

Most functions are not easy to picture. Those that are most readily presented in pictorial form to the mind are real functions defined on an interval of the real line; we imagine a graph. Some graphs are distinguished by their unbroken nature—it is possible to draw them without taking the pencil from the page. These graphs stand out from those others that are made up of several parts. Mathematicians try to reflect this unbroken nature in the mathematical concept of *continuity*. The attempt to do so centres around the idea that, given any point z in the domain of a function f, unbrokenness at z means that $f(x)$ does not stray far from $f(z)$ provided x is kept sufficiently close to z in the domain. The precise formulation of this idea captures with brilliance the intuitive notion of unbrokenness for real functions defined on an interval. It also draws into the net of *continuous functions* many, but not all, of those functions—such as the tangent function—with graphs that are broken only because their domains are broken, *disconnected*, subsets of the real line. Inevitably, in the more abstract setting of arbitrary metric spaces, it draws in also many strange functions, some of which challenge any intuitive notion of continuity we might have.

8.1 Local Continuity

Let us recall the concept of *continuity at a point* for a real function f defined on the interval $[0,1]$. The rather elaborate standard definition is that f is continuous at $a \in [0,1]$ if, and only if:

- For every $\epsilon \in \mathbb{R}^+$, there exists $\delta \in \mathbb{R}^+$ such that, for all $x \in [0,1]$ with $|x - a| < \delta$, we have $|f(x) - f(a)| < \epsilon$.

Neither ϵ nor δ is necessary here; we can state exactly the same condition as:

- For every open interval V centred at $f(a)$, there is an open interval U centred at a such that $f(U \cap [0,1]) \subseteq V$.

Moreover, looking at the diagram, it is not difficult to believe that this statement is equivalent to:

- For every open subset V of $\mathbb{R}$ with $f(a) \in V$, there is an open subset U of $[0,1]$ with $a \in U$ such that $f(U) \subseteq V$.

We want to define the concept of continuity at a point for an arbitrary function $f \colon X \to Y$, where (X, d) and (Y, e) are metric spaces. We replace $[0,1]$ and $\mathbb{R}$ by X and Y, respectively, in the formulations given above. The standard epsilon–delta definition can then be modified by replacing $|x - a|$ by $d(x, a)$ and $|f(x) - f(a)|$ by $e(f(x), f(a))$. The second formulation can be modified by replacing intervals by balls. The third statement needs no modification at all and has the further advantage that it can be used in more abstract areas of mathematics where neither distance nor balls are available; we therefore adopt it as our standard definition (8.1.2). But first we show that, for a function between metric spaces, it is equivalent to a number of other statements, any one of which can therefore be thought of as characterizing continuity at a point.

Theorem 8.1.1 (Criteria for Continuity at a Point)

Suppose (X, d) and (Y, e) are metric spaces, $z \in X$ and $f \colon X \to Y$. The following statements are equivalent:

(i)   (EPSILON–DELTA CRITERION) For every $\epsilon \in \mathbb{R}^+$, there exists $\delta \in \mathbb{R}^+$ such that, for all $x \in X$ with $d(x, z) < \delta$, we have $e(f(x), f(z)) < \epsilon$.

(ii) (EPSILON–DELTA BALL CRITERION) For every $\epsilon \in \mathbb{R}^+$, there exists $\delta \in \mathbb{R}^+$ such that $f(\flat[z\,;\delta]) \subseteq \flat[f(z)\,;\epsilon]$.

(iii) (OPEN BALL CRITERION) Each open ball of Y that contains $f(z)$ includes the image under f of an open ball of X that contains z.

(iv) (OPEN SET CRITERION) For each open subset V of Y with $f(z) \in V$, there exists an open subset U of X with $z \in U$ such that $f(U) \subseteq V$.

(v)   (CONVERGENCE CRITERION) For every sequence (x_n) of X that converges in X to z, the sequence $(f(x_n))$ of Y converges in Y to $f(z)$.

Proof

The epsilon–delta conditions are merely rewordings of each other and, since every open ball that contains $f(z)$ includes a ball centred at $f(z)$ (5.2.2), (ii) implies (iii). Since every open subset of Y that contains $f(z)$ includes an open ball centred at $f(z)$ (5.2.2) and every open ball of X is an open subset of X, (iii) implies (iv).

To show that (iv) implies (v), we proceed as follows. We assume (iv) holds. Suppose that (x_n) is a sequence in X that converges in X to z. Let V be any open subset of Y with $f(z) \in Y$. By hypothesis, there exists an open subset U of X with $z \in U$ such that $f(U) \subseteq V$. Since (x_n) con-

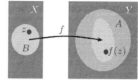

Open ball criterion. $f(z)$ is in ball A, and B is a ball that contains z and has $f(B) \subseteq A$.

verges to z, U includes a tail of (x_n) (6.1.3); therefore $f(U)$, and hence also V, includes a tail of $(f(x_n))$. But V is an arbitrary open subset of Y with $f(z) \in V$, so $(f(x_n))$ converges to $f(z)$ by definition (6.1.3).

Finally, we show that (v) implies (i). Suppose (i) is not satisfied. Then there exists $\epsilon \in \mathbb{R}^+$ such that for every $\delta \in \mathbb{R}^+$ there is some $x \in X$ with $d(x,z) < \delta$ and $e(f(x), f(z)) \geq \epsilon$. In particular, for each $n \in \mathbb{N}$, the set

$$A_n = \{x \in X \mid d(x,z) < 1/n \text{ and } e(f(x), f(z)) \geq \epsilon\}$$

is non-empty. Choose a sequence (a_n) in X with $a_n \in A_n$ for each $n \in \mathbb{N}$ (B.19.1). Clearly (a_n) converges to z, whereas $(f(a_n))$ does not converge to $f(z)$. So (v) is not satisfied. Thus the desired implication holds. $\square$

Definition 8.1.2

Suppose X and Y are metric spaces, $z \in X$ and $f : X \to Y$. We say that f is *continuous at z in X* if, and only if, for each open subset V of Y with $f(z) \in V$, there exists an open subset U of X with $z \in U$ such that $f(U) \subseteq V$.

Question 8.1.3

Suppose (X, d) and (Y, e) are non-empty metric spaces, $z \in X$ and $f : X \to Y$. Rephrased, the epsilon–delta ball criterion for continuity of f at z is that, for each $\epsilon \in \mathbb{R}^+$, the set $S_{z,\epsilon} = \{\delta \in \mathbb{R}^+ \mid f(\flat[z\,;\delta)) \subseteq \flat[f(z)\,;\epsilon)\}$ is not empty. Once the existence of some δ in $S_{z,\epsilon}$ is established, it is clear that $(0, \delta] \subseteq S_{z,\epsilon}$. This means that every positive number smaller than δ would satisfy the requirement for continuity just as well as δ itself. Is there a maximum value that δ can have? In other words, is $\sup S_{z,\epsilon} \in S_{z,\epsilon}$? The supremum can be infinite, in which case the answer to the question is *no*. But, if $x \in X$ and $d(x,z) < \sup S_{z,\epsilon}$, then $d(x,z) < \delta$ for some $\delta \in S_{z,\epsilon}$ (B.6.6), so that $e(f(x), f(z)) < \epsilon$. It follows that if the supremum is finite, it is a member of $S_{z,\epsilon}$.

A further question now arises. The function $\epsilon \mapsto \sup S_{z,\epsilon}$ is certainly decreasing in that if $\mu \in (0\,,\epsilon)$, then $S_{z,\mu} \subseteq S_{z,\epsilon}$, so that $\sup S_{z,\mu} \leq \sup S_{z,\epsilon}$.

Can we say anything about the ratio of ϵ to $\sup S_{z,\epsilon}$ as ϵ tends to 0? The question is put partly out of curiosity and partly because we think that, in some very special cases when f is a real differentiable function, there ought to be a relationship between the derivative and this ratio. We shall come back to this idea in the next chapter (9.5.1); for the moment, we give some examples to illustrate that, even for real functions, all sorts of things can happen to the ratio.

- If f is constant, then $\sup S_{z,\epsilon}$ is infinite for all ϵ and all $z \in X$.

- If f is not constant but is constant for all x sufficiently close to z, then $\sup S_{z,\epsilon}$ is larger than some fixed positive number for all ϵ, so that $\epsilon / \sup S_{z,\epsilon} \to 0$ as $\epsilon \to 0$.

- If f is an isometry, then, saving exceptional cases, $\sup S_{z,\epsilon} = \epsilon$ for all ϵ, so that $\epsilon / \sup S_{z,\epsilon} \to 1$ as $\epsilon \to 0$.

- If $X = [-1, 1]$ and $Y = \mathbb{R}$ and $f(x) = \sqrt{1 - x^2}$ for all $x \in [-1, 1]$, then $\epsilon / \sup S_{1,\epsilon} \to \infty$ as $\epsilon \to 0$. Note also that $|f'(x)| \to \infty$ as $x \to 1$ in $[-1, 1]$.

- If f is the modified step function defined on $[0, 1]$ by

$$
f(x) = \begin{cases} \dfrac{2^n + 1}{2} x - \dfrac{2^n - 1}{2^n}, & \text{if } n \in \mathbb{N} \quad \text{and} \quad \dfrac{2}{2^n + 1} \le x \le \dfrac{1}{2^{n-1}}, \\[2mm] \dfrac{1}{2^n}, & \text{if } n \in \mathbb{N} \quad \text{and} \quad \dfrac{1}{2^n} < x < \dfrac{2}{2^n + 1}, \\[2mm] 0, & \text{if } x = 0, \end{cases}
$$

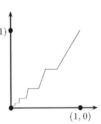

then f is continuous at every point of $[0, 1]$ and differentiable at all points other than those where $x = 0$ or $x = 2/(2^n + 1)$ or $x = 1/2^n$ for some $n \in \mathbb{N}$. The values of $\epsilon / \sup S_{0,\epsilon}$ range between $1/2$ and 1 as $\epsilon \to 0$. However, for each $n \in \mathbb{N}$ and $\epsilon \in [0, 1/2^{n+1}]$, we have $\epsilon / \sup S_{2^{-n}, \epsilon} = 2^n + 1/2$.

8.2 Limits of Functions

There is another way of expressing the epsilon–delta condition for continuity at a point that is sometimes useful. It is based on the idea of a limit.

Definition 8.2.1

Suppose (X, d) and (Y, e) are metric spaces and f is a function from a subset of X into Y. Suppose $y \in Y$ and $z \in \mathrm{acc}_X(\mathrm{dom}(f))$. We say that $f(x)$ *tends to*

y *as x tends to z*, written $f(x) \to y$ as $x \to z$, if, and only if, for every $\epsilon \in \mathbb{R}^+$, there exists $\delta \in \mathbb{R}^+$ such that, for all $x \in \operatorname{dom}(f) \setminus \{z\}$ with $d(x, z) < \delta$, we have $e(f(x), y) < \epsilon$. If such y exists, it is certainly unique (Q 8.1). In this case, we shall call y the *limit* of $f(x)$ as x tends to z and write $\lim_{x \to z} f(x) = y$.

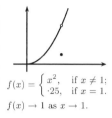

$$f(x) = \begin{cases} x^2, & \text{if } x \neq 1; \\ \cdot 25, & \text{if } x = 1. \end{cases}$$

$f(x) \to 1$ as $x \to 1$.

Theorem 8.2.2

Suppose (X, d) and (Y, e) are metric spaces and f is a function from a subset of X into Y. Suppose $z \in \operatorname{dom}(f)$. Then f is continuous at z if, and only if, either z is isolated in $\operatorname{dom}(f)$ or $\lim_{x \to z} f(x) = f(z)$.

Proof

If $z \in \operatorname{iso}(\operatorname{dom}(f))$, then $U = \{z\}$ is open in $\operatorname{dom}(f)$ (4.1.8). Moreover, for every open subset V of Y with $f(z) \in V$, we have $f(U) = \{f(z)\} \subseteq V$. So f is continuous at z.

If, on the other hand, $z \notin \operatorname{iso}(\operatorname{dom}(f))$, then $z \in \operatorname{acc}(\operatorname{dom}(f))$ (2.6.4). Then Definition 8.2.1 tells us that $\lim_{x \to z} f(x) = f(z)$ if, and only if, f satisfies the epsilon–delta criterion of 8.1.1 at z; that is, if, and only if, f is continuous at z.   □

Example 8.2.3

Consider the function f defined on $[-1, 1]$ by $f(x) = |x|$ if $x \neq 0$ and $f(0) = 1$. This function is continuous at every non-zero point of its domain. Moreover, $f(x) \to 0$ as $x \to 0$, but, since $f(0) \neq 0$, f is not continuous at 0.

Example 8.2.4

Consider the function $f \colon \mathbb{R}^2 \to \mathbb{R}$ given by $f(x) = x_1 x_2 / (x_1^2 + x_2^2)$ for each $x = (x_1, x_2) \in \mathbb{R}^2 \setminus \{0\}$ and $f(0) = \alpha$ for some $\alpha \in \mathbb{R}$. We assume that $\mathbb{R}^2$ has its Euclidean metric. It is easy to check continuity of f at every point of $\mathbb{R}^2 \setminus \{0\}$. Is f continuous at 0? The easiest way to see that it is not is to examine what happens to $f(x)$ as $x \to 0$ along different *curves* in $\mathbb{R}^2$. On the line $\{(a, a) \mid a \in \mathbb{R}\}$, f has the value $1/2$; on the line $\{(a, 2a) \mid a \in \mathbb{R}\}$, however, f has the value $2/5$. So there can be no unique limit for $f(x)$ as $x \to 0$. Therefore f is not continuous at 0 whatever the value of α.

8.3   Global Continuity

A function that is continuous at every point of its domain is called a *continuous* function. But there are various other ways of describing such global continuity. The definition we shall adopt, loosely stated, is that a continuous function is one that *pulls back* open subsets of the codomain to open subsets of the domain. Other equivalent formulations are explored in 8.3.1 below, amongst them the fact that, in a metric space (though not necessarily in a more general context), a function is continuous if, and only if, it maps every convergent sequence of the domain to a convergent sequence in the range with the appropriate limit. For an explanation of the inverse notation $f^{-1}(B)$ in 8.3.1, see B.14.3.

Theorem 8.3.1 (Criteria for Continuity)

Suppose X and Y are metric spaces and $f\colon X \to Y$. The following statements are equivalent:

(i) (OPEN SET CRITERION) For each open subset V of Y, the inverse image $f^{-1}(V)$ is open in X.

(ii) (CLOSED SET CRITERION) For each closed subset F of Y, the inverse image $f^{-1}(F)$ is closed in X.

(iii) (OPEN BALL CRITERION[1]) For each open ball B of Y, the inverse image $f^{-1}(B)$ is open in X.

(iv) (LOCAL CRITERION) f is continuous at every point of X.

(v) (CONVERGENCE CRITERION) For each sequence (x_n) in X that converges in X, the sequence $(f(x_n))$ converges in Y to $f(\lim x_n)$.

Proof

(i) and (ii) are equivalent by 4.1.4 because, for each subset S of Y, we have $X \backslash f^{-1}(S) = f^{-1}(Y \backslash S)$. And (i) implies (iii) because every open ball of Y is an open subset of Y (5.1.7).

Suppose (iii) holds, and let $x \in X$ be arbitrary. Suppose V is an open subset of Y with $f(x) \in V$. By 5.2.2, there exists an open ball B of Y with $f(x) \in B \subseteq V$. By hypothesis, $f^{-1}(B)$ is open in X. But $x \in f^{-1}(B)$ and $f(f^{-1}(B)) \subseteq B \subseteq V$ (B.14.4). So f is continuous at x (8.1.2) and (iv) holds.

(iv) and (v) are equivalent by 8.1.1. Finally, we suppose that f satisfies (iv) and show that it also satisfies (i). Suppose V is an open subset of Y. Let G be the union of all open subsets U of X for which $f(U) \subseteq V$. We show that $G = f^{-1}(V)$. Certainly, $G \subseteq f^{-1}(V)$. Moreover, for each $x \in f^{-1}(V)$, we have $f(x) \in V$, so that, since f is continuous at x, there exists an open subset U of

---

[1] There is no corresponding closed ball criterion (Q 8.5).

X with $x \in U$ and $f(U) \subseteq V$. Then $x \in U \subseteq G$. Since x is arbitrary in $f^{-1}(V)$, it follows that $f^{-1}(V) \subseteq G$ and hence that $f^{-1}(V) = G$. But G, being a union of open subsets of X, is open in X (4.3.2), so $f^{-1}(V)$ is open in X. Since V is an arbitrary open subset of Y, this shows that f satisfies (i).   □

Definition 8.3.2

Suppose X and Y are metric spaces and $f \colon X \to Y$. f is said to be *continuous on X* if, and only if, for each open subset V of Y, $f^{-1}(V)$ is open in X.

It follows immediately from 8.3.1 that a function is continuous if, and only if, it satisfies any one of the criteria listed there.

Example 8.3.3

Suppose X and Y are metric spaces and $f \colon X \to Y$. If X is a discrete metric space (4.3.7), then f is continuous irrespective of the metric on Y. This is so because every subset of a discrete metric space is open, so, for each subset V of Y, $f^{-1}(V)$ is necessarily open in X. In particular, if $\mathbb{N}$ is endowed with its usual metric inherited from $\mathbb{R}$, or with the discrete metric, or indeed with the inverse metric $(m, n) \mapsto \left| m^{-1} - n^{-1} \right|$, then every function from $\mathbb{N}$ into a metric space is continuous—in other words, all sequences are continuous functions provided $\mathbb{N}$ is endowed with a suitable metric.

Example 8.3.4

Every constant function is universally continuous—it does not matter what metrics are placed on its domain and codomain. To see this, suppose X and Y are metric spaces and $f \colon X \to Y$ is constant with value $w \in Y$. Suppose B is an open ball of Y. Then $f^{-1}(B) = X$ if $w \in B$ and $f^{-1}(B) = \varnothing$ if $w \notin B$. Both X and $\varnothing$ are open in X, so f satisfies the open ball criterion for continuity and is therefore a continuous function.

Question 8.3.5

Are there any non-constant non-empty functions that are universally continuous? Students who pursue mathematics into the broader area of topological spaces will find that universal continuity there is not possible for any other function. But we are concerned only with metric spaces, and here there are many functions with the property. Suppose X and Y are metric spaces and $f \colon X \to Y$. If X is finite, then X is necessarily a discrete metric space (4.3.7), so that f is continuous irrespective of the metrics on X and Y (8.3.3).

Example 8.3.6

The reader is undoubtedly familiar with a host of continuous functions from real analysis—the exponential and logarithmic functions, all polynomial and rational functions, the trigonometric functions, the absolute-value function and many others. Furthermore, in the context of the real line, we know that sums, products, scalar multiples and quotients of continuous functions are continuous. Some of these functions do not, however, have the look of continuity when we draw their graphs: the graph of a quotient f/g of continuous functions f and g may be broken because the points where g has value 0 are not in its domain; the tangent function has a graph made up of an infinite number of disjoint parts, and so on. The crux of the matter is that a function is continuous if it is continuous *at every point of its domain*; note that the breaks in the graphs of quotient functions and of the tangent function occur at points outside their domains. For functions defined on a proper subset of $\mathbb{R}$, a more appropriate visual check for continuity can be made not on $\mathbb{R}$ itself but on the *connected components* (11.5.1) of the domain—in other words, on the maximal intervals in the domain. For the tangent function, these are the intervals $((2n-1)\pi/2\,,(2n+1)\pi/2)$ for all $n \in \mathbb{Z}$. But even this test does not always establish continuity. Great care is required because there are functions that are continuous on every such maximal interval yet fail to be continuous functions (8.3.7).

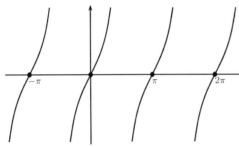

The graph of the tangent function.

Example 8.3.7

Define a function f on the subset $\{0\} \cup \bigcup \{(1/(n+1)\,,1/n)\mid n \in \mathbb{N}\}$ of $[0\,,1]$ as follows:

$$f(x) = \begin{cases} 1, & \text{if } n \in \mathbb{N} \text{ and } 1/(2n+1) < x < 1/2n; \\ -1, & \text{if } n \in \mathbb{N} \text{ and } 1/2n < x < 1/(2n-1); \\ 0, & \text{if } x = 0. \end{cases}$$

The function f is constant, and therefore certainly continuous, on each of the intervals that make up its domain, including the degenerate interval $\{0\}$. But f is not a continuous function; although it is continuous at every other point of its domain, it is not continuous at 0.

Question 8.3.8

Suppose (X, d) is a metric space. Are the point functions (1.2.1) continuous?
Suppose $z \in X$ and $a, b \in \mathbb{R}$ with $a < b$. The set $\{x \in X \mid d(x, z) < b\}$ is
 empty if $b \in \mathbb{R}^{\ominus}$ and equal to $\flat[z; b)$ otherwise and so is open;
also, $\{x \in X \mid d(x, z) \le a\}$ is closed, so that its complement
$\{x \in X \mid d(x, z) > a\}$ is open. It follows that $\delta_z^{-1}(a, b)$, which
equals $\{x \in X \mid d(x, z) < b\} \cap \{x \in X \mid d(x, z) > a\}$, is open by
4.3.2. So δ_z satisfies the open ball criterion for continuity and is
therefore a continuous function. Thus all point functions are continuous. It is a
particular consequence of this fact that every norm on a linear space, being the
point function δ_0 (1.7.8), is a continuous function with respect to the metric it
determines.

Question 8.3.9

Are metrics continuous? The question is strange; after all, metrics are functions
that are used to determine the continuity of other functions. Yet it is a valid
question, and its answer is not straightforward. Suppose (X, d) is a metric space.
Then d is a function defined not on X but on the Cartesian product $X \times X$, so
the metric that determines continuity of d is not d itself but whatever metric
is placed on $X \times X$. Let us suppose that $X \times X$ is given any product metric
(4.5.2). Suppose $((x_{n,1}, x_{n,2}))_{n \in \mathbb{N}}$ is a convergent sequence in $X \times X$ with limit
$z = (z_1, z_2)$. Then, by 6.5.1, $(x_{n,1})$ converges to z_1 in X and $(x_{n,2})$ converges to
z_2 in X. So $d(x_{n,1}, z_1) \to 0$ and $d(x_{n,2}, z_2) \to 0$ in $\mathbb{R}$ by 6.1.4. Then, by Q1.2,

$$|d(x_{n,1}, x_{n,2}) - d(z_1, z_2)| \le d(x_{n,1}, z_1) + d(x_{n,2}, z_2) \to 0,$$

and it follows from 6.1.4 that $d(x_{n,1}, x_{n,2}) \to d(z_1, z_2)$ in $\mathbb{R}$. Therefore d satisfies
the convergence criterion for continuity and is thus a continuous function. So
every metric is a continuous function provided only that its domain is endowed
with a product metric.

Question 8.3.10

Two functions we take entirely for granted are addition and multiplication
of real numbers. Are they continuous? As with metrics, we have to explain
what we mean by the question. Addition is the function $(a, b) \mapsto a + b$ and
multiplication is the function $(a, b) \mapsto ab$, both from $\mathbb{R}^2$ to $\mathbb{R}$. $\mathbb{R}$ has its usual
metric, and we suppose $\mathbb{R}^2$ to be endowed with a product metric. With this
proviso, both functions are continuous. Suppose $(a, b) \in \mathbb{R}^2$ and $((x_n, y_n))$ is
a sequence in $\mathbb{R}^2$ that converges to (a, b). By 6.5.1, $x_n \to a$ and $y_n \to b$ in $\mathbb{R}$.

Therefore, using 6.1.4,

$$|(x_n + y_n) - (a + b)| \leq |x_n - a| + |y_n - b| \to 0$$

and

$$|x_n y_n - ab| \leq |x_n - a|\,|b| + |y_n - b|\,|a| + |x_n - a|\,|y_n - b| \to 0,$$

and it follows from 6.1.4 that $(x_n + y_n)$ converges to $a + b$ and that $(x_n y_n)$ converges to ab in $\mathbb{R}$. This shows that addition and multiplication satisfy the convergence criterion for continuity at (a, b) and, since (a, b) is arbitrary in $\mathbb{R}^2$, are continuous functions with respect to the given metrics on $\mathbb{R}^2$ and $\mathbb{R}$.

The codomain features prominently in the various criteria for continuity. Is it a necessary part of the definition? After all, a codomain may be altered without essentially changing a function, the sole condition being that it must include the range of the function (B.14). Our question therefore can be formulated precisely as follows. Suppose (X, d) and (Y, e) are metric spaces and $f \colon X \to Y$ is a continuous function, and suppose Z is any metric superspace of $(f(X), e)$. Is $f \colon X \to Z$ continuous? Note that we do not assume that $Y \subseteq Z$. On the contrary, Z may contain some points of $Y \backslash f(X)$ and not contain others; moreover, we do not assume that the metric on Z coincides with e on any part of Z that lies in $Y \backslash f(X)$. Nonetheless, for this widest of possible interpretations of our question, the answer is *yes* (8.3.11).

Theorem 8.3.11

Suppose (X, d) and (Y, e) are metric spaces and $f \colon X \to Y$ is a continuous function. Suppose Z is any metric superspace of $(f(X), e)$. Then $f \colon X \to Z$ is continuous.

Proof

Suppose U is an open subset of Z. Then $U \cap f(X)$ is open in $(f(X), e)$ because Z is a metric superspace of $(f(X), e)$, and $U \cap f(X) = W \cap f(X)$ for some open subset W of (Y, e) because $(f(X), e)$ is a metric subspace of (Y, e). So $f^{-1}(U) = f^{-1}(U \cap f(X)) = f^{-1}(W \cap f(X)) = f^{-1}(W)$, which is open in X because $f \colon X \to Y$ is continuous. Since U is an arbitrary open subset of Z, this implies that $f \colon X \to Z$ is a continuous function.   □

Example 8.3.12

Suppose X is a metric space. Distinct points of X can be separated by disjoint open balls (6.2.1). This was the key to showing that a sequence in X has no more

than one limit. Further, arbitrary disjoint closed
subsets of X can be separated by disjoint open subsets (Q 5.9). We can now do even better. We can
separate non-empty disjoint closed subsets A and
B of X by a continuous function as follows. The
functions $x \mapsto \mathrm{dist}(x, A)$ and $x \mapsto \mathrm{dist}(x, B)$ are
both continuous (Q 8.6) and, moreover, the func-
tion $x \mapsto \mathrm{dist}(x, A) + \mathrm{dist}(x, B)$ is never zero by
3.6.10 because $\overline{A} \cap \overline{B} = \varnothing$. So

Disjoint closed sets A and B
separated by disjoint open sets.

$$f : x \mapsto \frac{\mathrm{dist}(x, B)}{\mathrm{dist}(x, A) + \mathrm{dist}(x, B)}$$

is defined and continuous on X. f maps every point of A to 1, every point
of B to 0 and every other point of X to some number in $(0, 1)$. Now let
$U = f^{-1}((1/2, 1])$ and $V = f^{-1}([0, 1/2))$. Since the intervals $(1/2, 1]$ and
$[0, 1/2)$ are open subsets of $[0, 1]$ and f is continuous, U and V are open
in X. They are disjoint by construction and $A \subseteq U$ and $B \subseteq V$.

8.4   Open and Closed Mappings

The open set criterion for continuity is expressed loosely by saying that a
continuous map is one that *pulls back* open subsets of the codomain to open
subsets of the domain. A similar statement can be made about the closed set
criterion. Maps that have the converse properties of mapping open sets onto
open sets and closed sets onto closed sets are called *open* and *closed* mappings,
respectively.

Definition 8.4.1

Suppose X and Y are metric spaces and $f: X \to Y$. We shall call f an *open
mapping* if, and only if, for each open subset U of X, the image $f(U)$ is open
in Y, and we shall call f a *closed mapping* if, and only if, for each closed subset
F of X, the image $f(F)$ is closed in Y.

Question 8.4.2

Is every continuous map open? Is every continuous map closed? Is every open
map continuous? Is every closed map continuous? Are open maps necessarily
closed? Are closed maps necessarily open? There is one answer to all of these
questions: *no*.

- The polynomial function $x \mapsto x^3 - x^2$ defined on $\mathbb{R}$ maps the open interval $(0,1)$ onto the half-open interval $[-4/27,0)$, which is not open in the codomain $\mathbb{R}$. This function is surjective because, for each $a \in \mathbb{R}$, the polynomial equation $x^3 - x^2 - a = 0$, having odd degree, has at least one real root, so $x \mapsto x^3 - x^2$ cannot be made into an open mapping by altering its codomain. But it is continuous.

- The function $1/x$ from $\mathbb{R}\backslash\{0\}$ to $\mathbb{R}$ is open and continuous but not closed because it maps the closed subset $\mathbb{R}^+$ of the domain to $\mathbb{R}^+$, which is not closed in the codomain.

- The function $g: \mathbb{R} \to \{0,1\}$ that maps each rational number to 1 and each irrational number to 0 is open and closed but not continuous since its codomain has the discrete metric. Change the codomain of g to $\mathbb{R}$ and the mapping is neither continuous nor open but remains closed.

8.5 Continuity of Compositions

Many of the functions we deal with are compositions of simpler functions. The function $x \mapsto e^{3x^2}$ may be regarded, for example, as the composition of three functions; it is $x \mapsto e^x$ after $x \mapsto 3x$ after $x \mapsto x^2$. The reader will on innumerable occasions have broken up functions such as e^{3x^2} into their component parts in order to differentiate them. It certainly makes the task much easier. We often use the same trick for testing continuity; the operative theorem states that a composite function is continuous if its component parts are continuous (8.5.1).

Theorem 8.5.1

Suppose X, Y and Z are metric spaces and $f: X \to Y$ and $g: Y \to Z$. If f and g are continuous, then $g \circ f$ is continuous.

Proof

Suppose f and g are continuous and W is an open subset of Z. Then $g^{-1}(W)$ is open in Y and so $f^{-1}(g^{-1}(W))$ is open in X, but this set is $(g \circ f)^{-1}(W)$. Since W is an arbitrary open subset of Z, $g \circ f$ is a continuous function. $\square$

Example 8.5.2

Given functions f and g as in 8.5.1, if $g \circ f$ is continuous, then g is continuous at least on that part of its domain that concerns us here; specifically, $g|_{f(X)}$

is continuous. But the general converse of 8.5.1 does not hold. In fact, it is very easy to construct continuous compositions from components that are not continuous. Let $f: \mathbb{R} \to \mathbb{R}$ be the function that maps each rational number to 1 and each irrational number to 0, and define $g: \mathbb{R} \to \mathbb{R}$ by $g(0) = 1$ and $g(x) = x$ if $x \in \mathbb{R} \backslash \{0\}$. Neither of these functions is continuous, but the composition $g \circ f$ is the constant function 1, which is certainly continuous.

8.6 Continuity of Restrictions and Extensions

Restrictions of continuous functions are continuous. However, there is many a function that is continuous on a subset of a metric space but cannot be extended continuously to the closure of its domain.

Theorem 8.6.1

Suppose X and Y are metric spaces and $f: X \to Y$ is continuous. Suppose S is a metric subspace of X. Then $f|_S$ is a continuous function.

Proof

Suppose V is open in Y. Since continuity of f ensures openness of $f^{-1}(V)$ in X, the set $S \cap f^{-1}(V)$ is open in S (4.4.1). But $f|_S^{-1}(V) = S \cap f^{-1}(V)$. Therefore, because V is an arbitrary open subset of Y, $f|_S$ is continuous. $\qquad\square$

Example 8.6.2

We shall be looking at methods for continuously extending certain continuous functions to larger domains in 10.9. But such extension is not always possible. Consider the continuous function $f: \mathbb{R}^+ \to \mathbb{R}$ given by $x \mapsto \ln x$. Its domain is a dense subset of $\mathbb{R}^\oplus$, but there is no way of extending f to $\mathbb{R}^\oplus$ and retaining continuity. Let us be more precise. Suppose Y is any metric superspace of $\mathbb{R}$. Extend f to $\mathbb{R}^\oplus$ by setting $f(0) = z \in Y$. If $z \in \mathbb{R}$, we certainly do not have $\ln x \to z$ as $x \to 0$, so that f is not continuous at z. If $z \notin \mathbb{R}$, then $z \notin \mathrm{Cl}_Y(\mathbb{R})$ (4.6.2), so that no sequence in $\mathbb{R}$ converges to z (6.6.2), although there are many sequences in $\mathbb{R}^+$ that converge to 0. Therefore f does not satisfy the convergence criterion for continuity at 0.

Example 8.6.3

Openness of a mapping, unlike continuity, depends very much on the codomain.

Restrictions of open mappings may not be open, and even making a restriction surjective does not guarantee its openness. Consider the open mapping $x \mapsto x^2$ from $\mathbb{R}$ onto $\mathbb{R}^{\oplus}$; its restriction to $(\mathbb{R}^{\oplus} \cap \mathbb{Q}) \cup (\mathbb{R}^{-} \backslash \mathbb{Q})$ has the same range $\mathbb{R}^{\oplus}$, but the image of the open subset $\mathbb{R}^{-} \backslash \mathbb{Q}$ of its domain is not open in $\mathbb{R}^{\oplus}$.

8.7 Continuity on Unions

A function f is continuous if, and only if, it is continuous at every point of its domain; the same applies to restrictions of f. Let us suppose that the domain of f is split up into several constituent parts and that the restriction of f to each of those parts is continuous; in other words, each restriction of f is continuous at every point of the appropriate constituent part. Does it follow that f is continuous at every point of its domain and is therefore a continuous function? It is important to know that it does not, even if the constituent parts are closed in the domain and mutually disjoint (8.3.7). However, 8.7.1 gives a sufficient condition for the truth of the implication.

Theorem 8.7.1

Suppose (X, d) and (Y, e) are metric spaces. Suppose $\mathcal{C}$ is a non-empty collection of mutually disjoint non-empty subspaces of X and $f \colon \bigcup \mathcal{C} \to Y$. Suppose that, for each $A \in \mathcal{C}$, we have $A \cap \mathrm{Cl}(\bigcup(\mathcal{C} \backslash \{A\})) = \varnothing$ and $f|_A$ continuous. Then f is continuous on $\bigcup \mathcal{C}$.

Proof

Suppose $a \in \bigcup \mathcal{C}$, and let $A \in \mathcal{C}$ be such that $a \in A$. If $\mathcal{C} = \{A\}$, the result is trivial, so we suppose otherwise. Let $\epsilon \in \mathbb{R}^{+}$. Since $f|_A$ is continuous, it is continuous at a, so there exists $\gamma \in \mathbb{R}^{+}$ such that, for all $x \in A$ for which $d(x, a) < \gamma$, we have $e(f(x), f(a)) < \epsilon$. Since $A \cap \mathrm{Cl}(\bigcup(\mathcal{C} \backslash \{A\})) = \varnothing$, 3.6.10 gives $\eta = \mathrm{dist}(a, \bigcup(\mathcal{C} \backslash \{A\})) \neq 0$, and, because A is not the only member of $\mathcal{C}$, $\eta \in \mathbb{R}^{+}$. Let $\delta = \min\{\gamma, \eta\}$. Then, for each $x \in \bigcup \mathcal{C}$ with $d(x, a) < \delta$, we have $x \in A$, so that $e(f(x), f(a)) < \epsilon$. Because ϵ is arbitrary in $\mathbb{R}^{+}$, f is continuous at a; and because a is arbitrary in $\bigcup \mathcal{C}$, f satisfies the local criterion for continuity (8.3.1) on $\bigcup \mathcal{C}$, so f is continuous on $\bigcup \mathcal{C}$. $\square$

Question 8.7.2

In 8.3.7 and 8.7.1, we considered only disjoint subsets of a domain. Let us look now at overlapping parts of a domain. Suppose X and Y are metric spaces and A and B are subsets X with $A \cap B \neq \varnothing$. Suppose $f: A \cup B \to Y$ has continuous restrictions to A and B. Does f have to be continuous? The answer is *no*. Consider the real function defined on $\mathbb{C}$ by

$$z \mapsto \begin{cases} 0, & \text{if } \Re z \geq 0 \text{ and } \Im z \geq 0; \\ |\Re(z)|, & \text{if } \Re z < 0 \text{ and } \Im z \geq 0; \\ |\Im(z)|, & \text{if } \Im z < 0. \end{cases}$$

This function is continuous on $\{z \in \mathbb{C} \mid \Im z \geq 0\}$ and is also continuous on $\{z \in \mathbb{C} \mid \Re z \geq 0 \text{ or } \Im z < 0\}$, but, being discontinuous at every point of the negative part of the real line, is not continuous on their union $\mathbb{C}$.

8.8   Continuity of Mappings into Product Spaces

Every finite product of metric spaces comes equipped with natural projections onto the coordinate spaces (1.6). These projections are continuous provided only that the product is endowed with a product metric. Moreover, a function that maps into the product is continuous if, and only if, its compositions with the natural projections are all continuous.

Theorem 8.8.1

Suppose $n \in \mathbb{N}$ and, for each $i \in \mathbb{N}_n$, (X_i, τ_i) is a metric space. Endow $P = \prod_{i=1}^{n} X_i$ with a product metric. Then, for each $j \in \mathbb{N}_n$, the natural projection $\pi_j: P \to X_j$ is continuous.

Proof

Suppose $j \in \mathbb{N}_n$ and V is open in X_j. Then $\pi_j^{-1}(V) = \{x \in P \mid x_j \in V\}$, which can be expressed as $\prod_{i=1}^{n} U_i$, where $U_j = V$ and $U_i = X_i$ for all $i \in \mathbb{N}_n \backslash \{j\}$. This is certainly a member of the product topology (4.5.2) because V is open in X_j and X_i is open in X_i for all $i \in \mathbb{N}_n \backslash \{j\}$. Therefore π_j is continuous. $\square$

Theorem 8.8.2

Suppose $n \in \mathbb{N}$ and, for each $i \in \mathbb{N}_n$, (X_i, τ_i) is a metric space. Endow $P = \prod_{i=1}^{n} X_i$ with a product metric. Suppose Z is a metric space and $f: Z \to P$. Then f is continuous if, and only if, $\pi_i \circ f$ is continuous for all $i \in \mathbb{N}_n$.

Proof

Certainly, if f is continuous, then so are all the compositions, by 8.5.1 and 8.8.1. For the converse, suppose that $\pi_i \circ f$ is continuous for every $i \in \mathbb{N}_n$. Suppose $z \in Z$ and (x_m) is a sequence in Z that converges to z in Z. Then $(\pi_i(f(x_m)))_{m \in \mathbb{N}}$ converges to $\pi_i(f(z))$ in X_i for each $i \in \mathbb{N}_n$ by the convergence criterion for continuity of $\pi_i \circ f$ (8.3.1), so that $(f(x_m))$ converges to $f(z)$ in P by 6.5.1. Since (x_m) is an arbitrary sequence in Z that converges to z in Z and z is arbitrary in Z, f satisfies the convergence criterion for continuity (8.3.1) and so is continuous.   □

Example 8.8.3

When we want to produce from the graph of an injective function $f \colon \mathbb{R} \to \mathbb{R}$ the graph of its inverse f^{-1}, we simply reflect the graph of f in the line $\{(a, a) \mid a \in \mathbb{R}\}$; in other words, we apply to the graph of f the mapping $(x_1, x_2) \mapsto (x_2, x_1)$. Let us label this function ψ. Is ψ a continuous mapping when its domain and codomain $\mathbb{R}^2$ have the usual Euclidean metric? Since the Euclidean metric is a product metric, 8.8.2 reduces the question to asking whether or not the two functions $\pi_1 \circ \psi$ and $\pi_2 \circ \psi$ are continuous. But $\pi_1 \circ \psi = \pi_2$ and $\pi_2 \circ \psi = \pi_1$, and both these maps are continuous by 8.8.1. So reflection in the line $\{(a, a) \mid a \in \mathbb{R}\}$ is a continuous mapping.

Example 8.8.4

Suppose X is a metric space, $f \colon X \to \mathbb{R}$ and $g \colon X \to \mathbb{R}$ are continuous functions and $\lambda \in \mathbb{R}$. One proof that the functions $f + g$, fg and λf are all continuous goes as follows. The map $x \mapsto (f + g)(x)$ is the composition of addition after $x \mapsto (f(x), g(x))$; addition is continuous (8.3.10, Q 8.12), and the latter map is continuous by 8.8.2. Then the composition is continuous by 8.5.1. Continuity of $x \mapsto f(x)g(x)$ is proved similarly. Finally, the map $x \mapsto \lambda f(x)$ is the composition of multiplication after $x \mapsto (\lambda, f(x))$; multiplication is continuous (8.3.10, Q 8.12), the constant function $x \mapsto \lambda$ is continuous (8.3.4) and so $x \mapsto (\lambda, f(x))$ is continuous by 8.8.2. Then the composition is continuous by 8.5.1.

Example 8.8.5

We know that if the metric on a product $P = \prod_{i=1}^{n} X_i$ of non-empty metric spaces is a conserving metric, then, for each $j \in \mathbb{N}_n$ and each $a \in P$, the copy $X_{j,a}$, in P, of X_j is isometric to X_j (1.6.4). We cannot hope for anything so nice with an arbitrary product metric. What we do get, however, is that the natural isomorphism $x \mapsto x_j$ from $X_{j,a}$ to X_j *preserves the topology* in that it identifies the open subsets of the domain with the open subsets of the range. To demonstrate that this is true, call the mapping ϕ. It is a restriction of the natural projection π_j and so is certainly continuous and therefore pulls back open sets to open sets. We see next that its inverse is also continuous. Suppose $w \in X_j$. If $(x_{m,j})$ is a sequence in X_j that converges to w, then,

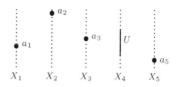

The marked open subset U of X_4 is identified with the open subset $\{a_1\} \times \{a_2\} \times \{a_3\} \times U \times \{a_5\}$ of $X_{4.a}$.

since the constant sequences $(a_i)_{m \in \mathbb{N}}$ are convergent in X_i for each $i \in \mathbb{N}_n \backslash \{j\}$ (8.3.4), the sequence $(\phi^{-1}(x_{m,j}))$ of $X_{j,a}$ converges to $z \in X_{j,a}$, where $z_i = a_i$ for all $i \in \mathbb{N}_n \backslash \{j\}$ and $z_j = w$ (6.5.1)—but this is $\phi^{-1}(w)$. Since w is arbitrary in X_j, this shows that ϕ^{-1} is a continuous function. So ϕ not only pulls back open subsets of its range to open subsets of its domain but also maps open subsets of its domain onto open subsets of its range. The bijective function ϕ therefore completely identifies the topology of X_j with the topology of $X_{j,a}$. Such a mapping is called a *homeomorphism* (see 13.6.1).

Note 8.8.6

In order for a function f to be continuous, the inverse image $f^{-1}(V)$ of every open subset V of its codomain must be open in its domain. In general, however, the topology on the domain may be much larger than is necessary to make f continuous—there may be many open subsets that are not inverse images under f of open subsets of the codomain or unions of such sets. The situation that obtains for natural projections is as sharp as possible: when a product of metric spaces is endowed with a metric that makes all the natural projections continuous, the topology generated by that metric necessarily includes the product topology (4.5.2). To put the matter briefly, the product topology is the smallest topology on the product that ensures that all the natural projections are continuous (Q 8.14).

In this book, we have given no general definition of an arbitrary product of sets. However, the reader who goes on to study infinite products of topological spaces will learn to define the product topology on such a product by this property: it is the smallest topology that makes all the natural projections continuous. Here we merely give a warning that, when the number of coordinate spaces is infinite, this definition generally gives a topology that is smaller than the most obvious infinite analogue of the topology we have given for finite products in 4.5.2.

8.9   Spaces of Continuous Functions

We have studied the metric spaces $B(X, Y)$ of bounded functions from a non-empty set X into a metric space Y. In this section, we assume X to be endowed with a metric and look at their much more important subspaces of bounded continuous functions. We introduced the most fundamental example, the space $\mathcal{C}([0, 1])$ of real continuous bounded functions on $[0, 1]$, in 7.7.5.

Definition 8.9.1

Suppose (X, d) and (Y, e) are metric spaces. We denote by $\mathcal{C}(X, Y)$ the metric space of continuous bounded functions from X to Y with the supremum metric given by $(f, g) \mapsto \sup\{e(g(x), f(x)) \mid x \in X\}$.

Suppose X is a metric space. The notation $\mathcal{C}(X, \mathbb{R})$ and $\mathcal{C}(X, \mathbb{C})$ for the spaces of continuous bounded real and complex functions, respectively, on X may be abbreviated to $\mathcal{C}(X)$ if it is clear from the context what codomain is intended. $\mathcal{C}(X, \mathbb{R})$ is a metric subspace of $B(X, \mathbb{R})$, but $B(X, \mathbb{R})$ has algebraic structure as well—it is an algebra (7.4.3). It follows from 8.8.4 that $\mathcal{C}(X, \mathbb{R})$ is algebraically closed (B.20.5) under addition, multiplication and scalar multiplication and is thus a subalgebra of $B(X, \mathbb{R})$. Similarly, $\mathcal{C}(X, \mathbb{C})$ is a subalgebra of $B(X, \mathbb{C})$.

If X and Y are metric spaces, is $\mathcal{C}(X, Y)$ closed in $B(X, Y)$? The answer to this question is not only pleasing (8.9.3) but quite extraordinary (8.9.4).

Theorem 8.9.2

Suppose (X, d) and (Y, e) are metric spaces and (f_n) is a sequence of continuous functions from X to Y that converges uniformly to a function $g\colon X \to Y$. Then g is continuous.

Proof

Let $z \in X$ and $\epsilon \in \mathbb{R}^+$. Because (f_n) converges uniformly to g, there exists $k \in \mathbb{N}$ such that $e(f_k(x), g(x)) < \epsilon/3$ for all $x \in X$. Because f_k is continuous at z, there exists $\delta \in \mathbb{R}^+$ such that, for all $x \in \flat_X[z\,; \delta)$, we have $e(f_k(x), f_k(z)) < \epsilon/3$. Then, for all $x \in \flat_X[z\,; \delta)$, we have

$$e(g(x), g(z)) \leq e(g(x), f_k(x)) + e(f_k(x), f_k(z)) + e(f_k(z), g(z)) < \epsilon.$$

Since ϵ is arbitrary in $\mathbb{R}^+$, g is continuous at z; but z is arbitrary in X, so that g satisfies the local criterion for continuity (8.3.1) and is therefore a continuous function.   □

Corollary 8.9.3

Suppose (X, d) and (Y, e) are metric spaces. Then $\mathcal{C}(X, Y)$ is closed in $B(X, Y)$.

Proof

Let s denote the supremum metric. Suppose that $g \in \mathrm{Cl}_{B(X,Y)}(\mathcal{C}(X,Y))$. By 6.6.2, there exists a sequence (f_n) in $\mathcal{C}(X,Y)$ such that $f_n \to g$. Then g is continuous by 8.9.2. $\square$

Aside 8.9.4

Something very nice is going on in 8.9.3. The space $B(X, Y)$ does not depend on a metric on X. In fact, X needs no metric; if we put a metric on X, we alter neither the set $B(X, Y)$ nor its metric. From 8.9.3, we know that $\mathcal{C}(X, Y)$ is closed in $B(X, Y)$. But the set $\mathcal{C}(X, Y)$ does depend on a metric on X— change the metric and the set may change. If the metric is the discrete metric, for example, then $\mathcal{C}(X, Y)$ is $B(X, Y)$ itself. For different metrics on X, we have different subspaces $\mathcal{C}(X, Y)$ of $B(X, Y)$, all with the supremum metric. Corollary 8.9.3 tells us that every one of these subspaces is closed in $B(X, Y)$.

8.10 Convergence as Continuity

Every sequence in a metric space is a continuous function provided its domain, usually $\mathbb{N}$, is endowed with a metric that makes it into a discrete metric space (8.3.3). By choosing a suitable metric for $\mathbb{N}$ and extending it appropriately to $\tilde{\mathbb{N}} = \mathbb{N} \cup \{\infty\}$ (B.7.1), we can ensure that continuity identifies precisely those sequences that converge in the space.

Theorem 8.10.1

Suppose X is a metric space. Endow $\tilde{\mathbb{N}}$ with the inverse metric of 1.1.12. Suppose $\tilde{x} : \tilde{\mathbb{N}} \to X$. For each $n \in \tilde{\mathbb{N}}$, denote the value of $\tilde{x}$ at n by x_n. Then the sequence $x = (x_n)_{n \in \mathbb{N}}$ is the restriction $\tilde{x}|_{\mathbb{N}}$ of $\tilde{x}$ to $\mathbb{N}$ and the following statements are equivalent:

(i) $\tilde{x}$ is continuous.

(ii) $\tilde{x}$ is continuous at ∞.

(iii) (x_n) converges in X to x_∞.

Proof

It is clearly the case that (x_n) is the stated restriction of $\tilde{x}$. By 8.3.1, (i) implies (ii). Suppose $\tilde{x}$ is continuous at ∞ and B is an open ball of X centred at x_∞. Then, by the open ball criterion of 8.1.1, there exists an open ball U of $\tilde{\mathbb{N}}$ with $\infty \in U$ such that $\tilde{x}(U) \subseteq B$. Because $\infty \in U$, U contains all except a finite number of members of $\mathbb{N}$ (4.3.9), so that B includes a tail of (x_n). Since B is an arbitrary open ball centred at x_∞, (x_n) converges to x_∞. So (ii) implies (iii). Last, suppose (x_n) converges in X to x_∞. Suppose V is an open subset of X. If $x_\infty \in V$, then V includes a tail of x_n, so $\tilde{x}^{-1}(V)$ has finite complement in $\tilde{\mathbb{N}}$. If $x_\infty \notin V$, then $\infty \notin \tilde{x}^{-1}(V)$. In either case, $\tilde{x}^{-1}(V)$ is open in $\tilde{\mathbb{N}}$ (4.3.9). Since V is an arbitrary open subset of X, $\tilde{x}$ is continuous. So (iii) implies (i).   □

Corollary 8.10.2

Suppose (X,d) is a metric space. For each convergent sequence $a = (a_n)$ in X, let $\tilde{a}$ denote the extension of a to $\tilde{\mathbb{N}}$ for which $a_\infty = \lim a_n$. Let $c(X)$ denote the space of convergent sequences in X with the supremum metric $(a,b) \mapsto \sup\{d(a_n, b_n) \mid n \in \mathbb{N}\}$. Then the map $a \mapsto \tilde{a}$ is an isometry from $c(X)$ onto $\mathcal{C}(\tilde{\mathbb{N}}, X)$, where $\tilde{\mathbb{N}}$ has the inverse metric of 1.1.12.

Proof

The mapping $a \mapsto \tilde{a}$ is surjective by 8.10.1 and is clearly injective. We verify as follows that it preserves the metric. Suppose $a, b \in c(X)$ and $\epsilon \in \mathbb{R}^+$. Then $\flat[a_\infty\,;\epsilon/2)$ includes a tail of (a_n) and $\flat[b_\infty\,;\epsilon/2)$ includes a tail of (b_n), so there exists $k \in \mathbb{N}$ such that $d(a_k, a_\infty) < \epsilon/2$ and $d(b_k, b_\infty) < \epsilon/2$, yielding $d(a_\infty, b_\infty) < d(a_k, b_k) + \epsilon \le \sup\{d(a_n, b_n) \mid n \in \mathbb{N}\} + \epsilon$. Since ϵ is arbitrary in $\mathbb{R}^+$, this gives $d(a_\infty, b_\infty) \le \sup\{d(a_n, b_n) \mid n \in \mathbb{N}\}$ and so also $\sup\{d(a_n, b_n) \mid n \in \tilde{\mathbb{N}}\} = \sup\{d(a_n, b_n) \mid n \in \mathbb{N}\}$. In other words, the mapping $a \mapsto \tilde{a}$ is an isometry.   □

Summary

We have considered local and global continuity and continuity of compositions and restrictions. We have contrasted continuous mappings with open and closed mappings. We have proved a necessary and sufficient condition for the continuity of a function into a product space and we have considered continuity on unions. Finally, we have introduced spaces of continuous functions and examined the relationship between convergence and continuity.

EXERCISES

†Q 8.1 Show that the y in Definition 8.2.1, if it exists, is unique.

Q 8.2 Define $f\colon \mathbb{R}^+ \to \mathbb{R}$ by $f(x) = \lfloor x \rfloor$ (B.6.9) for each $x \in \mathbb{R}^+$. At which points of its domain is f not continuous?

†Q 8.3 Find a function $f\colon \mathbb{R} \to \mathbb{R}$ that is not continuous at any point of $\mathbb{R}$ and a subset S of $\mathbb{R}$ such that both $f|_S$ and $f|_{\mathbb{R}\setminus S}$ are continuous.

Q 8.4 Suppose X and Y are metric spaces and $f\colon X \to Y$. Show that f is continuous at every isolated point of X.

†Q 8.5 Find a function $f\colon X \to Y$ between metric spaces X and Y that is not continuous but has the property that, for each closed ball B of Y, $f^{-1}(B)$ is closed in X.

Q 8.6 Suppose (X,d) is a metric space and A is a closed subset of X. Show that the function $x \mapsto \operatorname{dist}(x,A)$ defined on X is continuous.

Q 8.7 Show that every interval of $\mathbb{R}$ is a continuous image of $\mathbb{R}$ itself.

†Q 8.8 Suppose that S is a non-empty subset of $\mathbb{R}$. Show that every continuous real function with domain S has closed range if, and only if, S is closed in $\mathbb{R}$.

†Q 8.9 Suppose X and Y are metric spaces and $f\colon X \to Y$ is an injective open mapping (8.4.2). Show that $f^{-1}\colon f(X) \to X$ is continuous.

Q 8.10 Suppose X and Y are metric spaces and $f\colon X \to Y$. Suppose S is a non-empty subset of X. Is it true that f is continuous at every point of S if, and only if, the restriction of f to S is a continuous function?

Q 8.11 Show that the real function that maps each member of $c(\mathbb{R})$ to its limit in $\mathbb{R}$ is a continuous map.

†Q 8.12 Show that addition and multiplication of complex numbers are continuous functions when $\mathbb{C} \times \mathbb{C}$ is endowed with a product metric.

†Q 8.13 Suppose X and Y are metric spaces and endow the product $X \times Y$ with a product metric. Suppose $f\colon X \to Y$ is continuous. Show that the graph of f, namely the set $\{(x, f(x)) \mid x \in X\}$, is closed in $X \times Y$.

†Q 8.14 Suppose $n \in \mathbb{N}$ and, for each $i \in \mathbb{N}_n$, (X_i, τ_i) is a metric space. Endow $P = \prod_{i=1}^{n} X_i$ with a metric that makes all the natural projections continuous. Show that the topology on P includes the product topology.

Q 8.15 Suppose S is a non-empty closed subset of $\mathbb{R}$. Suppose $A \subseteq \mathbb{R}$ has the property that, for every $a \in A$, there is a unique nearest point $f(a)$ of S to a. Show that the function $a \mapsto f(a)$ from A to S is continuous.

Q 8.16 Find a metric space X and non-empty subsets A and S of X that have the property that for every $a \in A$ there is a unique nearest point $f(a)$ of S to a in X but for which the function f is not continuous.

Q 8.17 Let X and Y be metric spaces and $f \colon X \to Y$. Show that f is an open map if, and only if, for every open ball B of X, $f(B)$ is open in Y.

†Q 8.18 Find a surjective open mapping that is not a closed mapping. Find a surjective closed mapping that is not an open mapping.

†Q 8.19 Suppose X and Y are metric spaces and $f \colon X \to Y$ is bijective. Show that f is an open mapping if, and only if, f is a closed mapping.

Q 8.20 Suppose X and Y are metric spaces and $f \colon X \to Y$ is an open mapping. Suppose S is a subset of X and f is injective. Show that $f|_S$ is an open mapping if, and only if, $f(S)$ is open in Y.

Q 8.21 Suppose X and Y are metric spaces, $f \colon X \to Y$ is an open map, $S \subseteq X$ and $f(S)$ is open in Y. Suppose U is an open subset of S. Then there exists an open subset V of X such that $U = V \cap S$. Then $f(V) \cap f(S)$ is open in Y, but under what general condition can we say then that $f(U)$ is open in $f(S)$ and conclude that $f|_S$ is an open mapping?

Q 8.22 Suppose $n \in \mathbb{N}$ and, for each $i \in \mathbb{N}_n$, (X_i, τ_i) is a metric space. Endow $P = \prod_{i=1}^{n} X_i$ with a product metric. Show that each of the natural projections $\pi_i \colon P \to X_i$ is an open mapping.

Q 8.23 Let $\mathcal{C}^1([0,1])$ be the subspace of $\mathcal{C}([0,1])$ consisting of the functions that have a continuous derivative throughout $[0,1]$. Show that the mapping $f \mapsto f'$ from $\mathcal{C}^1([0,1])$ to $\mathcal{C}([0,1])$ is not continuous.

†Q 8.24 Consider a sequence (f_n) of continuously differentiable real functions defined on an interval $[a,b]$. Suppose that (f_n) converges pointwise to g. Suppose that (f_n') converges uniformly on $[a,b]$. Show that g is differentiable and that $g' = \lim f_n'$.

Q 8.25 In contrast to Q 8.24, find a sequence (f_n) of continuously differentiable real functions defined on $[0,1]$ for which (f_n) converges uniformly on $[0,1]$ to a differentiable function g and (f_n') converges pointwise on $[0,1]$ to a function that is not g'.

Q 8.26 Suppose V and W are real normed linear spaces and $f \colon V \to W$ satisfies $f(a+b) = f(a) + f(b)$ for all $a, b \in V$. Show that $f(\lambda a) = \lambda f(a)$ for all $a \in V$ and $\lambda \in \mathbb{Q}$. Deduce that, if f is continuous, f is linear.

9
Uniform Continuity

By and large it is uniformly true that in mathematics
there is a time lapse between a mathematical discovery
and the moment it becomes useful; and that this lapse
can be anything from 30 to 100 years, in some cases even more;
and that the whole system seems to function without any direction,
without any reference to usefulness, and without any desire
to do things which are useful.   *John von Neumann, 1903–1957*

In this chapter, we introduce uniform continuity, Lipschitz continuity and strong contraction. These are three more concepts of global continuity, all of them stronger than continuity and each one stronger than the one before it.

9.1 Uniform Continuity

The criterion for uniform continuity is very like the epsilon–delta criterion for continuity, but, in this case, for each $\epsilon \in \mathbb{R}^+$, there is a $\delta \in \mathbb{R}^+$ that serves the purpose of the definition right across a set. Uniform continuity is defined on a set; unlike continuity, it has no local counterpart.

Definition 9.1.1

Suppose (X, d) and (Y, e) are metric spaces and $f : X \to Y$. Suppose $S \subseteq X$. Then f is said to be *uniformly continuous* on S if, and only if, for every $\epsilon \in \mathbb{R}^+$ there exists $\delta \in \mathbb{R}^+$ such that, for every $x, z \in S$ for which $d(z, x) < \delta$, we have also $e(f(z), f(x)) < \epsilon$.

Theorem 9.1.2

Suppose (X, d) and (Y, e) are metric spaces, S is a subspace of X and $f : X \to Y$.

(i)   If f is uniformly continuous on S, then $f|_S$ is continuous.

(ii) If f is uniformly continuous on X, then $f|_S$ is uniformly continuous on S.

Proof

These facts follow immediately from Definition 9.1.1 and from the epsilon–delta criterion for continuity (8.3.1).   □

Example 9.1.3

Continuous functions that are not uniformly continuous abound. No polynomial function of degree greater than 1 is uniformly continuous on $\mathbb{R}$. The logarithmic function is not uniformly continuous on its domain; neither is the exponential function. Consider, for example, the exponential function $x \mapsto e^x$ defined on $\mathbb{R}$; there is no $\delta \in \mathbb{R}^+$ that guarantees $\left| e^a - e^b \right| < 1$ for all $a, b \in \mathbb{R}$ with $|a - b| < \delta$. Specifically, for all $a, b \in \mathbb{R}$ with $a < b$, we have $(e^b - e^a)/(b - a) > e^a$ (Q 9.1). Then, for any $\gamma \in (0, 1)$, pick $a = -\ln \gamma$ and $b = \gamma - \ln \gamma$, so that $e^b - e^a > (b - a)e^a = 1$.

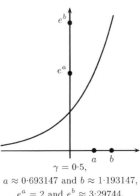

$\gamma = 0{\cdot}5$,
$a \approx 0{\cdot}693147$ and $b \approx 1{\cdot}193147$,
$e^a = 2$ and $e^b \approx 3{\cdot}29744$.

The importance of uniform continuity lies not so much in knowing which functions are uniformly continuous as in knowing on which sets continuity of a given function is uniform. The most useful theorem in this regard is that every continuous function is uniformly continuous on all *compact* subsets of its domain, a fact that follows immediately from 9.1.4 and 9.1.5. The definition of compactness given below is a convenient one for present purposes, but we shall give many more equivalent criteria, including the standard definition, when we come to discuss the concept in Chapter 12.

Definition 9.1.4

Suppose that X is a metric space. We shall say that X is a *compact metric space* if, and only if, X is bounded and has the nearest-point property.

Theorem 9.1.5

Suppose (X, d) and (Y, e) are metric spaces and the former has the nearest-point property. Suppose $f \colon X \to Y$ is continuous. Then f is uniformly continuous on every bounded subset of X.

Proof

Suppose S is a bounded subset of X. If $S = \varnothing$, the assertion is certainly true, so we suppose otherwise. Let $\epsilon \in \mathbb{R}^+$. For each $n \in \mathbb{N}$, set

$$A_n = \{x \in S \mid f(\flat[x\,;1/n)) \nsubseteq \flat[f(x)\,;\epsilon)\}\,.$$

We want to show that some A_n is empty. Suppose that, on the contrary, every A_n is non-empty. Choose a sequence (a_n) in S with $a_n \in A_n$ for each $n \in \mathbb{N}$ (B.19.1). Since S is bounded, the sequence (a_n) is bounded, and, because X has the nearest-point property, (a_n) has a subsequence (a_{m_n}) that converges in X (7.11.1). Let $z = \lim a_{m_n}$. By hypothesis, f is continuous at z, so, by the epsilon–delta ball criterion of 8.1.1, there exists $\delta \in \mathbb{R}^+$ such that

$$f(\flat[z\,;\delta)) \subseteq \flat[f(z)\,;\epsilon/2)\,.$$

We have $a_{m_n} \to z$, so we can pick $k \in \mathbb{N}$ that simultaneously satisfies $k > 2/\delta$ and $a_k \in \flat[z\,;\delta/2)$. It follows that $f(a_k) \in \flat[f(z)\,;\epsilon/2)$ and, from 5.1.10, we get $\flat[a_k\,;1/k) \subseteq \flat[z\,;\delta)$ and $\flat[f(z)\,;\epsilon/2) \subseteq \flat[f(a_k)\,;\epsilon)$. Therefore

$$f(\flat[a_k\,;1/k) \subseteq f(\flat[z\,;\delta)) \subseteq \flat[f(z)\,;\epsilon/2) \subseteq \flat[f(a_k)\,;\epsilon)\,,$$

which yields the contradiction that $a_k \notin A_k$. We must infer that A_m is empty for some $m \in \mathbb{N}$, so that, for all $a, b \in S$, $d(a, b) < 1/m \Rightarrow e(f(a), f(b)) < \epsilon$. Since ϵ is arbitrary, f is uniformly continuous on S. $\square$

Example 9.1.6

The exponential function and all polynomial functions, being continuous on $\mathbb{R}$, are uniformly continuous on every bounded subset of $\mathbb{R}$ because $\mathbb{R}$ has the nearest-point property (7.11.3).

Example 9.1.7

For each $n \in \mathbb{N}$, every closed bounded subset of $\mathbb{R}^n$ with a conserving metric has the nearest-point property by 7.13.2. So every continuous function on such a set is uniformly continuous by 9.1.5.

Example 9.1.8

The sine and cosine functions are uniformly continuous on the whole of $\mathbb{R}$. To see this, note that 9.1.7 ensures that they are both uniformly continuous on the interval $[-\pi\,,\pi]$. Then 2π-periodicity of the functions clinches the matter: let $\epsilon \in \mathbb{R}^+$ and let $\delta \in \mathbb{R}^+$ be such that, for all $a, b \in [-\pi\,,\pi]$ with $|b - a| < \delta$, we have $|\sin b - \sin a| < \epsilon$. Then for all $x, y \in \mathbb{R}$, there exist $a, b \in [-\pi\,,\pi]$

with $|a - b| \leq |x - y|$ and $\sin x = \sin a$ and $\sin y = \sin b$, from which the result follows easily.

Example 9.1.9

The tangent function is continuous but not uniformly continuous on the bounded interval $(-\pi/2, \pi/2)$ of $\mathbb{R}$. Theorem 9.1.5 does not apply here because the function is not continuous on any superset of $(-\pi/2, \pi/2)$ that has the nearest-point property. For $r \in (0, \pi/2)$, the tangent function is uniformly continuous on the interval $(-(\pi/2) + r, (\pi/2) - r)$ because it is continuous on the closed interval $[-(\pi/2) + r, (\pi/2) - r]$, which has the nearest-point property (7.13.2).

Example 9.1.10

Suppose (X, d) and (Y, e) are metric spaces and $f \colon X \to Y$. If d is the discrete metric, then f is uniformly continuous irrespective of the metric on Y. Specifically, for each $\epsilon \in \mathbb{R}^+$, and $a, b \in X$, $d(a, b) < 1 \Rightarrow a = b \Rightarrow e(f(a), f(b)) < \epsilon$. Note, in particular, that an infinite metric space with the discrete metric, such as S in 7.13.3, does not have the nearest-point property, so that it is not only on bounded subsets of spaces with the nearest-point property that continuous functions must be uniformly continuous.

It is important in this example that the metric is discrete and the space not merely a discrete space. There are continuous functions on discrete metric spaces that are not uniformly continuous. The subspace $S = \{1/n \mid n \in \mathbb{N}\}$ of $\mathbb{R}$ is a discrete metric space because each of its singleton sets is both open and closed in S. The function $1/n \mapsto n$ is continuous, as it must be (8.3.3), because S is discrete, but it is not uniformly continuous. Specifically, if $\delta \in \mathbb{R}^+$ and $m, n \in \mathbb{N}$ with $m > n > 2/\delta$, then $|1/m - 1/n| < \delta$ but $|m - n| \geq 1$.

9.2 Conservation by Uniformly Continuous Maps

Uniformly continuous functions have some very nice conserving properties. They map totally bounded sets onto totally bounded sets and Cauchy sequences onto Cauchy sequences (9.2.1). Most importantly, they map compact metric spaces onto compact metric spaces (it should not be overlooked that, on such spaces, every continuous function is uniformly continuous by 9.1.5). On the other hand, uniformly continuous functions need not preserve individually either boundedness or the nearest-point property.

Theorem 9.2.1

Suppose (X, d) and (Y, e) are metric spaces and $f \colon X \to Y$ is uniformly continuous. Then

(i) f maps every Cauchy sequence of X onto a Cauchy sequence of Y;

(ii) f maps every totally bounded subset of X onto a totally bounded subset of Y; and

(iii) f maps every compact subspace of X onto a compact subspace of Y.

Proof

Suppose (x_n) is a Cauchy sequence in X. Let $\epsilon \in \mathbb{R}^+$ and pick $\delta \in \mathbb{R}^+$ such that, for all $a, b \in X$ with $d(a, b) < \delta$, we have $e(f(a), f(b)) < \epsilon$. Since (x_n) is Cauchy, there is a ball B of X of radius less than $\delta/2$ that includes a tail of (x_n). Then $f(B)$ includes the corresponding tail of $(f(x_n))$. But $\operatorname{diam}(B) < \delta$ and the definition of δ ensures that $\operatorname{diam}(f(B)) < \epsilon$, so $f(B)$ is included in a ball C of Y of radius ϵ and C therefore includes a tail of $(f(x_n))$. Since ϵ is arbitrary in $\mathbb{R}^+$, $(f(x_n))$ is Cauchy. This proves (i).

Now suppose S is a totally bounded subset of X. Suppose (y_n) is any sequence in $f(S)$. For each $n \in \mathbb{N}$, the subset $S \cap f^{-1}(\{y_n\})$ of X is non-empty. We choose a sequence (x_n) with $x_n \in S \cap f^{-1}(\{y_n\})$ for each $n \in \mathbb{N}$ (B.19.1). Then $f(x_n) = y_n$ for each $n \in \mathbb{N}$. By the Cauchy criterion for total boundedness of S (7.8.2), (x_n) has a Cauchy subsequence (x_{m_n}). Then, by what we have just proved, $(f(x_{m_n}))$, that is (y_{m_n}), is a Cauchy subsequence of (y_n). Since (y_n) is an arbitrary sequence in $f(S)$, $f(S)$ satisfies the Cauchy criterion for total boundedness and so is totally bounded.

For the third part, suppose S is compact; that is, S is bounded and has the nearest-point property. Then S is totally bounded (7.11.1), whence $f(S)$ is totally bounded by (ii) and so bounded. Suppose (y_n) is an arbitrary sequence in $f(S)$ and, as in (ii), choose a sequence (x_n) in S such that $f(x_n) = y_n$ for every $n \in \mathbb{N}$ (B.19.1). Now (x_n) is a bounded sequence because S is bounded (9.1.4), so the convergence criterion of 7.11.1 ensures that (x_n) has a convergent subsequence. The image under f of such a subsequence is a convergent subsequence of (y_n) by 8.3.1 because f is continuous (9.1.2). Since (y_n) is arbitrary in $f(X)$, this means that $f(X)$ satisfies the convergence criterion of 7.11.1 and thus has the nearest-point property. So $f(X)$ is compact. $\square$

Example 9.2.2

Not all continuous functions enjoy the first two properties of 9.2.1. The function $x \mapsto 1/x$ defined on $(0, 1]$ is continuous, the inverse image of each open interval being also an open interval; it maps the Cauchy sequence $(1/n)$ onto the unbounded sequence (n); and it maps the totally bounded subset $(0\,,1]$ of $\mathbb{R}$ (7.8.5) onto the closed unbounded subset $[1\,,\infty)$ of $\mathbb{R}$.

Example 9.2.3

The tangent function maps the totally bounded interval $(-\pi/2\,,\pi/2)$ of $\mathbb{R}$ (7.8.5) onto the unbounded interval $(-\infty\,,\infty)$. It is not, as we already know, uniformly continuous on $(-\pi/2\,,\pi/2)$.

Question 9.2.4

Do any or all of the properties of 9.2.1 characterize uniform continuity? The answer is *no*. In fact every real function that is continuous on the whole of $\mathbb{R}$—for example, the exponential function—satisfies all of them. Such a function satisfies the first condition because all Cauchy sequences are convergent in $\mathbb{R}$ (7.13.2) and continuous functions map convergent sequences onto sequences that are convergent (8.3.1) and therefore Cauchy (6.8.2). It satisfies the second condition because, being continuous on $\mathbb{R}$, it is uniformly continuous on every bounded subset of $\mathbb{R}$ (9.1.5). Such sets are totally bounded (7.8.5) and so are mapped by the function onto totally bounded subsets of $\mathbb{R}$ by 9.2.1. It satisfies the third condition because it is uniformly continuous on closed bounded subsets of $\mathbb{R}$.

Example 9.2.5

Suppose S is a closed bounded subset of a finite-dimensional Euclidean space $\mathbb{R}^n$ and $f\colon S \to Y$ is a continuous function into some metric space Y. Then S is compact by 7.13.2 and f is uniformly continuous on S by 9.1.5, so that $f(S)$ is compact by 9.2.1.

Question 9.2.6

Do uniformly continuous functions map all bounded sets onto bounded sets? They do in familiar situations where the domain and codomain are subsets of $\mathbb{R}^n$, but that is because boundedness and total boundedness are the same thing in those spaces (7.10.4). It is not always so. Consider the identity function from $\mathbb{N}$ to $\mathbb{N}$, where the domain is given the discrete metric and the codomain the usual metric. The identity function is uniformly continuous because the metric on its domain is the discrete metric (9.1.10), but the domain is a bounded space and the range is not.

9.3   Uniform Continuity on Subsets of the Cantor Set

Subsets of the Cantor set (3.3.3) are totally bounded since they are bounded subsets of $\mathbb{R}$ (7.8.5). So every uniformly continuous image of a subset of the Cantor set is also totally bounded by 9.2.1. But who would guess that this is actually a characterization of totally bounded metric spaces (9.3.1)?

Theorem 9.3.1

Suppose (X, d) is a non-empty metric space. Then X is totally bounded if, and only if, there exists a bijective uniformly continuous function from a subset of the Cantor set $\mathcal{K}$ onto X.

Proof

Suppose X is totally bounded. For each $m \in \mathbb{N}$, choose a finite collection $\mathcal{B}_m$ of open balls of radius $1/m$ that covers X (B.19.1). All these balls together form a countable collection (B.17.4). By enumerating all the members of each $\mathcal{B}_m$ in turn, we form a sequence (U_n) of open balls in which, for each $m \in \mathbb{N}$, the balls of $\mathcal{B}_m$ precede those of $\mathcal{B}_{m+1}$.[1] Then (U_n) has the property that $\mathrm{diam}(U_n) \to 0$ as $n \to \infty$. For each $x \in X$, let $\alpha_n(x) = 2$ if $x \in U_n$ and $\alpha_n(x) = 0$ otherwise, and set $g(x) = \sum_{n=1}^{\infty} \alpha_n(x)/3^n$. Then $g(x) \in \mathcal{K}$ (Q 3.9). Note that there is an infinite number of values of n for which $\alpha_n(x) = 2$ because each $\mathcal{B}_m$ is a cover for X. It follows that g is injective because, for $x, z \in X$ with $x \neq z$, we have $\alpha_n(z) = 0$ whenever both $\alpha_n(x) = 2$ and $d(x, z) > \mathrm{diam}(U_n)$. Let $\phi = g^{-1}$. Then ϕ is a bijective map from the subset $g(X)$ of $\mathcal{K}$ onto X. We want to show that ϕ is uniformly continuous.

Let $\epsilon \in \mathbb{R}^+$. Let $p \in \mathbb{N}$ be such that $p > 2/\epsilon$. Then every member of $\mathcal{B}_p$ has diameter less than ϵ. Let $k \in \mathbb{N}$ be the largest subscript assigned to a member of $\mathcal{B}_p$ in the enumeration (U_n) of the covering balls. Suppose a and b are arbitrary members of $g(X)$ that satisfy $|a - b| < 1/3^k$. Let $\phi(a) = x$ and $\phi(b) = z$. Then $g(x) = a$ and $g(z) = b$, so that $\alpha_n(x) = \alpha_n(z)$ for all $n \in \mathbb{N}_k$—in other words, for all $n \in \mathbb{N}_k$, $x \in U_n$ if, and only if, $z \in U_n$. Since $\mathcal{B}_p$ covers X and all members of $\mathcal{B}_p$ occur in the first k terms of (U_n), there exists $q \in \mathbb{N}_k$ with $z \in U_q \in \mathcal{B}_p$, whence also $x \in U_q$. Then $d(\phi(a), \phi(b)) = d(x, z) \leq \mathrm{diam}(U_q) \leq 2/p < \epsilon$. Since ϵ is arbitrary in $\mathbb{R}^+$, the uniform continuity of ϕ follows. This proves the forward implication; the proof of the backward one is stated in the introduction to this section.   □

---

[1] Duplications are possible because radii are not well-defined, but this does not affect the argument.

We deduce from 9.3.1 that all totally bounded metric spaces are relatively small. Since they are all in one-to-one correspondence with a subset of $\mathbb{R}$, none has cardinality greater than $\mathbb{R}$ (B.17.2).

9.4 Lipschitz Functions

The type of global continuity that we habitually encounter amongst linear maps between normed linear spaces is Lipschitz continuity. It is stronger than uniform continuity and has the advantage that it preserves boundedness.

Definition 9.4.1

Suppose (X, d) and (Y, e) are metric spaces and $f: X \to Y$. If there exists $k \in \mathbb{R}^+$ such that $e(f(a), f(b)) \le kd(a, b)$ for all $a, b \in X$, then f is called a *Lipschitz function* on X with *Lipschitz constant* k.[2]

Theorem 9.4.2

Suppose (X, d) and (Y, e) are metric spaces, S is a subset of X and $f: X \to Y$.

(i) If f is a Lipschitz function on S with Lipschitz constant $k \in \mathbb{R}^+$, then f is uniformly continuous on S and δ in the definition of uniform continuity can be taken to be ϵ/k.

(ii) If f is a Lipschitz function on X with Lipschitz constant $k \in \mathbb{R}^+$, then $f|_S$ is a Lipschitz function with Lipschitz constant k.

Proof

For every $a, b \in X$, we have $d(a, b) < \epsilon/k \Rightarrow e(f(a), f(b)) \le kd(a, b) < \epsilon$, which proves (i). (ii) is obvious.   □

Theorem 9.4.3

Suppose (X, d) and (Y, e) are metric spaces, S is a bounded subset of X and $f: X \to Y$ is a Lipschitz function. Then $f(S)$ is bounded in Y.

Proof

Let $k \in \mathbb{R}^+$ be a Lipschitz constant for f and suppose $a, b \in S$. Then $e(f(a), f(b)) \le kd(a, b) \le k\operatorname{diam}(S)$, so that $\operatorname{diam}(f(S)) \le k\operatorname{diam}(S)$.   □

---

[2] Every real number larger than k is also a Lipschitz constant for f.

Example 9.4.4

We met one type of Lipschitz function at the start of this book, namely the isometry, with Lipschitz constant 1. But isometries behave much better than other Lipschitz functions, as they obey an equality rather than an inequality.

Example 9.4.5

Suppose X is a metric space. Then all the point functions δ_z are Lipschitz functions with Lipschitz constant 1 by 1.2.4; indeed all the pointlike functions (1.2.5) on X are Lipschitz functions with Lipschitz constant 1.

Example 9.4.6

Lipschitz functions occur quite naturally in every context where bounded functions do. Suppose X is a set and $\mathcal{S}$ is any non-empty set of functions defined on X. To each $x \in X$, there corresponds a point evaluation function $\hat{x}$ defined on $\mathcal{S}$ by the equations $\hat{x}(f) = f(x)$ for each $f \in \mathcal{S}$ (B.12.3). If $\mathcal{S}$ is a subset of $B(X, Y)$, where (Y, e) is some given metric space, then, for $x \in X$ and for each $f, g \in \mathcal{S}$, we have $e(\hat{x}(f), \hat{x}(g)) = e(f(x), g(x)) \leq s(f, g)$, where s denotes the usual supremum metric on $B(X, Y)$. It follows that $\hat{x}$ is a Lipschitz function with Lipschitz constant 1.

Example 9.4.7

Linear maps between normed linear spaces (B.20.12) have an extraordinary property that makes their continuity very much easier to handle than that of other maps: continuity at any one point of the domain implies Lipschitz continuity throughout the domain. Suppose X and Y are linear spaces endowed with norms, both of which we denote by $\|\cdot\|$. Suppose that $f \colon X \to Y$ is linear and that f is continuous at some point z of X. Then there exists $k \in \mathbb{R}^+$ such that, for all $x \in X$ with $\|x - z\| \leq 1/k$, we have $\|f(x) - f(z)\| < 1$. Suppose $a, b \in X$ with $a \neq b$. Set $\lambda = k^{-1}\|a - b\|^{-1}$. Then

$$\|(z + \lambda(a - b)) - z\| = \|\lambda(a - b)\| = \lambda\|a - b\| = 1/k,$$

whence $\|f(z + \lambda(a - b)) - f(z)\| < 1$. Then, applying linearity of f (B.20.12), we get $\|\lambda f(a - b)\| < 1$ and therefore $\|f(a) - f(b)\| < 1/\lambda = k\|a - b\|$. So f is a Lipschitz function on X with Lipschitz constant k.

Example 9.4.8

Consider the space $\mathcal{C}([0, 1])$ of continuous bounded real functions defined on $[0, 1]$ with its usual supremum metric. An important function defined on

$\mathcal{C}([0,1])$ is the area function, the function that measures
the area between the graph of a continuous function and
the horizontal axis, areas beneath the axis being computed as negative. Let us name this function A. Then
$A \colon \mathcal{C}([0,1]) \to \mathbb{R}$, and integration theory tells us that, for
each $f \in \mathcal{C}([0,1])$,

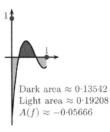

Dark area $\approx 0{\cdot}13542$
Light area $\approx 0{\cdot}19208$
$A(f) \approx -0{\cdot}05666$

$$A(f) = \int_0^1 f(x)\,dx$$

and $|A(f)| \leq \sup\{|f(x)| \mid x \in [0,1]\}$. So, if (f_n) is a sequence in $\mathcal{C}([0,1])$ that
converges to 0, then $(A(f_n))$ converges to 0 in $\mathbb{R}$. Since $A(0) = 0$, this tells
us that A is continuous at the zero function. But A is linear; therefore A is a
Lipschitz function, a fact that can be verified directly with ease.

Let us carry this a little further. Since $\mathcal{C}([0,1])$ is a metric space, we can
consider the space of real continuous bounded functions defined on it, namely
$\mathcal{C}(\mathcal{C}([0,1]))$. The members of this space all act on functions in the same way
that A does. A, however, is not a member of this space because, although it
is continuous, it is not bounded on $\mathcal{C}([0,1])$. Specifically, let $r \in \mathbb{R}^+$ and let
$g \colon [0,1] \to \mathbb{R}$ be the constant function $x \mapsto r$. Then, $g \in \mathcal{C}([0,1])$ and $A(g) = r$,
so, since r is arbitrary in $\mathbb{R}^+$, A is not bounded on $\mathcal{C}([0,1])$. In fact, the only
linear map between normed linear spaces that is a bounded function is the zero
map (Q9.11)—the reader who meets the common term *bounded linear map*
should be aware that it describes those linear maps that have restrictions that
are bounded, in the usual sense, on the unit ball. This property, for linear maps,
is equivalent to continuity.

Example 9.4.9

The modified step function of 8.1.3 is uniformly continuous on $[0,1]$ because
it is continuous and its domain is closed and bounded (9.1.7). But it is not
a Lipschitz function. Let $n \in \mathbb{N}$ and $a = 1/2^{n-1}$ and $b = 2/(2^n + 1)$.
Then $f(a) = 1/2^{n-1}$ and $f(b) = 1/2^n$, so that $a - b = 1/(2^{n-1}(2^n + 1))$ and
$f(a) - f(b) = 1/2^n$, whence $(f(a) - f(b))/(a - b) = (2^n + 1)/2$.

9.5   Differentiable Lipschitz Functions

Let us recall the ratio ϵ/δ that we discussed in 8.1.3. For each $\epsilon \in \mathbb{R}^+$, a uniformly continuous function admits a corresponding $\delta \in \mathbb{R}^+$, applicable now
across the whole of the domain, that enables the function to satisfy the condition for uniform continuity. But we know from 8.1.3 that, as smaller and

smaller values are taken for ϵ, there is no guarantee that admissible values of δ follow a regular pattern. For a Lipschitz function f, however, the ratio ϵ/δ need never exceed any Lipschitz constant for f (9.4.2). And our comment about differentiable functions in 8.1.3 is justified by 9.5.1.

Theorem 9.5.1

Suppose I is a non-degenerate interval of $\mathbb{R}$ and $f\colon I \to \mathbb{R}$ is differentiable on I. Then f is a Lipschitz function on I if, and only if, f' is bounded on I.

Proof

Suppose first that $k \in \mathbb{R}^+$ and that $|f'(x)| \le k$ for all $x \in I$. Suppose $a, b \in I$ and $a \ne b$. By the Mean Value Theorem, there exists $c \in I$ with c between a and b such that $f(b) - f(a) = (b - a)f'(c)$. This yields $|f(b) - f(a)| \le k|b - a|$. So f is a Lipschitz function with Lipschitz constant k. For the converse, suppose that f' is not bounded on I and let $r \in \mathbb{R}^+$ be arbitrary. Then there exist $a, b \in I$ such that $(f(b) - f(a))/(b - a) > r$, whence $|f(b) - f(a)| > r|b - a|$. So, since r is arbitrary in $\mathbb{R}^+$, f is not Lipschitz.   □

Example 9.5.2

The function $x \mapsto \sqrt{1 - x^2}$ defined on $[0, 1]$ (8.1.3) is differentiable on the interval $[0, 1)$; in fact, the derivative is continuous. But the derivative is bounded only on intervals $[0, \alpha]$ for $\alpha \in (0, 1)$; it is not bounded on $[0, 1)$. So this function is Lipschitz on every interval $[0, \alpha]$ with $\alpha \in (0, 1)$ but not Lipschitz on $[0, 1)$. It is, however, uniformly continuous on $[0, 1]$ simply because it is continuous on this closed bounded interval (9.1.7).

Example 9.5.3

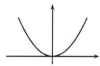

The function x^2 has bounded derivative $2x$ on every bounded interval of $\mathbb{R}$, so that, although x^2 is not even uniformly continuous on $\mathbb{R}$, it is Lipschitz on every bounded interval of $\mathbb{R}$.

9.6   Uniform and Lipschitz Continuity of Compositions

Compositions of uniformly continuous functions are uniformly continuous and compositions of Lipschitz maps are Lipschitz.

Theorem 9.6.1

Suppose (X, d), (Y, e) and (Z, m) are metric spaces, $f: X \to Y$ and $g: Y \to Z$.

(i)   If f and g are uniformly continuous on X and $f(X)$, respectively, then $g \circ f$ is uniformly continuous on X.

(ii) If f and g are Lipschitz functions with Lipschitz constants k and l on X and $f(X)$, respectively, then $g \circ f$ is a Lipschitz function on X with Lipschitz constant kl.

Proof

For (i), suppose the condition is satisfied and let $\epsilon \in \mathbb{R}^+$. Then there exist $\gamma, \delta \in \mathbb{R}^+$ such that $d(a, b) < \delta \Rightarrow e(f(a), f(b)) < \gamma \Rightarrow m(g(f(a)), g(f(b))) < \epsilon$ for all $a, b \in X$. For (ii), suppose the condition is satisfied. If $a, b \in X$, then $m(g(f(a)), g(f(b))) \le l\, e(f(a), f(b)) \le kl\, d(a, b)$, as required. $\square$

9.7   Uniform and Lipschitz Continuity on Unions

A function that is uniformly continuous on a number of disjoint closed sets may well not be uniformly continuous on their union even if the condition of 8.7.1 is satisfied (Q 9.4); for a sufficient condition for uniform continuity, we confine ourselves to finite unions (9.7.1). Lipschitz continuity has even less stability.

Theorem 9.7.1

Suppose (X, d) and (Y, e) are metric spaces and $\mathcal{C}$ is a finite collection of non-empty subsets of X such that $\mathrm{dist}(A, B) > 0$ for all $A, B \in \mathcal{C}$. Suppose $f: \bigcup \mathcal{C} \to Y$ has uniformly continuous restriction to each member of $\mathcal{C}$. Then f is uniformly continuous on $\bigcup \mathcal{C}$.

Proof

Let $\epsilon \in \mathbb{R}^+$. For each $A \in \mathcal{C}$, $f|_A$ is uniformly continuous by hypothesis. Let γ_A be such that, for all $u, v \in A$, $d(u, v) < \gamma_A \Rightarrow e(f(u), f(v)) < \epsilon$. The set $\{\mathrm{dist}(A, B) \mid A, B \in \mathcal{C}\} \cup \{\gamma_A \mid A \in \mathcal{C}\}$ is a finite subset of $\mathbb{R}^+$ and so has a minimum member $\delta \in \mathbb{R}^+$. Then, for all $u, v \in \bigcup \mathcal{C}$ with $d(u, v) < \delta$, there exists $A \in \mathcal{C}$ such that $u, v \in A$ and then, since $\delta \le \gamma_A$, we have $e(f(u), f(v)) < \epsilon$. $\square$

Example 9.7.2

The condition $\mathrm{dist}(A, B) > 0$ in 9.7.1 cannot in general be weakened to $\overline{A} \cap \overline{B} = \varnothing$. Consider the two subsets $A = \{(x, 1/x) \mid x \in \mathbb{R}\backslash\{0\}\}$ and $B = \{(x, 0) \mid x \in \mathbb{R}\}$ of $\mathbb{R}^2$ with the Euclidean metric. Both are closed in $\mathbb{R}^2$, the first by 3.3.4 and the second because it is an isometric copy of $\mathbb{R}$, which is universally closed (4.6.2). Define f to be 1 on A and 0 on B. Then f is uniformly continuous on each of A and B but not on $A \cup B$.

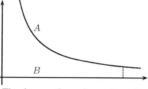

The larger the values of x, the closer the points of A are to B.

Specifically, let $\delta \in \mathbb{R}^+$. Then, for $x > 1/\delta$, the distance from $(x, 1/x)$ to $(x, 0)$ is less than δ and the distance between their images under f is 1.

Example 9.7.3

No theorem like 9.7.1 is possible for Lipschitz functions. Consider the function $f : \mathbb{N} \to \mathbb{N}$ given by $2n - 1 \mapsto 2n - 1$ and $2n \mapsto 4n$ for each $n \in \mathbb{N}$, where $\mathbb{N}$ is endowed, as domain and codomain, with its usual metric. The restriction of f to the odd natural numbers is the identity function, which is Lipschitz with Lipschitz constant 1; and the restriction of f to the even natural numbers is the doubling function, which is Lipschitz with Lipschitz constant 2. These two sets are a distance 1 apart. But f is not Lipschitz because for each $k \in \mathbb{R}^+$ and $n \in \mathbb{N}$ with $n \geq k/2$, we have $f(2n) - f(2n - 1) = 2n + 1 > k(2n - (2n - 1))$.

9.8   Uniform and Lipschitz Continuity on Products

Not every product metric ensures the uniform continuity of the natural projections (Q 9.6). We need to make restrictions in order to get theorems similar to 8.8.1 and 8.8.2.

Theorem 9.8.1

Suppose $n \in \mathbb{N}$ and, for each $i \in \mathbb{N}_n$, (X_i, τ_i) is a metric space. Endow $P = \prod_{i=1}^{n} X_i$ with a conserving metric e. Then all the natural projections $\pi_i : P \to X_i$ are Lipschitz maps with Lipschitz constant 1.

Proof

Suppose $a, b \in P$. Then, for each $i \in \mathbb{N}_n$, $\tau_i(\pi_i(a), \pi_i(b)) \leq e(a, b)$ because e is a conserving metric.   □

Theorem 9.8.2

Suppose $n \in \mathbb{N}$ and, for each $i \in \mathbb{N}_n$, (X_i, τ_i) is a metric space. Endow $P = \prod_{i=1}^{n} X_i$ with a conserving metric e. Suppose (Z, m) is a metric space and $f \colon Z \to P$. Then:

(i) f is uniformly continuous if, and only if, $\pi_i \circ f$ is uniformly continuous for all $i \in \mathbb{N}_n$.

(ii) f is a Lipschitz function if, and only if, $\pi_i \circ f$ is a Lipschitz function for all $i \in \mathbb{N}_n$.

Proof

The forward implications are immediate consequences of 9.8.1 and 9.6.1. For the backward implication in (i), suppose $\pi_i \circ f$ is uniformly continuous for each $i \in \mathbb{N}_n$. Let $\epsilon \in \mathbb{R}^+$ and, for each $i \in \mathbb{N}_n$, let γ_i be such that, for each $a, b \in Z$, we have $m(a, b) < \gamma_i \Rightarrow \tau_i(\pi_i(f(a)), \pi_i(f(b))) < \epsilon/n$. Let $\delta = \min\{\gamma_i \mid i \in \mathbb{N}_n\}$. Then $m(a, b) < \delta \Rightarrow \sum_{i=1}^{n} \tau_i(\pi_i(f(a)), \pi_i(f(b))) < \epsilon$ and, because e is a conserving metric, we have also $m(a, b) < \delta \Rightarrow e(f(a), f(b)) < \epsilon$, as required.

For the backward implication in (ii), suppose that, for each $i \in \mathbb{N}_n$, $\pi_i \circ f$ is a Lipschitz function with Lipschitz constant l_i. Let $k = \sum_{i=1}^{n} l_i$. Then, for each $a, b \in Z$, we have $\tau_i(\pi_i(f(a)), \pi_i(f(b))) \leq l_i m(a, b)$, whence $\sum_{i=1}^{n} \tau_i(\pi_i(f(a)), \pi_i(f(b))) \leq k m(a, b)$. Then, because e is a conserving metric, we get $e(f(a), f(b)) \leq k m(a, b)$, as required. $\square$

9.9 Strong Contractions

Any Lipschitz map between metric spaces that has Lipschitz constant 1 or less may be called a *contraction*; if it has Lipschitz constant less than 1, it is a *strong contraction*. A strong contraction from a metric space X into itself simultaneously pulls all the points of X closer to one another, and its iteration tends to pull all the points towards a single point. But without Cauchy sequences converging, we cannot guarantee the existence of such a point. This defect will be remedied in the chapter on completeness (Chapter 10).

Definition 9.9.1

Suppose (X, d) is a metric space. A map $f \colon X \to X$ is called a *strong contraction* on X if, and only if, there exists $k \in [0, 1)$ such that $d(f(x), f(z)) \leq k\, d(x, z)$ for all $x, z \in X$.

Theorem 9.9.2

Suppose X is a metric space and f is a strong contraction on X. Then

(i) f is a Lipschitz function with Lipschitz constant less than 1; and

(ii) f is uniformly continuous on its domain.

Proof

This is clear from Definition 9.9.1 and from 9.4.2.   □

Question 9.9.3

Suppose (X, d) is a metric space and $f \colon X \to X$ satisfies $d(f(a), f(b)) < d(a, b)$ for all $a, b \in X$. Is f necessarily a strong contraction? At first sight it might look as if it must be. But, on second thought, the condition that there exist $k \in [0, 1)$ such that $d(f(x), f(y)) \leq k\, d(x, y)$ is precisely the same as the condition that there exist $k' \in [0, 1)$ such that $d(f(x), f(y)) < k'\, d(x, y)$ because, if $k < 1$, then there exists k' with $k < k' < 1$ (B.6.11).

The answer to our question is *no*. The function f given by $x \mapsto x^2/2$ defined on $(0, 1)$ is not a strong contraction, despite the fact that it contracts the interval $(0, 1)$ to the smaller interval $(0, 1/2)$. Since $a^2/2 - b^2/2 = (a - b)(a + b)/2$ and $(a + b)/2 < 1$ for all $a, b \in (0, 1)$, this function satisfies the condition that $d(f(a), f(b)) < d(a, b)$ for all $a, b \in (0, 1)$, but it does not satisfy the condition to be a strong contraction because, for every $k \in [0, 1)$, there exist $a, b \in (0, 1)$ such that $(a + b)/2 > k$.

Theorem 9.9.4

Suppose (X, d) is a metric space and f is a strong contraction on X. For each $n \in \mathbb{N}$, let f^n denote the composition of n copies of f (B.13.2). Then

(i)   for each $a, b \in X$, the real sequence $(d(f^n(a), f^n(b)))$ converges to 0; and

(ii) for each $x \in X$, the sequence $(f^n(x))$ is a Cauchy sequence in X.

Proof

Let $k \in [0, 1)$ be a Lipschitz constant for f. For each $a, b \in X$, we have the inequality $d(f(a), f(b)) \leq kd(a, b)$, and it follows using induction that $d(f^n(a), f^n(b)) \leq k^n d(a, b)$ for all $n \in \mathbb{N}$. But because $k \in [0, 1)$, $k^n \to 0$, so that $d(f^n(a), f^n(b)) \to 0$ also. This proves (i).

For (ii), we invoke our knowledge of real series. Suppose $x \in X$. Let $\epsilon \in \mathbb{R}^+$ be arbitrary. Let $m \in \mathbb{N}$ be such that $k^m d(x, f(x)) < (1 - k)\epsilon$. Then, for all

$n \in \mathbb{N}$ we have, using induction and the triangle inequality,

$$d(f^m(x), f^{m+n}(x)) \quad \leq \quad k^m d(x, f^n(x)) \quad \leq \quad k^m \sum_{i=1}^{n} d(f^{i-1}(x), f^i(x))$$

$$\leq \quad k^m d(x, f(x)) \sum_{i=0}^{n-1} k^i$$

$$\leq \quad \frac{k^m}{1-k} d(x, f(x)) \quad < \quad \epsilon.$$

So the ball $\flat[f^m(x); \epsilon]$ includes the mth tail of $(f^n(x))$. Since ϵ is arbitrary in $\mathbb{R}^+$, this establishes that $(f^n(x))$ is a Cauchy sequence. $\square$

Summary

We have introduced strong forms of continuity. We have seen that uniformly continuous functions preserve total boundedness and Cauchy sequences and that Lipschitz functions preserve boundedness as well. We have shown that every continuous function defined on a bounded subset of a metric space with the nearest-point property is uniformly continuous. We have shown that every totally bounded metric space, and in particular every compact metric space, is a uniformly continuous injective image of a subset of the Cantor set. Last, we have examined briefly the properties of strong contractions.

EXERCISES

†Q 9.1   Show that for all $a, b \in \mathbb{R}$ with $a < b$, we have $(e^b - e^a)/(b - a) > e^a$.

Q 9.2   Suppose X is a metric space and C is a non-empty closed subset of X. Show that $x \mapsto \text{dist}(x, C)$ is uniformly continuous on X.

Q 9.3   Determine whether or not each of the following functions is uniformly continuous on the specified domain:

(i) $x \mapsto (1 + x^2)^{-1}$ on $[-1, 1]$.
(ii) $x \mapsto (1 + x^4)^{-1}$ on $\mathbb{R}$.
(iii) $x \mapsto x/(1 - x)^2$ on $(1, \infty)$.

†Q 9.4   Find a function that is uniformly continuous on an infinite number of closed sets, every two of which are of distance at least 1 from each other, but is not uniformly continuous on their union.

Q 9.5   Show that the function $f : x \mapsto x^2$ is uniformly continuous on the set $S = \bigcup\{[n, n + n^{-2}] \mid n \in \mathbb{N}\}$.

†Q 9.6   Show that not every product metric ensures the uniform continuity of the natural projections.

Q 9.7   Show that a uniformly continuous image of a metric space that has the nearest-point property need not have that property.

Q 9.8   Give an example to show that the image of an open set under a uniformly continuous map need not be open. Is the same true for Lipschitz maps? For contractions? For isometric maps?

Q 9.9   Suppose (X, d) and (Y, e) are metric spaces and $f \colon X \to Y$. Suppose there exists $k \in \mathbb{R}^+$ such that $e(f(a), f(b)) \geq kd(a, b)$ for all $a, b \in X$. Show that f is injective and that f^{-1} is a Lipschitz function. Show also that f is an open mapping if $f(X)$ is open in Y.

†Q 9.10 Suppose X is a non-empty set, (Y, e) is a metric space and $S \subseteq B(X, Y)$. For each $x \in X$, let $\hat{x}$ denote the function $f \mapsto f(x)$ defined on S (see 9.4.6). Show that $\{\hat{x} \mid x \in X\}$ is a bounded subset of $\mathcal{C}(S, Y)$ if, and only if, S is bounded in $B(X, Y)$.

†Q 9.11 Suppose X and Y are normed linear spaces and $f \colon X \to Y$ is a linear map. Show that if f is a bounded function, then $f = 0$.

Q 9.12   Suppose (X, d) and (Y, e) are metric spaces and (f_n) is a sequence of uniformly continuous functions from X to Y that converges uniformly to a function $g \colon X \to Y$. Show that g is uniformly continuous.

†Q 9.13 Suppose (X, d) and (Y, e) are metric spaces and (f_n) is a sequence of functions from X to Y that are Lipschitz with Lipschitz constant $k \in \mathbb{R}^+$. Suppose that (f_n) converges uniformly to $g \colon X \to Y$. Show that g is Lipschitz with Lipschitz constant k.

Q 9.14   Show that the function $x \mapsto x + x^{-1}$ defined on $[1, \infty)$ is a contraction that is not strong.

10
Completeness

Every problem in the theory of functions leads to certain questions
in the theory of sets, and it is to the degree that
these latter questions are resolved, that it is
possible to solve the given problem
more or less completely.   *René-Louis Baire, 1874–1932*

The Euclidean metric on $\mathbb{R}$ is derived from the absolute-value function, which in turn depends on the ordering of $\mathbb{R}$. The fact that the ordering of $\mathbb{R}$ is complete (B.6.7) was crucial in establishing, in 2.8.4 and 4.6.2, that $\mathbb{R}$ is universally closed. We have therefore called this property of universal closure *completeness* (4.6.3) and we have already shown that an arbitrary metric space X is complete if, and only if, every Cauchy sequence in X converges in X (6.11.3). Many metric spaces, with or without ordering, have this property. Moreover, any that does not can be realized as a dense subspace of one that does (10.12.2).

10.1 Virtual Points

Here we define what we shall call *virtual points*. They are those pointlike functions (1.5.1) that take on values close to 0 without attaining that value; in all other respects, virtual points behave exactly like the point functions δ_z of 1.2.1.

Definition 10.1.1

Suppose (X, d) is a metric space and $u \colon X \to \mathbb{R}$. Then u will be called a *virtual point* of X if, and only if, u satisfies the following three conditions:

- $u(a) - u(b) \leq d(a, b) \leq u(a) + u(b)$ for all $a, b \in X$.
- $\inf u(X) = 0$.
- $0 \notin u(X)$.

u measures the distance
from the vacant spot to
each point x of X.

We shall denote the set of virtual points of a metric space X by $\mathrm{vp}(X)$. It is immediate from their definition that virtual points are pointlike functions (1.2.5) and are not point functions. Theorem 10.1.2 below follows immediately from 9.4.5.

Theorem 10.1.2

Suppose X is a metric space. Every virtual point of X is a Lipschitz function with Lipschitz constant 1 when $\mathbb{R}$ is endowed with its usual metric.

10.2   Criteria for Completeness

In 10.2.1 below, we prove the equivalence of several properties, any one of which might be regarded as a characterization of completeness.

Theorem 10.2.1 (Criteria for Completeness)

Suppose (X, d) is a metric space. The following statements are equivalent:

(i)   (VIRTUAL POINT CRITERION) X has no virtual points.

(ii)   (UNIVERSAL CRITERION) X is closed in every metric superspace of X.

(iii)   (CAUCHY CRITERION) Every Cauchy sequence in X converges in X.

(iv)   (NEST CRITERION) Every nest $\mathcal{F}$ of non-empty closed subsets of X for which $\inf\{\mathrm{diam}(A) \mid A \in \mathcal{F}\} = 0$ has singleton intersection.

(v)   (NESTED SEQUENCE CRITERION) Every sequence (F_n) of non-empty closed subsets of X for which $F_{n+1} \subseteq F_n$ for each $n \in \mathbb{N}$ and $\mathrm{diam}(F_n) \to 0$ has non-empty intersection.

Proof

Since all the conditions are satisfied if $X = \varnothing$, we assume $X \neq \varnothing$. Suppose X has no virtual points and (Y, d) is a superspace of X and $w \in \partial_Y X$. The function $x \mapsto d(w, x)$ defined on X is pointlike and the infimum of its range is 0, so that, since it is not a virtual point of X, there exists $z \in X$ such that $d(w, z) = 0$. Then $w = z \in X$. Since w is arbitrary in $\partial_Y X$, it follows that X is closed in Y. So (i) implies (ii). That (ii), (iii) and (iv) are equivalent has already been proved (6.11.3, 4.7.2), and it is clear that (iv) implies (v).

Now we suppose that X satisfies (v) and that u is a pointlike function on X that satisfies $\inf u(X) = 0$. For each $n \in \mathbb{N}$, $A_n = u^{-1}([0, 1/n])$ is non-empty and $A_{n+1} \subseteq A_n$. Since u is continuous (10.1.2) and each interval $[0, 1/n]$ is closed, each A_n is closed in X (8.3.1). Moreover, for $n \in \mathbb{N}$ and $a, b \in A_n$, we have $u(a) \leq 1/n$ and $u(b) \leq 1/n$, so that $d(a,b) \leq u(a) + u(b) \leq 2/n$; therefore $\mathrm{diam}(A_n) \leq 2/n$. So $\mathrm{diam}(A_n) \to 0$. By hypothesis, $\bigcap\{A_n \mid n \in \mathbb{N}\}$ is not empty, but for z in this intersection, we have $u(z) < 1/n$ for all $n \in \mathbb{N}$ and therefore $u(z) = 0$. So $u \notin \mathrm{vp}(X)$. It follows that X has no virtual points. So (v) implies (i). $\qquad\square$

In 4.6.3, we defined completeness as universal closure;[1] now we can say that a metric space is complete if, and only if, it satisfies any one of the criteria listed in 10.2.1. The standard definition is that a metric space is complete if, and only if, every Cauchy sequence in it converges in the space.

Example 10.2.2

$\mathbb{R}$ with its usual metric is complete (4.6.2).

Example 10.2.3

Every metric space with the discrete metric is complete since in such a space every Cauchy sequence is eventually constant and therefore converges. Note, however, that discrete metric spaces need not be complete; an easy counterexample is the subspace $\{1/n \mid n \in \mathbb{N}\}$ of $\mathbb{R}$.

Example 10.2.4

Every metric space that has the nearest-point property is complete (7.11.1). In particular, every finite-dimensional normed linear space is complete (Q 7.23).

Example 10.2.5

Completeness is a necessary condition for a metric space to have the nearest-point property. But even a bounded complete metric space need not have the nearest-point property: every infinite set with the discrete metric is complete (10.2.3) and has diameter 1, but has no accumulation point, and therefore does not have the nearest-point property.

---

[1] A metric space X may be closed in some superspace and an isometric copy of X not closed in some enveloping superspace of its own. But universal closure of X—being closed in *every* metric superspace—is a property intrinsic to X and is therefore preserved by isometries. This fact is made clearer by the Cauchy criterion for completeness, which refers to nothing at all outside X.

10.3 Complete Subsets

The reader will notice that the property of universal closure, unlike that of closure, is independent of any particular enveloping metric space. So whether or not a metric space is being considered as a space in its own right or as a subspace of some larger space is irrelevant when we are talking about completeness.

Definition 10.3.1

Suppose (X, d) is a metric space and S is a subset of X. We say that S is a *complete subset* of X if, and only if, the metric subspace (S, d) of (X, d) is a complete metric space.

Theorem 10.3.2

Suppose X is a complete metric space and $S \subseteq X$. Then S is complete if, and only if, S is closed in X.

Proof

Certainly, if S is complete, S is closed in X by definition (4.6.3). For the converse, suppose S is closed in X. Then each nest $\mathcal{F}$ of closed subsets of S for which $\inf\{\operatorname{diam}(A) \mid A \in \mathcal{F}\} = 0$ is a nest of closed subsets of X (Q 4.6) satisfying the same condition and so has non-empty intersection (10.2.1). Therefore S satisfies the nest condition for completeness (10.2.1).   □

Example 10.3.3

Because $\mathbb{R}$ is complete, the complete subsets of $\mathbb{R}$ are its closed subsets; that is, precisely those that have the nearest-point property. In particular, every closed interval of $\mathbb{R}$ is complete, and, despite its obvious fragmentation, the Cantor set $\mathcal{K}$ is also complete.

Example 10.3.4

An open subset of a complete metric space is not complete unless it is also closed. However, open subsets can be made into complete spaces by judiciously altering the metric. Let's do it. Suppose (X, d) is a complete metric space and U is a non-empty proper open subset of X. For each $x \in U$, $\operatorname{dist}_d(x, U^c) \neq 0$ because U^c is closed in X; define $f(x) = 1/\operatorname{dist}_d(x, U^c)$, and let e denote the metric on U defined by $(a, b) \mapsto d(a, b) + |f(a) - f(b)|$ (Q 1.16).

  Suppose (x_n) is a Cauchy sequence in (U, e). Because $d \leq e$, (x_n) is Cauchy in (U, d) also and therefore converges in the complete space (X, d). Let z be its limit in (X, d). Because the distance function is continuous (Q 8.6), we now have $\text{dist}_d(x_n, U^c) \to \text{dist}_d(z, U^c)$ (8.3.1). If the latter quantity were zero, we should have $f(x_n) \to \infty$, giving $e(x_1, x_n) = d(x_1, x_n) + |f(x_1) - f(x_n)| \to \infty$, making (x_n) unbounded in (X, e), contradicting 7.6.1. So $\text{dist}_d(z, U^c) > 0$ and $z \in U$. Moreover, since $\text{dist}_d(x_n, U^c) \to \text{dist}_d(z, U^c)$ and the inverse function is continuous, we have $f(x_n) \to f(z)$ in $\mathbb{R}$ (8.3.1). Therefore we have $e(x_n, z) = d(x_n, z) + |f(x_n) - f(z)| \to 0$, so that $x_n \to z$ in (U, e) (6.1.4). Because (x_n) is an arbitrary Cauchy sequence in (U, e), (U, e) is complete.

10.4   Unions and Intersections of Complete Subsets

Completeness is universal closure and behaves under unions and intersections exactly as we might expect it to do from our knowledge of closed sets.

Theorem 10.4.1

Suppose X is a metric space and $\mathcal{U}$ is a set of complete subspaces of X. Then:

(i) $\bigcap \mathcal{U}$ is complete.

(ii)   If $\mathcal{U}$ is finite, then $\bigcup \mathcal{U}$ is complete.

Proof

For (i), suppose (x_n) is a Cauchy sequence in $\bigcap \mathcal{U}$. Then (x_n) is Cauchy in every member of $\mathcal{U}$ and so converges in each of them (10.2.1). Since limits in X are unique, the unique limit is in U for every $U \in \mathcal{U}$, so (x_n) converges in $\bigcap \mathcal{U}$. Thus $\bigcap \mathcal{U}$ satisfies the Cauchy criterion for completeness.

For (ii), suppose Y is a metric superspace of $\bigcup \mathcal{U}$. Then each member of $\mathcal{U}$, being complete, is closed in Y, and, if $\mathcal{U}$ is finite, $\bigcup \mathcal{U}$ is also closed in Y (4.3.2). Since Y is an arbitrary metric superspace of $\bigcup \mathcal{U}$, this shows that $\bigcup \mathcal{U}$ is universally closed and thus complete. $\square$

Example 10.4.2

The union of an infinite number of complete subsets of a metric space need not be complete. For each $n \in \mathbb{N}$, the singleton subset $\{1/n\}$ of $\mathbb{R}$ is complete. But $\{1/n \mid n \in \mathbb{N}\}$ is not complete.

10.5 Products of Complete Metric Spaces

A product metric on a finite product of complete metric spaces may not ensure that the product is complete (Q 10.3), but a conserving metric does.

Theorem 10.5.1

Suppose $n \in \mathbb{N}$ and, for each $i \in \mathbb{N}_n$, (X_i, τ_i) is a non-empty metric space. Suppose the product $P = \prod_{i=1}^{n} X_i$ is endowed with a conserving metric d. Then P is complete if, and only if, X_i is complete for all $i \in \mathbb{N}_n$.

Proof

Suppose first that P is complete, that $i \in \mathbb{N}_n$ and that $(x_{i,m})_{m \in \mathbb{N}}$ is a Cauchy sequence in X_i. For each $j \in \mathbb{N}_n \backslash \{i\}$, let $a_j \in X_j$. For each $m \in \mathbb{N}$, let $z_m \in P$ be given by $\pi_i(z_m) = x_{i,m}$ and $\pi_j(z_m) = a_j$ for each $j \in \mathbb{N}_n \backslash \{i\}$. Then (z_m) is Cauchy in P (6.10.1) and so converges in P (10.2.1). Then $(\pi_i(z_m))$—that is $(x_{i,m})$—converges in X_i (6.5.1). So X_i is complete by 10.2.1. This proves the forward implication; the backward implication is easier using the same tools. $\square$

Example 10.5.2

It follows from 4.6.2 and 10.5.1 that $\mathbb{R}^n$, for any $n \in \mathbb{N}$, is complete when endowed with any conserving metric. Of course, $\mathbb{C}^n$ with a conserving metric is similarly complete. The set $\mathcal{M}_{n \times n}(\mathbb{R})$ of $n \times n$ matrices with real entries is isomorphic to $\mathbb{R}^{n^2}$. Endowed with a similarly appointed metric, it also is a complete metric space. Closed subspaces of all these spaces are also complete by 10.3.2. Indeed, all these spaces have the nearest-point property (7.13.2).

10.6 Completeness and Continuity

We do not expect every continuous image of a complete metric space to be
complete because continuity does not always preserve the Cauchy property in sequences (9.2.2). Anyway, a counterexample is given by the exponential function, which maps the complete space $\mathbb{R}$ onto the incomplete space $\mathbb{R}^+$. Uniform continuity preserves the Cauchy property (9.2.1), but it does

0 is not in the range of the exponential function.

not preserve completeness any more than continuity does; indeed Lipschitz functions need not preserve completeness either

(Q 10.6). Not even bijective continuous functions with continuous inverse need preserve completeness (10.6.3); isometries, of course, preserve completeness just as they do every other metric property properly belonging to a metric space, but there are other functions that preserve completeness, as we see now in 10.6.1.

Theorem 10.6.1

Suppose that (X, d) and (Y, e) are metric spaces and that (X, d) is complete. Suppose there exists a bijective function $f \colon X \to Y$ such that f is continuous and f^{-1} is uniformly continuous. Then Y is complete.

Proof

Suppose (a_n) is a Cauchy sequence in Y. Then $(f^{-1}(a_n))$ is Cauchy in X by 9.2.1 and so converges in X because X is complete. Therefore (a_n), being the image of $(f^{-1}(a_n))$ under the continuous function f, also converges (8.3.1). Since (a_n) is an arbitrary Cauchy sequence in Y, Y is complete by 10.2.1. $\square$

Example 10.6.2

Let d denote the usual metric on $\mathbb{R}^{\oplus}$ and m denote the exponential metric $(a, b) \mapsto \left| e^a - e^b \right|$. $(\mathbb{R}^{\oplus}, d)$ is complete, being a closed subspace of $\mathbb{R}$ with its usual metric. The identity map from $(\mathbb{R}^{\oplus}, d)$ to $(\mathbb{R}^{\oplus}, m)$ is bijective; it is easily verified that continuity of the exponential function implies continuity of this identity map, and the identity function from $(\mathbb{R}^{\oplus}, m)$ to $(\mathbb{R}^{\oplus}, d)$ is a Lipschitz map because $|a - b| \leq \left| e^a - e^b \right|$ for all $a, b \in \mathbb{R}^{\oplus}$. So $\mathbb{R}^{\oplus}$ with the exponential metric is complete.

Example 10.6.3

When the closed interval $[1, \infty]$ is endowed with its usual metric, it is a complete metric space. Now endow it with the inverse metric $(a, b) \mapsto \left| a^{-1} - b^{-1} \right|$; this space is not complete because the sequence (n) is Cauchy but does not converge. Notice, however, that the identity map from $[1, \infty]$ with the Euclidean metric to $[1, \infty]$ with the inverse metric is continuous and has continuous inverse.

10.7 Completeness of the Hausdorff Metric

The collection $\mathcal{S}(X)$ of non-empty closed bounded subsets of a metric space (X, d) can be endowed with the Hausdorff metric (7.3.1). The resulting space includes a closed isometric copy of X itself (10.7.1), so that $\mathcal{S}(X)$ is not complete unless X is complete. As it happens, completeness of X is not only necessary but sufficient for completeness of $\mathcal{S}(X)$.

Theorem 10.7.1

Suppose (X, d) is a non-empty metric space. Let $\mathcal{S}(X)$ denote the collection of all non-empty closed bounded subsets of X. Let h be the Hausdorff metric (7.3.1) on $\mathcal{S}(X)$, namely

$$(A, B) \mapsto \max\{\sup\{\operatorname{dist}(b, A) \mid b \in B\}, \sup\{\operatorname{dist}(a, B) \mid a \in A\}\}.$$

Then $\{\{x\} \mid x \in X\}$ is an isometric copy of X in $\mathcal{S}(X)$ and is closed in $\mathcal{S}(X)$.

Proof

Let $X' = \{\{x\} \mid x \in X\}$. Then $X' \subseteq \mathcal{S}(X)$ and $h(\{a\}, \{b\}) = d(a, b)$ for all $a, b \in X$, so that X' is an isometric copy of X in $\mathcal{S}(X)$. To show that X' is closed in $\mathcal{S}(X)$, we proceed as follows. Suppose that $F \in \operatorname{Cl}_{\mathcal{S}(X)}(X')$. Then $\operatorname{dist}_h(F, X') = 0$ (3.6.10), so, for each $\epsilon \in \mathbb{R}^+$, there exists $x \in X$ such that $h(F, \{x\}) < \epsilon/2$, whence $d(x, z) < \epsilon/2$ for all $z \in F$, so that, since $F \neq \varnothing$, $\operatorname{diam}(F) \leq \epsilon$. Because ϵ is arbitrary in $\mathbb{R}^+$, we then get $\operatorname{diam}(F) = 0$. Therefore F is a singleton set; in other words, $F \in X'$. Since F is arbitrary in $\operatorname{Cl}(X')$, this establishes that X' is closed in $\mathcal{S}(X)$. $\qquad\square$

Theorem 10.7.2

Suppose (X, d) is a non-empty metric space. Let $\mathcal{S}(X)$ denote the collection of all non-empty closed bounded subsets of X. Let h be the Hausdorff metric on $\mathcal{S}(X)$. Then $(\mathcal{S}(X), h)$ is complete if, and only if, (X, d) is complete.

Proof

(This proof is based on notes of C.-H. Chu [4].) Suppose first that X is complete. Suppose (A_n) is a Cauchy sequence in $\mathcal{S}(X)$; we show that (A_n) converges in $\mathcal{S}(X)$. Let $m_0 = 1$ and, for each $n \in \mathbb{N}$, recursively define m_n to be the least integer greater than m_{n-1} such that the corresponding tail of (A_n) is included in an open ball of $(\mathcal{S}(X), h)$ of radius $1/2^{n+1}$. In particular, $h(A_{m_n}, A_{m_{n+1}}) < 1/2^n$ for each $n \in \mathbb{N}$. Let Z be the set

$$\{z \in X \mid z \text{ is the limit of a sequence } (a_n) \text{ where } a_n \in A_{m_n} \text{ for each } n \in \mathbb{N}\}.$$

If $\{n \in \mathbb{N} \mid A_n = \overline{Z}\}$ is infinite, then (A_n) has a constant subsequence and, being Cauchy, converges (6.8.3), so we suppose this set is finite.

Suppose $\epsilon \in \mathbb{R}^+$ and let $k \in \mathbb{N}$ be such that $1/2^{k-1} < \epsilon$ and $A_{m_n} \neq \overline{Z}$ for each $n \in \mathbb{N}$ with $n \geq k$. Suppose $w \in A_{m_k}$. Consider the sequence (C_n), where $C_1 = \{w\}$ and $C_{n+1} = A_{m_{k+n}}$ for each $n \in \mathbb{N}$. For each $n \in \mathbb{N}$ and each $u \in C_n$, there exists at least one $v \in C_{n+1}$ such that $d(u,v) < 1/2^{k+n-1}$ because $h(A_{m_{k+n-1}}, A_{m_{k+n}}) < 1/2^{k+n-1}$. So the range of the relation

$$\left\{ ((u,n),(v,n+1)) \mid n \in \mathbb{N},\ u \in C_n,\ v \in C_{n+1},\ d(u,v) < 1/2^{k+n-1} \right\}$$

is included in its domain. It follows, using B.19.2, that there exists a sequence (c_n) with $c_n \in C_n$ and $d(c_n, c_{n+1}) < 1/2^{k+n-1}$ for all $n \in \mathbb{N}$. Such a sequence (c_n) is certainly Cauchy in X and, because X is complete, it has a limit l in X. It follows that $l \in Z$ and, in particular, that $Z \neq \varnothing$. Using the triangle inequality and our knowledge of real series, we have, because $w = c_1$,

$$\text{dist}(w, \overline{Z}) \leq d(w, l) \leq \sum_{n=1}^{\infty} d(c_n, c_{n+1}) \leq \sum_{n=1}^{\infty} 1/2^{k+n-1} = 1/2^{k-1},$$

whence, since w is arbitrary in A_{m_k},

$$\sup\left\{ \text{dist}(x, \overline{Z}) \mid x \in A_{m_k} \right\} \leq 1/2^{k-1} < \epsilon.$$

Moreover, for each $u \in \overline{Z}$, there exists $z \in Z$ with $d(z,u) < \epsilon/4$ and there is a sequence (b_n) with $b_n \in A_{m_n}$ for each $n \in \mathbb{N}$ such that $b_n \to z$. So there exists $j > k$ such that $d(z, b_j) < \epsilon/4$. Since $j > k$, we have $h(A_{m_j}, A_{m_k}) < 1/2^k$, so there exists $v \in A_{m_k}$ such that $d(b_j, v) < 1/2^k < \epsilon/2$. It follows that $d(u,v) \leq d(u,z) + d(z,b_j) + d(b_j,v) < \epsilon$, and therefore that $\text{dist}(u, A_{m_k}) < \epsilon$, which, because u is arbitrary in $\overline{Z}$, yields

$$\sup\left\{ \text{dist}(x, A_{m_k}) \mid x \in \overline{Z} \right\} \leq \epsilon.$$

Since A_{m_k} is bounded, this establishes also that $\overline{Z}$ is bounded and therefore that $\overline{Z} \in \mathcal{S}(X)$. The last two displayed inequalities give $h(\overline{Z}, A_{m_k}) \leq \epsilon$ so that, since $A_{m_k} \neq \overline{Z}$, $\text{dist}_{\mathcal{S}(X)}\big(\overline{Z}, \{A_n \mid n \in \mathbb{N}\} \setminus \{\overline{Z}\}\big) \leq \epsilon$. Because ϵ is arbitrary in $\mathbb{R}^+$, this yields $\text{dist}_{\mathcal{S}(X)}\big(\overline{Z}, \{A_n \mid n \in \mathbb{N}\} \setminus \{\overline{Z}\}\big) = 0$. Since (A_n) is Cauchy, 6.8.4 ensures that (A_n) converges in $\mathcal{S}(X)$. So $\mathcal{S}(X)$ is complete.

For the converse, if $\mathcal{S}(X)$ is complete, then $\{\{x\} \mid x \in X\}$, being closed in $\mathcal{S}(X)$ by 10.7.1, is also complete (10.3.2). Then, since X is isometric to $\{\{x\} \mid x \in X\}$ (10.7.1), X also is complete. $\square$

10.8   Complete Spaces of Functions

Completeness of a space of bounded functions does not depend on any metric property of the domain; it depends entirely on completeness of the codomain. Because $\mathbb{R}$ and $\mathbb{C}$ are complete, this one fact provides us with a rich source of complete metric spaces.

Theorem 10.8.1

Suppose X is a non-empty set and (Y, e) is a metric space. Then the space $B(X, Y)$ of bounded functions from X into Y, with its usual supremum metric s, is a complete metric space if, and only if, Y is complete.

Proof

Suppose first that Y is complete and that (f_n) is a Cauchy sequence in $B(X, Y)$. Then, for each $x \in X$, the sequence $(f_n(x))$ is Cauchy in Y because $e(f_n(x), f_m(x)) \leq s(f_n, f_m)$ for all $m, n \in \mathbb{N}$, and, since Y is complete, $(f_n(x))$ converges in Y. We define a function $g \colon X \to Y$ by $g(x) = \lim f_n(x)$ for each $x \in X$. Since (f_n) is bounded (7.6.1), we have $g \in B(X, Y)$ by 7.7.4. We must now show that $f_n \to g$ in $B(X, Y)$.

   Suppose $r \in \mathbb{R}^+$. Because (f_n) is Cauchy, there exists $h \in B(X, Y)$ such that $\flat[h\,;r/3]$ includes a tail of (f_n). So, for sufficiently large $n \in \mathbb{N}$, we have $s(h, f_n) < r/3$ and therefore, for each $x \in X$, $e(h(x), f_n(x)) \leq s(h, f_n) < r/3$. Because $f_n(x) \to g(x)$ for all $x \in X$, it follows that $e(h(x), g(x)) \leq r/3$ for all $x \in X$, yielding $s(g, h) \leq r/3$. Since $s(g, h) \leq r/3$ and $\flat[h\,;r/3]$ includes a tail of (f_n), certainly $\flat[g\,;r]$ includes the same tail. But r is arbitrary in $\mathbb{R}^+$, so this proves that (f_n) converges to g in $B(X, Y)$. Since (f_n) is an arbitrary Cauchy sequence in $B(X, Y)$, $B(X, Y)$ is complete.

   Towards the converse, suppose that $B(X, Y)$ is complete and let K denote the set of all constant functions from X to Y. Certainly $K \subseteq B(X, Y)$. If $f \in B(X, Y) \setminus K$ and $a, b \in f(X)$ with $a \neq b$, then $\flat[f\,;e(a, b)/3]$ contains no member of K, so that $B(X, Y) \setminus K$ is open in $B(X, Y)$ (5.2.2). It follows that K is closed in $B(X, Y)$ and therefore complete (10.3.2). For each $y \in Y$, denote by $\tilde{y}$ the member of K with values equal to y. For all $a, b \in Y$, we have $s(\tilde{a}, \tilde{b}) = e(a, b)$, so that the map $y \mapsto \tilde{y}$ is an isometry of Y onto K. Being isometric to a complete metric space, Y itself is complete.   □

Example 10.8.2

For any non-empty set X, the spaces $B(X, \mathbb{R})$ and $B(X, \mathbb{C})$ are complete because $\mathbb{R}$ and $\mathbb{C}$ are complete. In particular, the spaces $\ell_\infty(\mathbb{R})$ and $\ell_\infty(\mathbb{C})$ of bounded sequences in $\mathbb{R}$ and $\mathbb{C}$ (7.4.4), being $B(\mathbb{N}, \mathbb{R})$ and $B(\mathbb{N}, \mathbb{C})$, respectively, are both complete.

Example 10.8.3

Suppose X is a metric space and Y is a complete metric space. Then, by 10.3.2, the metric space $\mathcal{C}(X, Y)$ of continuous bounded functions (8.9.1) from X to Y is a complete metric space when endowed with the supremum metric because it is a closed subspace of the complete space $B(X, Y)$ (8.9.3). In particular, the spaces $\mathcal{C}(X)$ of continuous bounded real or complex functions on any metric space X are complete.

Question 10.8.4

We have seen that $c(\mathbb{R})$, the space of convergent real sequences, is a subspace of ℓ_∞ (7.6.2). Is $c(\mathbb{R})$ closed in ℓ_∞ and therefore complete? The reader who cares to do so can prove directly that it is. For the more adventurous reader, we offer the following approach suggested by the comment in 8.9.4. First, $B(\tilde{\mathbb{N}}, \mathbb{R})$ is complete irrespective of the metric on $\tilde{\mathbb{N}}$ (10.8.1) because $\mathbb{R}$ is complete; next, its subspace $\mathcal{C}(\tilde{\mathbb{N}}, \mathbb{R})$ is closed in $B(\tilde{\mathbb{N}}, \mathbb{R})$ (8.9.3) and therefore complete, whatever metric is placed on $\tilde{\mathbb{N}}$. Finally, $c(\mathbb{R})$ is an isometric copy of $\mathcal{C}(\tilde{\mathbb{N}}, \mathbb{R})$ (8.10.2) when $\tilde{\mathbb{N}}$ has the inverse metric of 1.1.12. So $c(\mathbb{R})$ is also complete.

Example 10.8.5

It is easy to check (Q 10.5) that the space $c_0(\mathbb{R})$ of real sequences that converge to 0 with the supremum metric is closed in $c(\mathbb{R})$ and is therefore complete.

Example 10.8.6

Suppose X is a real normed linear space and U is its open unit ball. Then $B(U, \mathbb{R})$ is complete (10.8.1). Let X^* denote the set of all linear maps (B.20.12) from X to $\mathbb{R}$ whose restrictions to U are bounded, and let S denote the set of those bounded restrictions. It is not difficult to establish that the map $f \mapsto f|_U$ is bijective from X^* onto S. We use 6.6.3 to show that S is closed in $B(U, \mathbb{R})$. Suppose $g \in \mathrm{Cl}_{B(U,\mathbb{R})}(S)$ and (f_n) is a sequence in X^* such that $f_n|_U \to g$. Then for all $a, b \in U$ and $\lambda \in \mathbb{R}$ with $a + b \in U$ and $\lambda a \in U$, linearity of the f_n yields $g(a + b) = \lim f_n(a + b) = \lim f_n(a) + \lim f_n(b) = g(a) + g(b)$ and $g(\lambda a) = \lim f_n(\lambda a) = \lambda \lim f_n(a) = \lambda g(a)$, forcing g to be the restriction to U of a linear map and thus a member of S. So S is closed in $B(U, \mathbb{R})$ and therefore complete (10.3.2). We endow X^* with the metric determined by the bounds of its members on U, namely $(f, g) \mapsto \sup\{|f(x) - g(x)| \mid x \in U\}$. Then X^* is an isometric copy of S and is therefore also complete. X^* with this metric—its usual metric—is known as the *dual* of X. The algebraic structure of the dual need not concern us here; what is important is that X^* is a complete space whether or not X is complete. For practical examples, ℓ_1 can be identified as an isometric copy of c_0^* and ℓ_2 as an isometric copy of ℓ_2^*; it follows that ℓ_1 and ℓ_2 are complete (see 7.4.5). We omit the proofs; they can be found in [6].

10.9 Extending Continuous Functions

Continuous functions that cannot be extended continuously to the closure of their domains abound (8.6.2); this failure may occur even when we are allowed to extend the codomain of the function (10.9.2). Uniformly continuous functions behave much better. We see now that every uniformly continuous function can be continuously extended to the closure of its domain in some larger space without enlarging the codomain, provided only that the codomain is complete (10.9.1).

Theorem 10.9.1

Suppose (X, d) and (Y, e) are metric spaces and Y is complete. Suppose S is a dense subset of X and $f: S \to Y$ is a uniformly continuous function. Then there exists a uniformly continuous function $\tilde{f}: X \to Y$ such that $\tilde{f}|_S = f$. Moreover, f has no other continuous extension to X.

Proof

A function whose domain is a subset of a rectangle, each value of the function being represented by the intensity of the shade of grey. We ask the reader to pretend the domain is a dense subset of the rectangle.

For each $\epsilon \in \mathbb{R}^+$, let δ_ϵ be the largest real number in $(0, 1]$ such that for all $a, b \in S$ with $d(a, b) < \delta_\epsilon$ we have $e(f(a), f(b)) < \epsilon$ (compare 8.1.3). Then, for each $x \in X$, let $B_{x,\epsilon} = \flat_X[x; \delta_\epsilon/2]$. Suppose $z \in X$. Because $d(a, b) < \delta_\epsilon$ for all $a, b \in S \cap B_{z,\epsilon}$, we have also $e(f(a), f(b)) < \epsilon$, whence $\mathrm{diam}(f(B_{z,\epsilon})) \leq \epsilon$ and then also $\mathrm{diam}\left(\overline{f(B_{z,\epsilon})}\right) \leq \epsilon$ (3.6.11). Because S is dense in X, each of the balls $B_{z,\epsilon}$ has non-empty intersection with S (4.2.1), so each of the sets $f(B_{z,\epsilon})$ is non-empty. Because Y is complete, the nest $\left\{ \overline{f(B_{z,\epsilon})} \,\middle|\, \epsilon \in \mathbb{R}^+ \right\}$ of

non-empty closed subsets of Y has singleton intersection (10.2.1). Set $\tilde{f}(z)$ to be the sole member of that intersection; it is equal to $f(z)$ if $z \in S$, so, when this action is performed for each $z \in X$, $\tilde{f}$ is an extension of f to X. We must show that $\tilde{f}$ is uniformly continuous.

Let $\gamma \in \mathbb{R}^+$ and suppose that $u, v \in X$ satisfy $d(u, v) < \delta_{\gamma/3}/2$. Then the open set $B_{u,\gamma/3} \cap B_{v,\gamma/3}$ contains both u and v and is thus not empty; because S is dense in X, it contains some point a of S (4.2.1). Therefore $f(a) \in f(B_{u,\gamma/3}) \cap f(B_{v,\gamma/3})$. By definition, we have both $\tilde{f}(u) \in \overline{f(B_{u,\gamma/3})}$ and $\tilde{f}(v) \in \overline{f(B_{v,\gamma/3})}$. We have shown above that these sets have diameter not exceeding $\gamma/3$, forcing both $e(f(a), \tilde{f}(u)) \leq \gamma/3$ and $e(f(a), \tilde{f}(v)) \leq \gamma/3$ and yielding $e(\tilde{f}(u), \tilde{f}(v)) \leq 2\gamma/3 < \gamma$. So $\tilde{f}$ is uniformly continuous on X.

Towards uniqueness of $\tilde{f}$, we suppose that $g\colon X \to Y$ is continuous and $g|_S = f$. Suppose that $w \in X$ and that (s_n) is a sequence in S such that $s_n \to w$ in X. Since g and $\tilde{f}$ are both continuous on X, we have $g(s_n) \to g(w)$ and $\tilde{f}(s_n) \to \tilde{f}(w)$ in Y. But, for each $n \in \mathbb{N}$, $g(s_n) = f(s_n) = \tilde{f}(s_n)$, so that, because limits are unique in Y, $g(w) = \tilde{f}(w)$. And because w is arbitrary in X, this finally yields the conclusion that $g = \tilde{f}$. $\qquad\square$

The extension. We ask the reader to believe that apparently sharp contrasts are effected continuously.

Example 10.9.2

Continuous functions cannot always be extended continuously. The function $x \mapsto 1/x$ from $\mathbb{R}\backslash\{0\}$ to $\mathbb{R}\backslash\{0\}$ is continuous but not uniformly continuous on $\mathbb{R}\backslash\{0\}$. It can be extended to $\mathbb{R}$, but, whatever value we give it at 0, the resulting function is not continuous, even if the codomain is extended to $\mathbb{R}$.

There is another important extension theorem that applies to all members of the complete space $\mathcal{C}(S,\mathbb{R})$, where S is any closed subset of a metric space X: a bounded real continuous function defined on S can always be extended continuously to X (10.9.4) (even though, as we demonstrate in 10.9.5, there may be no continuous extension for a continuous bounded function defined on a dense subset of X). This theorem is known as Tietze's Extension Theorem; we place the bulk of the work in the following lemma, where the reader may detect echoes of 8.3.12.

Lemma 10.9.3

Suppose X is a metric space and S is a closed subset of X. Suppose $f \in \mathcal{C}(S,\mathbb{R})$ is not constant. Let $a = \inf f(S)$ and $b = \sup f(S)$. Let $k = (b-a)/3$ and $A = f^{-1}(-\infty, a+k]$ and $B = f^{-1}[b-k, \infty)$. Then the assignment

$$x \mapsto a + k + \frac{k\,\mathrm{dist}(x,A)}{\mathrm{dist}(x,A) + \mathrm{dist}(x,B)}$$

yields a function $f^{\dagger} \in \mathcal{C}(X,\mathbb{R})$ that has the properties $f^{\dagger}(X) \subseteq [a+k, b-k]$ and $\inf(f - f^{\dagger})(S) = -k$ and $\sup(f - f^{\dagger})(S) = k$.

Proof

Because $k > 0$ and $\inf f(S) = a$ and $\sup f(S) = b$, A and B are non-empty; because f is continuous, they are both closed in S (8.3.1) and therefore also closed in X (Q 4.5); and because $a + k < b - k$, $A \cap B = \varnothing$. So, for each $x \in X$, $\operatorname{dist}(x, A) + \operatorname{dist}(x, B) > 0$ (3.6.10) and $f^\dagger$ is well defined on X by

$$f^\dagger(x) = a + k + \frac{k \operatorname{dist}(x, A)}{\operatorname{dist}(x, A) + \operatorname{dist}(x, B)} \quad \text{for each } x \in X.$$

This function $f^\dagger$ is a composition of continuous functions and so is continuous (compare 8.3.12). By construction, the minimum and maximum values of $f^\dagger$ are $a + k$ and $b - k$ and are attained at all points of A and B, respectively. Because $\inf f(S) = a$ and $f(S \backslash A) \subseteq (a + k, \infty)$, we have also $\inf f(A) = a$; because $\sup f(S) = b$ and $f(S \backslash B) \subseteq (-\infty, b - k)$, we have also $\sup f(B) = b$; and since $f^\dagger(A) = \{a + k\}$ and $f^\dagger(B) = \{b - k\}$, we deduce that $\inf(f - f^\dagger)(A) = -k$ and $\sup(f - f^\dagger)(B) = k$. But all other values of $f - f^\dagger$ on S lie between these two values. Specifically, for $s \in S \backslash (A \cup B)$, we have $a + k < f(s) < b - k$ and $a + k < f^\dagger(s) < b - k$, so that $\left| f(s) - f^\dagger(s) \right| < b - a - 2k = k$. So $\inf(f - f^\dagger)(S) = -k$ and $\sup(f - f^\dagger)(S) = k$, as required.   □

Theorem 10.9.4 (Tietze's Extension Theorem)

Suppose X is a metric space and S is a closed subset of X. Suppose $f \in \mathcal{C}(S, \mathbb{R})$. Then there exists $\tilde{f} \in \mathcal{C}(X, \mathbb{R})$ such that $\tilde{f}|_S = f$ and $\inf \tilde{f}(X) = \inf f(S)$ and $\sup \tilde{f}(X) = \sup f(S)$.

Proof

If f is constant, the result is easy, so we suppose otherwise. Let $a = \inf f(S)$ and $b = \sup f(S)$, and set $k = (b - a)/3$. Using the notation of 10.9.3, define $h_1 = f^\dagger$ and, for each $n \in \mathbb{N}$, recursively define $h_{n+1} = h_n + (f - h_n)^\dagger$ (B.19). Using 10.9.3 and applying induction—the details are left to the reader (Q 10.7)—it is easily verified that $h_n \in \mathcal{C}(X, \mathbb{R})$ and

$$\inf(f - h_n)(S) = -\frac{2^{n-1}k}{3^{n-1}} \quad \text{and} \quad \sup(f - h_n)(S) = \frac{2^{n-1}k}{3^{n-1}} \quad \text{and}$$

$$h_n(X) \subseteq \left[a + \frac{2^{n-1}k}{3^{n-1}}, b - \frac{2^{n-1}k}{3^{n-1}} \right] \text{ and } (h_{n+1} - h_n)(X) \subseteq \left[-\frac{2^{n-1}k}{3^n}, \frac{2^{n-1}k}{3^n} \right]$$

for all $n \in \mathbb{N}$. For $m, n \in \mathbb{N}$ with $m < n$ and all $x \in X$, we then have

$$|h_n(x) - h_m(x)| \leq \sum_{i=m}^{n-1} \frac{2^{i-1}k}{3^i} < \frac{2^{m-1}k}{3^{m-1}},$$

so that (h_n) is a Cauchy sequence in the complete space $\mathcal{C}(X, \mathbb{R})$ (10.8.3) and so converges to some $\tilde{f} \in \mathcal{C}(X, \mathbb{R})$. Moreover, since $h_n(X) \subseteq [a, b]$ for all $n \in \mathbb{N}$, we have $\tilde{f}(X) \subseteq [a, b]$. And because, for all $s \in S$ and all $n \in \mathbb{N}$, $|f(s) - h_n(s)| \leq 2^{n-1}k/3^{n-1} \to 0$, it follows that $\tilde{f}(s) = f(s)$, as required.   □

Example 10.9.5

There are continuous bounded functions that cannot be continuously extended. One example is the function $f : x \mapsto \sin(1/x)$ defined on $\mathbb{R}^+$ that is pictured in 3.3.5. Its domain $\mathbb{R}^+$ is not closed in $\mathbb{R}^\oplus$, so Tietze's Theorem cannot be invoked to extend it to $\mathbb{R}^\oplus$ or further, but it is instructive to see where the theorem fails for this function. The problem lies in the lemma. In this case, the sets A and B of 10.9.3 are closed in $S = \mathbb{R}^+$ but not in $X = \mathbb{R}^\oplus$ and, although $A \cap B = \varnothing$, the set $\mathrm{Cl}_X(A) \cap \mathrm{Cl}_X(B)$ is not empty, so that $\mathrm{dist}(x,A) + \mathrm{dist}(x,B)$ takes the value 0 and the function $f^\dagger$ is not well defined.

Example 10.9.6

Some real bounded continuous functions defined on non-closed sets can be extended continuously. Suppose (X,d) is any metric space, S is any subset of X and $f \colon S \to \mathbb{R}$ is bounded and uniformly continuous. Then f can be regarded as a function from S into the bounded interval $I = [\inf f(S), \sup f(S)]$, which is complete by 10.3.2. Then, by 10.9.1, f can be extended continuously to $\overline{S}$ in such a way that the range of the extension lies in I. Then, since the extension is bounded, it in turn can be extended by Tietze's Theorem (10.9.4) to a continuous function from X into I.

Example 10.9.7

It follows immediately from the definition of compactness (9.1.4) and the Cauchy criterion for having the nearest-point property (7.11.1) that a metric space is compact if, and only if, it is complete and totally bounded. Moreover, we have seen that every totally bounded non-empty metric space is a uniformly continuous image of a subset A of the Cantor set (9.3.1). Suppose now that X is a non-empty compact metric space. By 9.3.1, because X is totally bounded, there exists a subset A of the Cantor set $\mathcal{K}$ and a uniformly continuous bijective mapping $\phi \colon A \to X$. By 10.9.1, because X is complete, ϕ can be uniformly continuously extended to $\overline{A}$; we denote the unique extension also by ϕ, merely noting that it is not injective unless A is closed. We can now make a further extension: ϕ has a uniformly continuous extension to the whole of the Cantor set $\mathcal{K}$. The procedure is as follows.

For each $x \in \mathcal{K}$, let $\alpha(x) = x - \mathrm{dist}\left(x, \overline{A}\right)$ and $\beta(x) = x + \mathrm{dist}\left(x, \overline{A}\right)$. Since $\overline{A}$ has the nearest-point property (7.12.2), there is a nearest point of $\overline{A}$ to each point x of $\mathcal{K}$ (7.11.1); the only possibilities are $\alpha(x)$ and $\beta(x)$, so either $\alpha(x) \in \overline{A}$ or $\beta(x) \in \overline{A}$. If $x \notin \overline{A}$ and both $\alpha(x)$ and $\beta(x)$ are in $\overline{A}$, then, since $x = (\alpha(x) + \beta(x))/2$, we have $x \in \{k/3^n \mid n \in \mathbb{N}, \ k \in \mathbb{N}_{3^n}, \ k/3 \notin \mathbb{N}\}$ by Q3.9. With this fact in mind, we define $n \colon \mathcal{K} \to \overline{A}$ by setting $n(x) = x$ if $x \in \overline{A}$ and, if $x \notin \overline{A}$,

$$n(x) = \begin{cases} \alpha(x) \text{ if } \alpha(x) \in \overline{A} \text{ and either } \beta(x) \notin \overline{A} \text{ or } x \in \{k/3^n \mid n \in \mathbb{N}, k \equiv 1(\text{mod } 3)\}; \\ \beta(x) \text{ if } \beta(x) \in \overline{A} \text{ and either } \alpha(x) \notin \overline{A} \text{ or } x \in \{k/3^n \mid n \in \mathbb{N}, k \equiv 2(\text{mod } 3)\}. \end{cases}$$

It is not difficult to see that the only points at which n might not be continuous are those where there is more than one nearest point. But, if x is such a point of $\mathcal{K}$, then $x = k/3^n$ for some $n \in \mathbb{N}$ and $k \in \mathbb{N}_{3^n}$, where k is not divisible by 3. If

0 $\alpha(x)$ x $\beta(x)$ 1

Approximations of an example $\overline{A}$ (above) and $\mathcal{K}$ (below). $x = 7/9 \in \mathcal{K}\backslash\overline{A}$, $\alpha(x) = 2/3 \in \overline{A}$ and $\beta(x) = 8/9 \in \overline{A}$. So $n(x) = \alpha(x)$ and $n(z) = \alpha(x)$ for all $z \in \mathcal{K} \cap [\alpha(x), x]$. Thus n is continuous at x because no point of $\mathcal{K}$ lies in $(x, \beta(x))$.

$k \equiv 1(\text{mod } 3)$, then $(x, x + 1/3^n) \cap \mathcal{K} = \varnothing$ by Q3.8, so that n has the constant value $\alpha(x)$ on the set $\mathcal{K} \cap [\alpha(x), x + 1/3^n)$; in particular, n is continuous at x. A similar argument can be made if $k \equiv 2(\text{mod } 3)$. Then the composition $\phi \circ n$ is a continuous function from $\mathcal{K}$ onto X (8.5.1); it is uniformly continuous by 9.1.5 because its domain $\mathcal{K}$ is compact (7.12.2).

10.10 Banach's Fixed-Point Theorem

Strong contractions on a metric space, when iterated, tend to pull all the points of the space together into a single point (10.10.3). Banach's Theorem, also called the *Banach Contraction Principle*, is that such a *fixed point* must exist if the space is complete and can, in that case, be computed by iteration. This theorem is invaluable for developing algorithmic procedures and generally for computing solutions to equations.

Definition 10.10.1

Suppose X is a non-empty set and $f: X \to X$. A point $x \in X$ is called a *fixed point* for f if and only if $f(x) = x$.

Example 10.10.2

Every continuous function from $[0, 1]$ to $[0, 1]$ has at least one fixed point, though it may have many. At least, the following assertion is intuitively true: suppose $f: [0, 1] \to [0, 1]$ is continuous, $f(0) = \alpha \in (0, 1]$ and $f(1) = \beta \in [0, 1)$. Then the graph of f joins $(0, \alpha)$ continuously to $(1, \beta)$ and must cross the line $\{z \in \mathbb{R}^2 \mid z_1 = z_2\}$ somewhere along the way. Can you prove that it crosses the line (Q10.10)?

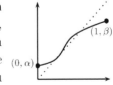

Theorem 10.10.3 (Banach's Fixed-Point Theorem)

Suppose (X, d) is a non-empty complete metric space and $f\colon X \to X$ is a strong contraction on X with Lipschitz constant $k \in (0, 1)$. Then f has a unique fixed point in X and, for each $w \in X$, the sequence $(f^n(w))$ converges to this point.

Proof

Suppose that $w \in X$. By 9.9.4, $(f^n(w))$ is Cauchy in X and, because X is complete, it converges in X. Let $z = \lim f^n(w)$. Because f is continuous (9.9.2), the convergence criterion for continuity of f at z (8.1.1) yields $f(z) = \lim f^{n+1}(w)$, and, because $(f^{n+1}(w))$ is a subsequence of $(f^n(w))$, 6.7.1 then gives $\lim f^{n+1}(w) = \lim f^n(w)$, so that $f(z) = z$, as required. The function f can have no other fixed point for, if $a \in X$ were such a point, then we should have $d(a, z) = d(f(a), f(z)) \leq k d(a, z)$, which forces $d(a, z) = 0$ and thus $a = z$ because $k < 1$.   □

Example 10.10.4

Suppose $f\colon \mathbb{R} \to \mathbb{R}$ is a differentiable function and there exists $k \in [0, 1)$ such that $|f'(x)| \leq k$ for all $x \in \mathbb{R}$. Then, for each $x, y \in \mathbb{R}$ with $x < y$, it follows from the Mean Value Theorem that there exists $c \in (x, y)$ with $f(y) - f(x) = (y - x)f'(c)$. From this we get $|f(y) - f(x)| \leq k|y - x|$, so that f is a strong contraction. Then Banach's Theorem tells us that f has a unique fixed point.

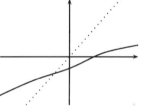

A graph with a shallow slope meets the dotted line in a unique point.

Example 10.10.5

Let $\mathcal{S}$ denote the set of non-empty closed bounded subsets of $\mathbb{R}$ endowed with the Hausdorff metric h. This space is complete (10.7.2). For each $A \in \mathcal{S}$, define $f(A) = \{a/3 \mid a \in A\} \cup \{(a + 2)/3 \mid a \in A\}$. Then $f(A)$, being a union of two closed bounded subsets of $\mathbb{R}$, is in $\mathcal{S}$. So $f\colon \mathcal{S} \to \mathcal{S}$. Suppose now that $A, B \in \mathcal{S}$. For each $a \in A$, we have $\operatorname{dist}(a/3, f(B)) \leq |a/3 - b/3| = |a - b|/3$ for all $b \in B$, so that $\operatorname{dist}(a/3, f(B)) \leq \operatorname{dist}(a, B)/3$. Similarly we have $\operatorname{dist}((a + 2)/3, f(B)) \leq |(a + 2)/3 - (b + 2)/3| = |a - b|/3$ for all $b \in B$, so that $\operatorname{dist}((a + 2)/3, f(B)) \leq \operatorname{dist}(a, B)/3$. The two inequalities together give us $\sup\{\operatorname{dist}(z, f(B)) \mid z \in f(A)\} \leq \frac{1}{3}\sup\{\operatorname{dist}(a, B) \mid a \in A\}$. A similar calculation gives $\sup\{\operatorname{dist}(w, f(A)) \mid w \in f(B)\} \leq \frac{1}{3}\sup\{\operatorname{dist}(b, A) \mid b \in B\}$. Therefore $h(f(A), f(B)) \leq \frac{1}{3}h(A, B)$, making f a strong contraction. There is no prize for guessing what the unique fixed point of this contraction is.

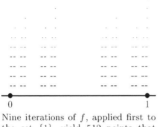

Nine iterations of f, applied first to the set $\{1\}$, yield 512 points that already look like the Cantor set.

What is amazing is that it can be obtained by iteration starting with any non-empty closed bounded subset of $\mathbb{R}$. If we start with the interval $[0,1]$, f extracts the middle third to give $[0,1/3] \cup [2/3,1]$. Its application again and again is the standard construction performed in 3.3.3. But we can equally well start with any member of $\mathcal{S}$. If we start, for example, with the set $\{1\}$; then $(f^n(\{1\}))$ is a sequence of finite subsets of $[0,1]$—an increasing sequence in this case—that converges in $(\mathcal{S},h)$ to the fixed point of f, which is, of course, the Cantor set.

Example 10.10.6

The graph of the cosine function clearly crosses the line $y = x$ somewhere between $x = 0$ and $x = \pi/2$, and it does so exactly once. In other words, the cosine function restricted to $[0,\pi/2]$ has a unique fixed point. What is that unique number $z \in [0,\pi/2]$ such that $\cos z = z$? It turns out that we can apply Banach's Theorem to discover it despite the fact that the cosine function is not a strong contraction on $[0,\pi/2]$. Let us see why this is so. The function $x \mapsto \cos(\cos x)$ is a strong contraction by 10.10.4 because its derivative $x \mapsto \sin(\cos x)\sin x$ is bounded by $\sin 1 \in (0,1)$. So

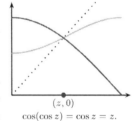

$(z,0)$

$\cos(\cos z) = \cos z = z.$

iteration of the cosine function will lead us to the fixed point of $\cos \circ \cos$. This is the same as the fixed point of $\cos$ itself (Q10.9). Starting at 0 and iterating the cosine function, we get 0, 1, 0·54030230586814, 0·85755321584639, 0·65428979049778, 0·79348035874257, leading eventually to the approximate value of 0·73908513321516 for z.

Example 10.10.7

The Cantor set is an example of a *fractal*.[2] On the right is a picture of another fractal, *Sierpiński's triangle*, this time in $\mathbb{R}^2$ rather than $\mathbb{R}$. The picture was obtained by iterating the function
$$A \mapsto \{(\cdot 5a_1 + c_1, \cdot 5a_2 + c_2) \mid a \in A, c \in \{(1,1),(1,50),(50,50)\}\}$$
on the space of non-empty closed bounded subsets of $\mathbb{R}^2$ starting at a single point.

---

[2] The interested reader is referred to [1] and [7] for more information about fractals.

Another well-known fractal in $\mathbb{R}^2$ is *Barnsley's fern*, which is an idealized version of *Asplenium adiantum-nigrum*. One version of it looks not unlike the real fern pictured here. The function given by Barnsley [1] to produce by iteration a picture of *Barnsley's fern* is $A \mapsto \bigcup\{\{M_i a + C_i \mid a \in A\} \mid i \in \mathbb{N}_4\}$, where

$$M_1 = \begin{pmatrix} 0 & 0 \\ 0 & 0{\cdot}16 \end{pmatrix}, \ M_2 = \begin{pmatrix} 0{\cdot}85 & 0{\cdot}04 \\ -0{\cdot}04 & 0{\cdot}85 \end{pmatrix}, \ M_3 = \begin{pmatrix} 0{\cdot}2 & -0{\cdot}26 \\ 0{\cdot}23 & 0{\cdot}22 \end{pmatrix},$$

$$M_4 = \begin{pmatrix} -0{\cdot}15 & 0{\cdot}28 \\ 0{\cdot}26 & 0{\cdot}24 \end{pmatrix}, \ C_1 = \begin{pmatrix} 0 \\ 0 \end{pmatrix}, \ C_2 = \begin{pmatrix} 0 \\ 0{\cdot}16 \end{pmatrix}, \ C_3 = \begin{pmatrix} 0 \\ 0{\cdot}16 \end{pmatrix}, \ C_4 = \begin{pmatrix} 0 \\ 0{\cdot}44 \end{pmatrix}.$$

Example 10.10.8

Let $n \in \mathbb{N}$. A system of n linear equations in n real unknowns can be represented as $Ax = c$, where A is an $n \times n$ matrix, and c and x, represented as $n \times 1$ column matrices, are members of $\mathbb{R}^n$, c being fixed and x having entries that are the n unknowns. This system of equations has a unique solution for x if, and only if, A is invertible; in that case, the solution is given by $x = A^{-1}c$. This might not, however, be the most efficient way of solving the system. Banach's Fixed-Point Theorem sometimes provides an alternative method. Define $f \colon \mathbb{R}^n \to \mathbb{R}^n$ to be $x \mapsto x - Ax + c$. The solutions to the original system are precisely the fixed points of f. If f happens to be a strong contraction when $\mathbb{R}^n$ is endowed with some complete metric, then not only do we know that there is a unique fixed point, but we can get the solution simply by starting anywhere in $\mathbb{R}^n$ and iterating the function f. We know that all conserving metrics make $\mathbb{R}^n$ complete (10.5.1); in particular, this is true of the metrics μ_1, μ_2 and μ_∞ (1.6.3). Set $(w_{i,j})$ to be the matrix $I - A$, where I is the identity $n \times n$ matrix. A little calculation reveals that, if we set

$$k_1 = \max\left\{\sum_{i=1}^{n} |w_{i,j}| \ \middle| \ j \in \mathbb{N}_n\right\}, \quad k_2 = \sqrt{\sum_{i,j=1}^{n} |w_{i,j}|^2}, \quad k_\infty = \max\left\{\sum_{j=1}^{n} |w_{i,j}| \ \middle| \ i \in \mathbb{N}_n\right\},$$

then, for all $a, b \in \mathbb{R}^n$ and $i \in \{1, 2, \infty\}$, we have $\mu_i(f(a), f(b)) \le k_i \mu_i(a, b)$. So, if any of the k_i is less than 1—and the reader can check that one of them may be and the others not—iteration provides a method of solution. Note that fixed points and the process of iteration are purely algebraic phenomena that have nothing to do with the metric, but a suitable metric both describes what happens when we iterate and provides an explanation of why it happens.

10.11 Baire's Theorem

Complete metric spaces have the wonderful property that every countable intersection of dense open subsets is non-empty; this property is crucial for many further developments in modern analysis. In general, intersections of dense subsets of a metric space may be empty, as $\mathbb{Q} \cap (\mathbb{R}\backslash\mathbb{Q})$ is empty. But we are persuaded to think a little deeper when we consider intersections of open dense subsets. If the number of sets is non-zero and finite, then the intersection is both open and dense (10.11.1), and if the number of sets is uncountable, the intersection may be empty, even in a complete space (10.11.2). The theorem that countable intersections are non-empty, indeed dense, in a complete metric space (10.11.4) is due to Baire and has historically been known as the Baire Category Theorem.[3]

Theorem 10.11.1

Suppose (X, d) is a metric space and $\mathcal{U}$ is a non-empty finite collection of open dense subsets of X. Then $\bigcap \mathcal{U}$ is an open dense subset of X.

Proof

Because $\mathcal{U}$ is finite, $\bigcap \mathcal{U}$ is open in X by 4.3.2. Each individual member of $\mathcal{U}$ is dense in X; let $\mathcal{V}$ be a subset of $\mathcal{U}$ such that $\bigcap \mathcal{V}$ is dense in X and suppose that there is no proper superset of $\mathcal{V}$ in $\mathcal{U}$ that has dense intersection. Suppose $\mathcal{V} \neq \mathcal{U}$ and let $A \in \mathcal{U}\backslash\mathcal{V}$. Then $A \cap \bigcap \mathcal{V}$ is not dense in X by hypothesis, so there exists a non-empty open subset B of X such that $B \cap A \cap \bigcap \mathcal{V} = \varnothing$ (4.2.1). But A is dense in X and $B \cap \bigcap \mathcal{V}$ is open, so that $B \cap \bigcap \mathcal{V} = \varnothing$ by 4.2.1, which contradicts the assumption that $\bigcap \mathcal{V}$ is dense in X, again by 4.2.1. We conclude that $\mathcal{V} = \mathcal{U}$.   □

Example 10.11.2

The intersection of an infinite collection of open dense sets need not be open by 4.3.3. It need not be dense either. Consider $\{\mathbb{R}\backslash\{r\} \mid r \in \mathbb{R}\}$. Each member of this set is open and dense in $\mathbb{R}$, but the intersection is empty. Even a countable collection of dense open subsets of a metric space need not have dense intersection. For example, the set $\{\mathbb{Q}\backslash\{q\} \mid q \in \mathbb{Q}\}$ is a countable set of subsets of $\mathbb{Q}$, each open and dense in $\mathbb{Q}$, yet $\bigcap\{\mathbb{Q}\backslash\{q\} \mid q \in \mathbb{Q}\} = \varnothing$. Note that $\mathbb{Q}$ is not complete.

---

[3] Baire's categories have nothing to do with modern category theory. For Baire, a subset of a metric space is of the *first category* if it is a countable union of *nowhere dense* sets (Q 4.19) and is of the *second category* otherwise. His theorem can then be expressed by saying that, in a complete metric space, every non-empty open subset is of the second category (see Q 10.8).

Definition 10.11.3

Suppose X is a metric space. X is called a *Baire metric space* if, and only if, every intersection of a non-empty countable collection of dense open subsets of X is dense in X.

Theorem 10.11.4 (Baire's Theorem)

Every complete metric space is a Baire space.

Proof

Suppose that X is a complete metric space and that $\mathcal{C}$ is a non-empty countable collection of dense open subsets of X. If $\mathcal{C}$ is finite, we invoke 10.11.1 to establish that $\bigcap \mathcal{C}$ is dense in X. Otherwise, we suppose that $\mathcal{C}$ is infinite and proceed as follows.

Suppose W is an arbitrary non-empty open subset of X. We want to show that $W \cap \bigcap \mathcal{C}$ is non-empty. This is certainly so if $W \cap \operatorname{iso}(X) \neq \varnothing$ because $\operatorname{iso}(X) \subseteq \bigcap \mathcal{C}$ (Q 4.18), so we assume that W contains no isolated point of X.

Let (U_n) be an enumeration of the members of $\mathcal{C}$, set $V_0 = X$ and, for each $n \in \mathbb{N}$, set $V_n = \bigcap \{U_i \mid i \in \mathbb{N}_n\}$. Let $\mathcal{S}$ denote the collection of open balls of X that are included in W; $\mathcal{S}$ is non-empty because W is non-empty and open in X (5.2.2). Note also that every member of $\mathcal{S}$ has positive diameter because W contains no isolated point of X. If any member of $\mathcal{S}$ is included in V_n for all $n \in \mathbb{N}$, then certainly $W \cap \bigcap \mathcal{C} \neq \varnothing$, so we suppose otherwise.

For each $A, B \in \mathcal{S}$, in what follows we shall write $B \prec A$ if, and only if, $\overline{B} \subseteq A$ and there exists $n \in \mathbb{N}$ such that $B \subseteq V_n$ and $A \not\subseteq V_n$. Consider the relation

$$\rho = \{(A, B) \mid A, B \in S,\ B \prec A,\ \operatorname{diam}(B) \leq \operatorname{diam}(A)/2\}.$$

The domain of ρ is $\mathcal{S}$ for the following reasons. For any $A \in \mathcal{S}$, there exists $n \in \mathbb{N}$ such that $A \subseteq V_{n-1}$ and $A \not\subseteq V_n$, by hypothesis. But $A \cap U_n \neq \varnothing$ (4.2.1) because A is open and U_n is dense in X and, because A and U_n are both open in X, $A \cap U_n$ is open in X (4.3.2). So, since $\operatorname{diam}(A) > 0$, Q 5.2 ensures that there exists $B \in \mathcal{S}$ with $\operatorname{diam}(B) \leq \operatorname{diam}(A)/2$ such that $\overline{B} \subseteq A \cap U_n$, yielding $B \subseteq V_n$ and therefore also $B \prec A$.

So $\operatorname{ran}(\rho) \subseteq \operatorname{dom}(\rho)$. Therefore, by B.19.2, there exists a sequence (B_n) in $\mathcal{S}$ such that $(B_n, B_{n+1}) \in \rho$ for all $n \in \mathbb{N}$. Then $\operatorname{diam}(B_n) \to 0$ and, for each $n \in \mathbb{N}$, $B_{n+1} \prec B_n$. This yields, by induction, $B_{n+1} \subseteq V_n$ for each $n \in \mathbb{N}$.

Since X is complete, the nested sequence criterion of 10.2.1 ensures that $\bigcap \{\overline{B_n} \mid n \in \mathbb{N}\} \neq \varnothing$. Moreover, because, for each $n \in \mathbb{N}$, we have from above

$B_{n+1} \prec B_n$, and therefore $\overline{B_{n+1}} \subseteq B_n$, and $B_{n+1} \subseteq V_n$, we now get

$$\bigcap \{\overline{B_n} \mid n \in \mathbb{N}\} \subseteq \bigcap \{B_n \mid n \in \mathbb{N}\} \subseteq \bigcap \{V_n \mid n \in \mathbb{N}\} = \bigcap \mathcal{C}.$$

In other words, $\bigcap \mathcal{C}$ includes the non-empty set $\bigcap \{\overline{B_n} \mid n \in \mathbb{N}\}$. Thus $W \cap \bigcap \mathcal{C}$ is non-empty, as required. Since W is an arbitrary non-empty open subset of X, $\bigcap \mathcal{C}$ is dense in X by 4.2.1. So X is a Baire space.   □

Example 10.11.5

Every closed subset of a complete metric space is complete (10.3.2) and so is a Baire space. We shall see a little later (13.6.3) that every open subset of a complete metric space is also a Baire space, showing that not all Baire spaces are complete.

Question 10.11.6

All finite metric spaces are complete. Countable infinite metric spaces may or may not be complete: $\mathbb{N}$ is complete, whereas $\{1/n \mid n \in \mathbb{N}\}$ is not. Finite spaces consist of isolated points. Countable spaces may or may not have isolated points: $\mathbb{N}$ consists of isolated points, whereas $\mathbb{Q}$ has none. $\mathbb{Q}$ is not complete. This array of facts leads to a question: is there any countable complete metric space without an isolated point? The answer is *no*. Suppose X is a countable metric space that has no isolated points. Then, for each $x \in X$, the open set $X \backslash \{x\}$ is also dense. Since the intersection of all these sets is empty, it follows that X is not a Baire space and therefore not complete.

Example 10.11.7

That $\mathbb{R}$ is uncountable we accepted long ago, and we can prove it without recourse to advanced techniques. But Baire's Theorem, dependent on a wealth of theory and, of course, on B.19.1, gives a strikingly simple proof: $\mathbb{R}$ is complete and has no isolated point and is therefore uncountable (see 10.11.6).

Example 10.11.8

One of the most far-reaching implications of Baire's Theorem can be presented as an inverse mapping theorem: suppose X is a real or complex linear space and $\|\cdot\|'$ and $\|\cdot\|''$ are norms on X for which the spaces $(X, \|\cdot\|')$ and $(X, \|\cdot\|'')$ are both complete. Then the identity map from $(X, \|\cdot\|')$ to $(X, \|\cdot\|'')$ is continuous if, and only if, its inverse is also continuous. Although we have all the tools to prove this now, its proof is rather tricky and we omit it.

10.12   Completion of a Metric Space

Not all metric spaces are complete, so we are not always permitted to use Banach's Fixed-Point Theorem or the Baire Category Theorem. Before 6.11.3, we entertained the hope that every metric space X might have a minimal complete superspace. Now we transform the hope into reality. Actually, we shall work not with X itself but with the isometric copy made up of its point functions (1.2.1); we realize the completion of this space by appending to it all the virtual points of X. The resulting complete space is a subspace of the space of 1.5.1.

Definition 10.12.1

Suppose (X, d) is a metric space. A metric space (Y, e) is called a *completion* of (X, d) if, and only if, (Y, e) is complete and (X, d) is isometric to a dense subspace of (Y, e).

Theorem 10.12.2

Suppose (X, d) is a metric space. Let $\tilde{X} = \delta(X) \cup \mathrm{vp}(X)$, where $\delta(X)$ denotes the set of point functions (1.2.1) and $\mathrm{vp}(X)$ the set of virtual points (10.1.1) of X. Endow $\tilde{X}$ with the metric s given by $(u, v) \mapsto \sup\{|u(x) - v(x)| \mid x \in X\}$ (1.5.1). Then $(\tilde{X}, s)$ is a completion of (X, d).

Proof

The subspace $(\delta(X), s)$ of $(\tilde{X}, s)$ is an isometric copy of (X, d) by 1.5.1. For the density of $\delta(X)$, we proceed as follows. Suppose $u \in \tilde{X}$ and $r \in \mathbb{R}^+$. Then there exists $x \in X$ such that $u(x) < r$ and, for every $a \in X$, because u is pointlike,

$$|u(a) - \delta_x(a)| = |u(a) - d(x, a)| \leq u(x) < r,$$

so that $s(u, \delta_x) \leq r$ and thus $\mathrm{dist}_s(u, \delta(X)) \leq r$. Since r is arbitrary in $\mathbb{R}^+$, it follows that $\mathrm{dist}_s(u, \delta(X)) = 0$, so that $u \in \mathrm{Cl}_{\tilde{X}}(\delta(X))$. But u is arbitrary in $\tilde{X}$, so $\delta(X)$ is dense in $(\tilde{X}, s)$, as claimed.

   We want to show that $\tilde{X}$ is complete. Suppose α is a virtual point of $\tilde{X}$. Since $\delta(X)$ is dense in $\tilde{X}$ and α is continuous, we have $\inf \alpha(\delta(X)) = \inf \alpha(\tilde{X}) = 0$. So $\alpha|_{\delta(X)}$ is a virtual point of $\delta(X)$. Then the corresponding map v on the isometric copy X of $\delta(X)$, given by $v(x) = \alpha(\delta_x)$ for all $x \in X$, is a virtual point of X. So $v \in \tilde{X}$. For all $x \in X$, we have $\alpha(v) - \alpha(\delta_x) \leq s(v, \delta_x)$ and $\alpha(\delta_x) = v(x) = v(x) - \delta_x(x) \leq s(v, \delta_x)$, so that, by addition, $\alpha(v) \leq 2s(v, \delta_x)$. Since x is arbitrary, this gives $\alpha(v) \leq 2\mathrm{dist}_s(v, \delta(X)) = 0$, so that $\alpha(v) = 0$. This contradicts the assumption that α is a virtual point of $\tilde{X}$. So $\tilde{X}$ has no virtual points and is thus complete (10.2.1). $\square$

Note 10.12.3

The intuition that no virtual point of a metric space (X, d) is also a point of X can be justified from the axioms of set theory. It then follows from 10.12.2 that the superset $X \cup \mathrm{vp}(X)$ of X endowed with the appropriate extension of d is a completion of X that, unlike $\delta(X) \cup \mathrm{vp}(X)$, is also a genuine superspace of X.

Example 10.12.4

Suppose (X, d) and (Y, e) are metric spaces and $f \colon S \to Y$ is a uniformly continuous function on a dense subset S of X. Then 10.9.1 and 10.12.3 imply that f has a uniformly continuous extension $\tilde{f} \colon X \to Y \cup \mathrm{vp}(Y)$, where $Y \cup \mathrm{vp}(Y)$ has the appropriate metric to make it a completion of Y.

Theorem 10.12.5

Suppose that (X, d) is a metric space and (X', m) and (X'', s) are completions of X and that $\psi \colon X \to X'$ and $\phi \colon X \to X''$ are isometries onto dense subspaces of X' and X'', respectively. Then there is an isometry from X' to X'' that maps $\psi(X)$ onto $\phi(X)$.

Proof

The map $\phi \circ \psi^{-1}$ is an isometry from the dense subspace $\psi(X)$ of X' onto the dense subspace $\phi(X)$ of X''. Since X'' is complete, $\phi \circ \psi^{-1}$ has a unique continuous extension f to X' (10.9.1), and this extension is isometric (Q 10.14). Because f is an isometry, its range is complete, so is closed in X''. Since this range includes $\phi(X)$, it includes $\overline{\phi(X)}$, which is X''. $\square$

Summary

In this chapter, we have examined the concept of completeness and have shown how to complete any metric space. We have demonstrated the completeness of some important spaces. We have shown that completeness of the Hausdorff metric depends only on completeness of the original metric. We have also proved two theorems, Baire's and Banach's, that have far-reaching consequences both for analysis and for problem solving in many other areas of application.

EXERCISES

Q 10.1   Let X be the set of continuous functions $\mathcal{C}([0\,,1])$ equipped with the integral metric, $(f,g) \mapsto \int_0^1 |g(t) - f(t)|\,dt$. Show that X is not complete.

Q 10.2   Suppose X is a complete metric space. Show that the space $c(X)$ of convergent sequences in X is complete.

†Q 10.3   Show that a product metric on a finite product of complete metric spaces need not make the product complete.

Q 10.4   Find a set X and two metrics d and m on X such that the Cauchy sequences of (X,d) and (X,m) are identical and the identity map from (X,d) to (X,m) is continuous but not uniformly continuous.

†Q 10.5   Show that $c_0(\mathbb{R})$ is closed in $c(\mathbb{R})$ and is therefore complete.

Q 10.6   Find a Lipschitz function $f \colon \mathbb{R} \to \mathbb{R}$ that has incomplete range.

†Q 10.7   Supply the inductive argument left aside in the proof of 10.9.4.

†Q 10.8   Suppose X is a non-empty metric space. Show that the following are equivalent:

(i) X is a Baire space.

(ii)   Every countable union of closed nowhere dense (Q 4.19) subsets of X has dense complement.

(iii)   Every countable union of nowhere dense subsets of X has empty interior.

(iv)   No non-empty countable union of nowhere dense subsets of X is open in X.

Q 10.9   Suppose X is a complete metric space and $f \colon X \to X$ is such that $f \circ f$ is a strong contraction. Must f have a fixed point?

†Q 10.10   Give a rigorous justification of the statement made in 10.10.2.

Q 10.11   Find simultaneous solutions to the equations $3\sin a - 2\cos b = 6a - 12$ and $\cos a + 3\sin b = 6b + 6$.

Q 10.12   Let $r \in (0\,,1)$ and $\alpha \in \mathbb{R}$. Define $\phi \colon \mathcal{C}([0\,,r]) \to \mathcal{C}([0\,,r])$ by setting

$$(\phi(f))(x) = \alpha + \int_0^x f(t)\,dt$$

for each $f \in \mathcal{C}([0\,,r])$ and $x \in [0\,,r]$. Show that ϕ is a strong contraction on $\mathcal{C}([0\,,r])$ and find its unique fixed point.

Q 10.13   Show that the function $x \mapsto x^2 + x^{-1}$ (Q 9.14) has no fixed point in the complete metric space $[1\,,\infty)$.

†Q 10.14 Suppose (X, d) and (Y, e) are metric spaces and Y is complete. Suppose S is a dense subset of X and $f : S \to Y$ is an isometric map. Show that the unique continuous extension of f to X is also an isometry.

Q 10.15 When an incomplete normed linear space is completed, is the metric on the completion necessarily determined by a norm that extends the original norm?

†Q 10.16 Suppose C is a closed non-empty convex subset of ℓ_2 and $w \in \ell_2$. Show that there exists a unique $z \in C$ such that $\|z - w\|_2 = \text{dist}(w, C)$. You may use the fact that ℓ_2 is complete (10.8.6).

11
Connectedness

No matter how correct a mathematical theorem
may appear to be, one ought never to be
satisfied that there was not something
imperfect about it until it also gives
the impression of being beautiful.   George Boole, 1815–1864

We have mentioned *connectedness* a number of times already in this book and we have characterized connected spaces as those metric spaces that have no proper non-empty subset with empty boundary (3.2.2). Now is the time to make a formal definition and to explore its most immediate consequences. Let the reader note, however, that there are many different concepts of connectedness, each one important in some area of study, and that we discuss only three.

11.1 Connected Metric Spaces

A *disconnected metric space*, like that displayed in 3.2.2, is one that can be expressed as a union of two disjoint non-empty open subsets. A connected space is, as we might expect, one that is not disconnected. There are, however, several other ways of formulating the concept, as we show now in 11.1.1.

Theorem 11.1.1 (Criteria for Connectedness)

Suppose X is a metric space. The following statements are equivalent:

(i)   (BOUNDARY CRITERION) Every proper non-empty subset of X has non-empty boundary in X.

(ii) (OPEN–CLOSED CRITERION) No proper non-empty subset of X is both open and closed in X.

(iii) (OPEN UNION CRITERION) X is not the union of two disjoint non-empty open subsets of itself.

(iv) (CLOSED UNION CRITERION) X is not the union of two disjoint non-empty closed subsets of itself.

(v) (CONTINUITY CRITERION) Either $X = \varnothing$ or the only continuous functions from X to the discrete space $\{0, 1\}$ are the two constant functions.

Proof

Suppose X satisfies (i). Then no non-empty proper subset of X both includes its boundary and is disjoint from its boundary; in other words, X satisfies (ii) (4.1.1 and 4.1.2).

Suppose X satisfies (ii). Suppose U is a non-empty open subset of X. Then U^c is closed and so, by hypothesis, is either empty or not open. Therefore X satisfies (iii).

Suppose X satisfies (iii) and X is the disjoint union of two closed subsets A and B of X. Then $A^c = B$ and $B^c = A$ are disjoint and open in X and X is their union, so that, by hypothesis, either $B = \varnothing$ or $A = \varnothing$. Therefore X satisfies (iv).

Suppose X satisfies (iv) and $f : X \to \{0, 1\}$ is continuous. Then $f^{-1}(\{0\})$ and $f^{-1}(\{1\})$ are disjoint and closed in X (8.3.1) and X is their union. By hypothesis, at least one of these sets is empty, so that f is constant or $X = \varnothing$. So X satisfies (v).

Suppose X satisfies (v) and $\varnothing \neq A \subseteq X$ with $\partial A = \varnothing$. Then $\partial(A^c) = \varnothing$ also (3.1.2). So A and A^c are both open in X. Define $f : X \to \{0, 1\}$ by setting $f(x) = 1$ if $x \in A$ and $f(x) = 0$ if $x \in A^c$. The open subsets of $\{0, 1\}$ are $\varnothing$, $\{0\}$, $\{1\}$ and $\{0, 1\}$ and, by construction, the inverse image of each of these sets is open in X, so f is continuous (8.3.1). By hypothesis, f is constant, so either $A = \varnothing$ or $A^c = \varnothing$, the latter yielding $A = X$. Therefore X satisfies (i). $\qquad \square$

Definition 11.1.2

A metric space X is called a *connected metric space* if, and only if, X cannot be expressed as the union of two disjoint non-empty open subsets of itself.

Theorem 11.1.1 can now be interpreted as saying that a metric space is connected if, and only if, it satisfies any one of the criteria listed there.

Example 11.1.3

$\mathbb{R}$ with its usual metric is connected. To see this, suppose U is a non-empty proper subset of $\mathbb{R}$ and let $a \in \mathbb{R} \backslash U$. If the set $(a, \infty) \cap U$ is not empty, let b be its infimum. Then $\mathrm{dist}(b, U) = 0$ (2.2.5). Either $b = a$ or $[a, b) \subseteq \mathbb{R} \backslash U$ and, in either case, $\mathrm{dist}(b, \mathbb{R} \backslash U) = 0$, so $b \in \partial U$. A similar argument shows that ∂U is not empty in the case when $(-\infty, a) \cap U \neq \varnothing$.

11.2 Connected Subsets

A connected subset of a metric space is a subset that, endowed with the subspace metric, is a connected metric space. Connectedness is a property that belongs properly to a metric space; it is not relative to any metric superspace that the space may sit inside. Moreover, the closure of every connected subset of a metric space is also connected.

Definition 11.2.1

A subset S of a metric space X is called a *connected subset* of X if, and only if, the subspace S of X is a connected metric space.

Theorem 11.2.2

Suppose X is a metric space, Z is a metric subspace of X and $S \subseteq Z$. Then S is a connected subset of X if, and only if, S is a connected subset of Z.

Proof

The topology on S is the same whether it is viewed as a subspace of X or of Z (Q 4.13), so, whether or not S can be disconnected with respect to this topology is independent of the viewpoint. □

Theorem 11.2.3

Suppose X is a metric space, S is a connected subset of X and $S \subseteq A \subseteq \overline{S}$. Then A is connected.

Proof

If S is empty, this is trivial, so we suppose otherwise. Suppose $f \colon A \to \{0,1\}$ is continuous. Then $f|_S$ is continuous (8.6.1) and therefore constant because S is connected (11.1.1). For each $z \in A$, we have $z \in \overline{S}$, so that there exists a sequence (x_n) in S that converges to z (6.6.2). Then, because f is continuous, the sequence $(f(x_n))$ converges to $f(z)$ (8.3.1). But $(f(x_n))$ is a constant sequence, so that f has the same value at z as it has on S. As z is arbitrary in A, this proves that f is constant on A. Because f is an arbitrary continuous function from A to $\{0,1\}$, A is connected by the continuity criterion (11.1.1). □

Question 11.2.4

Every singleton subset of a metric space is connected; so is the empty subset. Certainly no other finite set is connected. Are there any other countable connected subsets of a metric space? Strange as it may seem, in the more abstract world of topological spaces, infinite connected countable spaces do exist (see [6]). There are no such metric spaces, however. Suppose (X, d) is a countable metric space with more than one element, and let $a, b \in X$ with $a \neq b$. Since the interval $(0, d(a, b))$ is uncountable (B.17), there exists $s \in (0, d(a, b))$ such that no point of X is of distance s from a. Then $\flat[a\,;s)$ equals $\flat[a\,;s]$ and so is both open and closed in X. But $a \in \flat[a\,;s)$ and $b \notin \flat[a\,;s)$, so that $\flat[a\,;s)$ is a proper non-empty subset of X. So X is not connected.

No point of X is distant s from a.

11.3 Connectedness and Continuity

The Intermediate Value Theorem says that each continuous real function defined on $[0, 1]$ takes on every value that lies between its supremum and its infimum values. It is a very special case of 11.3.1.

Theorem 11.3.1

Every continuous image of a connected metric space is connected.

Proof

Suppose X and Y are metric spaces and $g: X \to Y$ is continuous. Suppose $g(X)$ is not a connected space. Then there exists a continuous function f from $g(X)$ onto $\{0, 1\}$. This yields $(f \circ g)(X) = f(g(X)) = \{0, 1\}$, and, since $f \circ g$ is continuous (8.5.1), X does not satisfy the continuity criterion for connectedness and is therefore not connected.   □

Corollary 11.3.2

Suppose S is a non-empty subset of $\mathbb{R}$. Then S is connected if, and only if, S is an interval.

Proof

First, if S is not an interval, then there exists $w \in \mathbb{R}$ with $\inf S < w < \sup S$ such that $w \notin S$ (B.8.2). So $S \cap (-\infty, w)$ and $S \cap (w, \infty)$ are non-empty open subsets of S (4.4.1) and their union is S. Therefore S is not connected. For the converse, it has been established in 11.1.3 that $\mathbb{R}$ is connected. By 11.3.1, every continuous image of $\mathbb{R}$ is connected, so by Q 8.7, every interval is connected. □

Corollary 11.3.3 (Intermediate Value Theorem)

Suppose X is a connected metric space and $f\colon X \to \mathbb{R}$ is continuous. Suppose $\alpha \in (\inf f(X), \sup f(X))$. Then there exists $z \in X$ such that $f(z) = \alpha$.

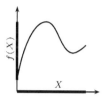

Proof

By 11.3.1, $f(X)$ is connected; therefore, by 11.3.2, $f(X)$ is an interval, whence $\alpha \in f(X)$. □

11.4   Unions, Intersections and Products of Connected Sets

Every finite product of connected subsets endowed with a product metric is connected. Under certain conditions, a union of connected subsets is also connected. An intersection of two connected subsets of a metric space need not be connected.

Theorem 11.4.1

Suppose X is a metric space and $\mathcal{S}$ is a chained collection of connected subsets of X (B.15.1). Then $\bigcup \mathcal{S}$ is also connected.

Proof

Suppose $f\colon \bigcup \mathcal{S} \to \{0,1\}$ is continuous. Since every member of $\mathcal{S}$ is connected, $f|_C$ is constant for each $C \in \mathcal{S}$. Suppose $x, y \in \bigcup \mathcal{S}$ and let $A, B \in \mathcal{S}$ be such that $x \in A$ and $y \in B$. If $A = B$, then, since $f|_A$ is constant, we have $f(x) = f(y)$. If $A \neq B$, there exist $n \in \mathbb{N}\backslash\{1\}$ and a chain $(U_1, \ldots, U_n)$ from A to B in $\mathcal{S}$. For each $i \in \mathbb{N}_n\backslash\{1\}$, pick $u_i \in U_{i-1} \cap U_i$. Because $f|_{U_i}$ is constant for each $i \in \mathbb{N}_n$, we have $f(x) = f(u_2)$ and $f(u_{i-1}) = f(u_i)$ for each $i \in \mathbb{N}_n\backslash\{1,2\}$ and $f(u_n) = f(y)$, which together yield $f(x) = f(y)$. Since x and y are arbitrary in $\bigcup \mathcal{S}$, this shows that f is constant and that $\bigcup \mathcal{S}$ is connected (11.1.1). □

Corollary 11.4.2

Suppose X is a metric space and $\mathcal{S}$ is a non-empty collection of connected subsets of X for which $\bigcap \mathcal{S} \neq \varnothing$. Then $\bigcup \mathcal{S}$ is connected.

Proof

$\mathcal{S}$ is chained because, for each $A, B \in \mathcal{S}$, we have $A \cap B \neq \varnothing$. So $\bigcup \mathcal{S}$ is connected by 11.4.1. $\square$

Example 11.4.3

A non-empty intersection of connected subsets of a metric space need not be connected. Consider, for a counterexample, the unit circle $\{z \in \mathbb{C} \mid |z| = 1\}$ and the ellipse $\left\{z \in \mathbb{C} \mid 4(\Re(z))^2 + (\Im(z))^2 = 1\right\}$. Both are connected subsets of $\mathbb{C}$, but their intersection is the two-point set $\{i, -i\}$.

Theorem 11.4.4

Suppose $n \in \mathbb{N}$ and, for each $i \in \mathbb{N}_n$, (X_i, τ_i) is a non-empty metric space. Endow the product $P = \prod_{i=1}^{n} X_i$ with a product metric. Then P is connected if, and only if, X_i is connected for all $i \in \mathbb{N}_n$.

Proof

The natural projections π_i of P onto the coordinate spaces X_i are all continuous (8.8.1) because the metric is a product metric, so, by 11.3.1, if P is connected, then the coordinate spaces $X_i = \pi_i(P)$ are also connected.

For the converse, suppose that all the coordinate spaces are connected. For each $j \in \mathbb{N}_n$ and $a \in P$, denote the subset $\{x \in P \mid x_i = a_i \text{ for all } i \in \mathbb{N}_n \backslash \{j\}\}$ of P by $X_{j,a}$. Then $X_{j,a}$, being an isometric copy of X_j (1.6.4), is connected. The collection $\{X_{j,a} \mid j \in \mathbb{N}_n, a \in P\}$ is chained and its union is P, by B.15.2, so P is connected by 11.4.1. $\square$

11.5 Connected Components

We mentioned *connected components* in 8.3.6, noting that we cannot hope to identify *unbrokenness* and *continuity* with each other on disconnected subsets of the domain of a function. We also gave an example (8.3.7) of a function that is continuous on all the maximal connected subsets of its domain but is not continuous. Here we lay out the basic facts about connected components. They *partition* the space in that they are mutually disjoint and their union is the whole space; they are always closed, but need not be open.

Definition 11.5.1

Suppose X is a metric space. A subset U of X is called a *connected component* of X if, and only if, U is connected and there is no proper superset of U in X that is connected.

Example 11.5.2

A metric space X is connected if, and only if, its only connected component is X. In a discrete metric space, every singleton set is both open and closed and so has no proper superset that is connected. Therefore discrete metric spaces have the property that their connected components are their singleton subsets. In an arbitrary metric space, there may be any number of singleton connected components, but every other connected component must be uncountable (11.2.4).

Theorem 11.5.3

Suppose X is a metric space. Then

(i)   the connected components of X are mutually disjoint;

(ii) the connected components of X are all closed in X; and

(iii) X is the union of its connected components.

Proof

If U and V are connected components of X and $U \cap V \neq \varnothing$, then $U \cup V$ is connected (11.4.2), so that, by the definition of a connected component, we have $U = U \cup V = V$. This proves (i), and (ii) is an immediate consequence of 11.2.3. For (iii), we must show that each point of X belongs to some connected component. Towards this, suppose $x \in X$ and let $\mathcal{C}$ be the collection of all connected subsets of X that contain x. Since $\{x\}$ is connected, $\mathcal{C} \neq \varnothing$ and $x \in \bigcup \mathcal{C}$. Moreover, $\bigcup \mathcal{C}$ is connected by 11.4.2 because each member of $\mathcal{C}$ contains x. Suppose Z is a connected superset of $\bigcup \mathcal{C}$ in X. Then $x \in Z$ so that $Z \in \mathcal{C}$ and therefore $Z \subseteq \bigcup \mathcal{C}$, yielding $Z = \bigcup \mathcal{C}$. So $\bigcup \mathcal{C}$ is a connected component of X that contains x. $\qquad\square$

Question 11.5.4

Connected components have to be closed. Do they have to be open? The open–

closed criterion for connectedness persuades us to ask this
question; the answer is *yes* if there is only a finite number of connected components (Q 11.4), but *no* in general.
Consider the metric subspace $X = \{1/n \mid n \in \mathbb{N}\} \cup \{0\}$ of
$\mathbb{R}$. In this case, the set $\{0\}$ is a connected component of
X; it is closed in X, but it is not open because every open
interval that contains 0 contains $1/n$ for some $n \in \mathbb{N}$.

Each of the components
is the complement of the
closed union of the other
two, so is open.

11.6 Totally Disconnected Metric Spaces

In a discrete metric space, the singleton subsets are the connected components.
Such spaces are the simplest of the totally disconnected spaces.

Definition 11.6.1

A metric space in which the connected components are all singleton sets is said
to be *totally disconnected*.

Example 11.6.2

All discrete spaces are totally disconnected (11.5.2), but totally disconnected
spaces need not be discrete. Consider, for example, $\mathbb{Q}$. The open subsets of $\mathbb{Q}$
are of the form $U \cap \mathbb{Q}$, where U is open in $\mathbb{R}$. None of these is a singleton set,
so $\mathbb{Q}$ is not discrete. But $\mathbb{Q}$ is totally disconnected because every connected
component of $\mathbb{Q}$ is a connected subset of $\mathbb{R}$ (11.2.2) and is therefore an interval
(11.3.2), and the only intervals that are subsets of $\mathbb{Q}$ are the degenerate ones.

Example 11.6.3

The Cantor set $\mathcal{K}$ is totally disconnected. The reasoning is as follows: between
any two members a, b of $\mathcal{K}$ with $a < b$, there is some $x \in \mathbb{R}\backslash\mathcal{K}$, so that a and b
do not belong to the same connected component of $\mathcal{K}$.

11.7 Paths

The words *path* and *curve* are used in a variety of ways in different branches
of mathematics. We shall define a *path* to be any continuous function defined

on $[0,1]$. Thus a path carves out any stretched or twisted
unbroken image of that interval; this ought to correspond
to our intuitive idea of a *curve*, but we shall, for the most
part, avoid the term *curve* and leave it to geometers to
give a precise definition that suits their discipline. We
shall see presently that our idea actually gives us many
more curves than we bargained for (11.8.7).

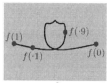

$f: [0,1] \to \mathbb{R}^2$ may move
back and forth or stand
still on a subinterval.

Definition 11.7.1

Suppose X is a metric space. Every continuous function $f: [0,1] \to X$ from the
closed interval $[0,1]$ into X is called a *path* in X from the point $f(0)$ to the
point $f(1)$; the points $f(0)$ and $f(1)$ of X are called the *endpoints* of the path.

Example 11.7.2

By strict definition, a function *is* its graph (B.14), so what we see when we
look at the graph of a function $f: [0,1] \to \mathbb{R}$ is a picture
of the path itself, which is, indeed, the *curve* traced out
in $\mathbb{R}^2$ by the closely related function $x \mapsto (x, f(x))$ from
$[0,1]$ to $\mathbb{R}^2$. Of course, there are many curves in $\mathbb{R}^2$ that
are not determined in this way by a real-valued function.
The image of the function $f: [0,1] \to \mathbb{R}$ itself, the curve
it traces out, is in $\mathbb{R}$ and is represented in the diagram by the black line. It is

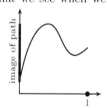

easy to see why one might ask for the term *curve* to be restricted to images of
paths that are differentiable and have derivative that is never zero.

Theorem 11.7.3

Every path is a uniformly continuous function, and its image is connected and
compact.

Proof

Since $[0,1]$ is compact (7.13.2), the first assertion follows from 9.1.5. Because
a path is continuous and its domain $[0,1]$ is compact and connected (7.12.2),
its image is compact by 9.2.1 and connected by 11.3.1.   □

11.8 Pathwise Connectedness

The idea that any two points in a metric space can be joined by an unbroken curve in the space is perhaps a more intuitive idea of connectedness than the one we have adopted—at least it might be if we were not aware of *space-filling curves* (11.8.7). It is, however, a stronger concept than connectedness (11.8.3).

Definition 11.8.1

Suppose X is a metric space. Then X is said to be *pathwise connected* if, and only if, for each $a, b \in X$, there is a path in X with endpoints a and b.

Example 11.8.2

Every normed linear space is pathwise connected. The continuous function $t \mapsto ta + (1 - t)b$ defined on $[0, 1]$ is a path with endpoints a and b.

Example 11.8.3

The closure in $\mathbb{R}^2$ of the graph $\Gamma = \{(x, \sin(1/x)) \mid x \in \mathbb{R}^+\}$ of the function $\sin(1/x)$ of 3.3.5 is an example of a connected metric space that is not pathwise connected. No point of $J = \{(0, y) \in \mathbb{R}^2 \mid y \in [-1, 1]\}$ is connected by a path to any point in Γ (see Q 11.8). However, it is easy to check that Γ itself is connected, and it follows from 11.2.3 that its closure $\Gamma \cup J$ is also connected.

Theorem 11.8.4

Every pathwise connected metric space is connected.

Proof

Suppose X is a pathwise connected metric space. If X is empty, the result is trivial. We suppose otherwise and let $a \in X$. Let $\mathcal{C}$ be the collection of all images of paths in X that have a as an endpoint. For each $x \in X$, there is a path in X with endpoints x and a by hypothesis, so that $\bigcup \mathcal{C} = X$. Each member of $\mathcal{C}$ is connected (11.7.3) and $a \in \bigcap \mathcal{C}$. By 11.4.2, $\bigcup \mathcal{C}$ is connected. $\square$

Theorem 11.8.5

Every continuous image of a pathwise connected metric space is also pathwise connected.

Proof

Suppose X and Y are metric spaces and $f: X \to Y$ is continuous. Suppose X is pathwise connected. For each $a, b \in f(X)$, there exist $x, z \in X$ such that $f(x) = a$ and $f(z) = b$. Since X is pathwise connected, there exists a continuous mapping $g: [0, 1] \to X$ such that $g(0) = x$ and $g(1) = z$. Then $f \circ g: [0, 1] \to Y$ is continuous and $(f \circ g)(0) = a$ and $(f \circ g)(1) = b$. Since a and b are arbitrary in $f(X)$, it follows that $f(X)$ is pathwise connected. $\qquad \square$

Example 11.8.6

The image of a path is always connected (11.7.3); in fact, it is always, as we might expect, pathwise connected: the interval $[0, 1]$ is clearly pathwise connected, so that every continuous image of $[0, 1]$ is pathwise connected by 11.8.5.

Example 11.8.7

Paths are not nearly as straightforward as they may seem. In many mathematical applications, we ask that the defining function have better properties than continuity—that it be *continuously differentiable* with non-zero derivative or *rectifiable*, for example. Some of the strangest paths are those whose images are called *space-filling curves*, or *Peano curves*. We have seen that the image of a path is always a pathwise connected compact space (11.7.3, 11.8.6). What

A space-filling curve in $\mathbb{R}^2$.

is amazing is that every such metric space is the image of some path. The actual functions may be elusive, but we can prove that there are such functions as follows. Suppose S is a non-empty pathwise connected compact metric space. Then there exists a continuous surjective function $\phi: \mathcal{K} \to S$ from the Cantor set onto S; this was shown in 10.9.7. This function can be extended continuously to the interval $[0, 1]$ as follows. The connected components of $[0, 1] \setminus \mathcal{K}$ are the open intervals that are removed in the construction of $\mathcal{K}$ (3.3.3, Q 11.5). There is a countable number of them, so we can enumerate them as (a_n, b_n) for each $n \in \mathbb{N}$. Note that $a_n, b_n \in \mathcal{K}$ for each $n \in \mathbb{N}$. Since S is pathwise connected, for each $n \in \mathbb{N}$ there exists a continuous function $\psi_n: [0, 1] \to S$ with $\psi_n(0) = \phi(a_n)$ and $\psi_n(1) = \phi(b_n)$. We can choose a sequence (ψ_n) of such functions by B.19.1. Then we can extend ϕ to $[0, 1]$ by setting $\phi(x) = \psi_n((x - a_n)/(b_n - a_n))$ for each $x \in (a_n, b_n)$ and $n \in \mathbb{N}$. It is easy to see that the extended function ϕ is continuous. The extended function ϕ has domain $[0, 1]$ and so is, by definition, a path. Its image is S. Therefore S, despite all appearances, is the image of a path—it is an example of what we are hesitating to call a *curve*.

11.9 Polygonal Connectedness

In a linear space, there are other forms of connectedness available to us. The simplest and strongest is convexity (5.4.1). But subsets of a linear space that are not convex may still be connected in a way that is, in general, stronger than pathwise connectedness (11.9.2). They may be polygonally connected.

Definition 11.9.1

Suppose X is a linear space, S is a subset of X and $a, b \in S$. For any $n \in \mathbb{N}$,

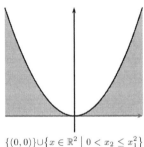

an n-tuple $(c_1, \ldots, c_n)$ of points of S will be called a *polygonal connection* from a to b in S if, and only if, $c_1 = a$ and $c_n = b$ and, for each $i \in \mathbb{N}_n \backslash \{1\}$, the line segment $\{(1-t)c_{i-1} + tc_i \mid t \in [0,1]\}$ is included in S. S is said to be *polygonally connected* if, and only if, for each $a, b \in S$, there exists a polygonal connection from a to b in S.

Example 11.9.2

All convex subsets of a linear space are polygonally connected and, for each pair of points, one line segment only is needed to effect the connection. The non-convex subset of $\mathbb{C}$ given by $S = \{z \in \mathbb{C} \mid |z - 1| \le 1\} \cup \{z \in \mathbb{C} \mid |z + 1| \le 1\}$ is polygonally connected because any two points that cannot be joined by a line segment in S can be joined by two such line segments meeting at the origin. The set $\{z \in \mathbb{C} \mid |z| = 1\}$ is pathwise connected but not polygonally connected.

$\{(0,0)\} \cup \{x \in \mathbb{R}^2 \mid 0 < x_2 \le x_1^2\}$ is pathwise connected, but not polygonally connected (Q 11.9).

Theorem 11.9.3

Suppose X is a normed linear space. Every polygonally connected subset of X is pathwise connected and therefore connected.

Proof

Suppose S is a polygonally connected subset of X and $a, b \in S$. Let $(c_1, \ldots, c_n)$ be a polygonal connection from a to b in S. Define a function $f: [0,1] \to X$ by $f((i - 2 + t)/(n - 1)) = (1 - t)c_{i-1} + tc_i$ for each $t \in [0,1)$ and $i \in \mathbb{N}_n \backslash \{1\}$ and $f(1) = c_n$. It is easy to check that f is continuous and that $f(0) = a$ and $f(1) = b$. Since a and b are arbitrary in S, S is pathwise connected. $\square$

We now have three types of connectedness. A connected metric space need not be pathwise connected and a pathwise connected subset of a linear space need not be polygonally connected. Incidentally, polygonal connectedness of a subset of a linear space is quite independent of any metric or norm that might be placed on the space, but we do need a norm or a suitable metric to ensure that functions such as $t \mapsto ta + (1-t)b$ are continuous and to show that a polygonally connected subset is pathwise connected. How closely are these different concepts of connectedness related to each other? We see next that all differences disappear when we are dealing with an open subset of a normed linear space (11.9.4). This fact is used in the calculus of several variables to show that a function that has zero derivative on an open connected subset of $\mathbb{R}^n$ is necessarily constant.

Theorem 11.9.4

Suppose $(X, \|\cdot\|)$ is a normed linear space and S is an open connected subset of X. Then S is polygonally connected and therefore also pathwise connected.

Proof

Suppose $z \in S$. Let U be the set of all those points of S that have a polygonal connection with z. Suppose $u \in U$ and let $(c_1, \ldots, c_n)$ be a polygonal connection from z to u in S. Since S is open in X, there exists an open ball B of X that is centred at u and included in S (5.2.2). This ball is convex (5.4.3), so that, for each $w \in B$, the line segment from u to w is included in S. So $(c_1, \ldots, c_n, w)$ is a polygonal connection from z to w, giving $w \in U$. Since w is arbitrary in B, it follows that $B \subseteq U$ and that $u \in U^\circ$ (5.2.1). Since u is arbitrary in U, U is open in X and therefore also in S.

© Eoghan Ó Searcóid, 2006

Suppose $p \in S \backslash U$. Since S is open in X, there is a ball D of X centred at p and included in S. It is convex, so, if there were any polygonal connection $(v_1, \ldots, v_n)$ in S from z to any point v_n of D, then $(v_1, \ldots, v_n, p)$ would be a polygonal connection from z to p in S, giving the contradiction $p \in U$. So no such connection exists and $D \subseteq S \backslash U$. Therefore $S \backslash U$ is open in X and then also in S. So U is closed in S.

U is not empty because $z \in U$, so, since S is connected and U is both open and closed in S, we have $U = S$. In other words, there is a polygonal connection in S from z to each point of S. Since z is arbitrary in S, this proves that S is polygonally connected.   □

Example 11.9.5

A closed connected subset of a normed linear space need not be even pathwise connected. An example is the closure in $\mathbb{R}^2$ of the graph of the function $x \mapsto \sin(1/x)$ defined on $\mathbb{R}^+$ (11.8.3).

Summary

In this chapter, we have introduced three different types of connectedness, each stronger than the previous one, and have shown that they are the same for open subsets of a normed linear space. We have discussed connected components and total disconnectedness. We have begun a discussion on paths and have demonstrated the existence of space-filling curves.

EXERCISES

Q 11.1   Show that a ball of a connected metric space need not be connected.

Q 11.2   Suppose X is a metric space. Show that X is disconnected if, and only if, there is a non-empty proper subset S of X such that $\overline{S} \cap \overline{S^c} = \varnothing$.

Q 11.3   Give an example to show that the interior of a connected subset of a metric space need not be connected.

†Q 11.4   Suppose X is a metric space and the number of connected components is finite. Show that each of them is both open and closed in X.

†Q 11.5   Suppose S is a proper closed subset of $[0, 1]$ and $\{0, 1\} \subseteq S$. Show that each connected component of $[0, 1] \setminus S$ is an open interval (a, b) with $a, b \in S$.

Q 11.6   Is there any injective continuous function from $\mathbb{R}^2$ to $\mathbb{R}$?

Q 11.7   Suppose X is a metric space with just one element. Show that X is both connected and totally disconnected.

†Q 11.8   Let $S = \{(x, \sin(1/x)) : 0 < x \leq 1\} \cup \{(0, y) : -1 \leq y \leq 1\}$. Show that S is connected but not pathwise connected.

†Q 11.9   Show that the subset $S = \{(0, 0)\} \cup \{x \in \mathbb{R}^2 \mid 0 < x_2 \leq x_1^2\}$ of $\mathbb{R}^2$ is pathwise connected but not polygonally connected. (See the diagram in 11.9.2.)

Q 11.10   Suppose X is a normed linear space and $\mathcal{C}$ is a chained collection (B.15.1) of convex subsets of X. Show that $\bigcup \mathcal{C}$ is polygonally connected.

12

Compactness

I write slowly.
This is chiefly because I am never satisfied until
I have said as much as possible in a few words,
and writing briefly takes far more time than
writing at length.   Carl Friedrich Gauss, 1777–1855

Finite quantities are much easier to deal with than infinite ones. Similarly, metric concepts are often easier to investigate on compact metric spaces than on arbitrary metric spaces. For example, we usually first examine the notion of continuity by studying functions defined on the interval $[0, 1]$. This interval, like all closed bounded intervals of the real line, is compact; it is, in fact, the prototype for compact subsets of arbitrary metric spaces.

12.1 Compact Metric Spaces

We have already defined compact metric spaces as those spaces that are bounded and have the nearest-point property (9.1.3). We have seen that they are those metric spaces that are complete and totally bounded (10.9.7), and we have, without proof, identified them as those metric spaces in which every nest of non-empty closed subsets has non-empty intersection (4.7.3). There are several other characterizations of compactness that we list with proof in 12.1.3. First we make some definitions: an *open cover* is simply a cover (B.11.1) that consists of open sets, and a collection of subsets of a metric space has the *finite intersection property* if, and only if, the collection of complements of its members includes no finite cover for the space.

Definition 12.1.1

Suppose X is a metric space and S is a subset of X. A collection $\mathcal{C}$ of open subsets of X is called an *open cover* for S in X if, and only if, $S \subseteq \bigcup \mathcal{C}$. If $\mathcal{C}$ is an open cover for S, then any subset $\mathcal{A}$ of $\mathcal{C}$ that also covers S is called an *open subcover* of $\mathcal{C}$ for S in X.

Definition 12.1.2

Suppose X is a set and $\mathcal{S}$ is a collection of subsets of X. We say that $\mathcal{S}$ has the *finite intersection property* if, and only if, for every non-empty finite subset $\mathcal{F}$ of $\mathcal{S}$, we have $\bigcap \mathcal{F} \neq \varnothing$.

Theorem 12.1.3 (Criteria for Compactness)

Suppose X is a non-empty metric space. The following statements are equivalent:

(i)   (OPEN COVER CRITERION) Every open cover for X has a finite subcover.

(ii) (BALL COVER CRITERION) Every cover of X by open balls has a finite subcover.

(iii) (FINITE INTERSECTION CRITERION) Every collection of closed subsets of X with the finite intersection property has non-empty intersection.

(iv) (CANTOR'S CRITERION) Every nest of non-empty closed subsets of X has non-empty intersection.

(v)   (CANTOR'S SEQUENCE CRITERION) Every sequence (F_n) of non-empty closed subsets of X for which $F_{n+1} \subseteq F_n$ for each $n \in \mathbb{N}$ has non-empty intersection.

(vi) (CONVERGENCE CRITERION) Every sequence in X has a subsequence that converges in X.

(vii) (SPATIAL CRITERION) X is complete and totally bounded.

(viii)(BBW CRITERION) X is bounded and has the nearest-point property.

(ix) (CANTOR SET CRITERION) X is a continuous image of the Cantor set.

(x)   (ATTAINED BOUND CRITERION) Every real continuous function defined on X is bounded and attains its bounds (7.5.1).

Proof

That (i) implies (ii) is clear. Suppose that X satisfies (ii) and that $\mathcal{F}$ is a collection of closed subsets of X for which $\bigcap \mathcal{F} = \varnothing$. Then $X = \bigcup \{F^c \mid F \in \mathcal{F}\}$. So the collection $\{b[a\,;\mathrm{dist}(a\,,F)) \mid F \in \mathcal{F},\, a \in F^c\}$ of balls of X covers X; by hypothesis, it has a finite subcover for X, and therefore there is a finite non-empty subset $\mathcal{C}$ of $\mathcal{F}$ such that $\bigcup \{F^c \mid F \in \mathcal{C}\} = X$. Then $\bigcap \mathcal{C} = \varnothing$, so that $\mathcal{F}$ does not have the finite intersection property. So (ii) implies (iii).

Now suppose that X satisfies (iii). Suppose $\mathcal{N}$ is a nest of non-empty closed subsets of X. The intersection of any non-zero finite number of members of $\mathcal{N}$ is equal to the smallest of them and so is non-empty. Therefore $\mathcal{N}$ has the finite intersection property. By hypothesis, $\bigcap \mathcal{N} \neq \varnothing$. So X satisfies (iv).

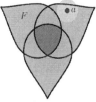

No ball $\flat[a\,;\mathrm{dist}(a\,,F))$ contains any point of the intersection.

If X satisfies (iv), then X clearly satisfies (v).

Suppose that X satisfies (v). Suppose $x = (x_n)$ is a sequence in X. Then, by hypothesis, the nest $\left\{\overline{\mathrm{tail}_n(x)} \mid n \in \mathbb{N}\right\}$ has non-empty intersection, so that (x_n) has a convergent subsequence by 6.7.2. It follows that X satisfies (vi).

Suppose X satisfies (vi). Then every sequence in X has a convergent subsequence, which is therefore Cauchy, so X is totally bounded by 7.8.2. Also, every Cauchy sequence in X has a convergent subsequence and so converges by 6.8.3; therefore X is complete. So X satisfies (vii).

That (vii) and (viii) are equivalent follows from 7.11.1 and was already noted in 10.9.7, where it was proved that they imply (ix).

Suppose that X satisfies (ix) and that $g \colon \mathcal{K} \to X$ is a continuous function from the Cantor set onto X. Suppose $f \colon X \to \mathbb{R}$ is continuous. Then $f \circ g$ is continuous (8.5.1). Its domain $\mathcal{K}$ is a closed bounded subset of $\mathbb{R}$ (4.1.11), so that its range $f(X)$ is also a closed bounded subset of $\mathbb{R}$ (9.2.5) and f attains its bounds (7.5.2). Therefore X satisfies (x).

Suppose that X satisfies (x). Then any given point function of X is bounded, so that X is bounded, and every pointlike function attains its minimum value, so that X has the nearest-point property by 7.11.1. These two facts, together with the Cauchy criterion of 7.11.1, ensure that X is totally bounded. Let $\mathcal{U}$ be an open cover for X. We show that $\mathcal{U}$ has a finite subcover. Define $u \colon X \to \mathbb{R}^{\oplus}$ by setting $u(x) = \sup\{\mathrm{dist}(x\,,U^c) \mid U \in \mathcal{U}\}$ for each $x \in X$. It is easy to check that u is continuous. By hypothesis, u attains its minimum value at some $w \in X$. Set $u(w) = r$. If $r = 0$, we should have $\mathrm{dist}(w\,,U^c) = 0$ and therefore $w \in U^c$ for all $U \in \mathcal{U}$ because such U^c are all closed in X, yielding the contradiction $w \notin \bigcup \mathcal{U}$. So $r > 0$. Since r is the minimum value for u, it follows that, for each $x \in X$, there exists $U \in \mathcal{U}$ such that $\flat[x\,;r) \subseteq U$. Since X is totally bounded, a finite collection of such balls covers X, so that a corresponding finite subset of $\mathcal{U}$ also covers X. So (x) implies (i) and the proof is complete.

We defined compactness in 9.1.4 using the BBW criterion of 12.1.3. Following this theorem, we can say that a metric space is compact if, and only if, it is empty or it satisfies any one of the criteria listed in 12.1.3. The standard formal definition is that a metric space S is compact if, and only if, every open cover for X has a finite subcover. This is a formulation that is applicable in more abstract areas of mathematics; moreover, since it is couched in terms of open sets, it makes clear that compactness is preserved when the metric is replaced by any other metric that produces the same topology.

Example 12.1.4

Closed bounded intervals $[a, b]$ of the real line are compact: they are complete (10.3.3) and totally bounded (7.8.5). Unbounded intervals $[a, \infty)$ or $(-\infty, b]$ are not compact because they are not bounded. Closed discs $\flat[z\,;r]$ of the complex plane are compact: they are complete (10.3.3) and totally bounded (7.10.4). Of course, the Cantor set is compact. The real line is complete but not compact; the set $\{1/n \mid n \in \mathbb{N}\}$ is totally bounded but not compact.

Question 12.1.5

We have seen that every metric space can be completed, a fact that is useful because of the nice properties that complete spaces enjoy. It might be desirable to be able to do something similar with compactness. After all, every compact metric space is already complete, and compact spaces have even nicer properties than complete ones. Can all metric spaces be compactified? Can they be extended in some judicious way so that the extension is a compact metric space? The answer is nearly always *no*,[1] for the very simple reason that compact metric spaces are small: since each totally bounded metric space is in one-to-one correspondence with a subset of the Cantor set (9.3.1), it can have cardinality no greater than that of $\mathbb{R}$ (B.17.2). So metric spaces that have cardinality greater than that of $\mathbb{R}$ are not metric subspaces of any compact metric space.

12.2 Compact Subsets

A subset of a metric space is said to be a compact subset if it is a compact space in its own right. Thus, like completeness and connectedness, compactness is an intrinsic property of a metric space and is independent of any superspace in which it is being considered to reside. It is useful to know, however, that the open cover criterion can be tested using covers by open subsets of any such metric superspace rather than by open subsets of the subspace itself (12.2.2).

---

[1] The reader who goes on to study general topological spaces will find that this situation can be very easily remedied if we drop the condition that the topology on the enveloping space be determined by a metric.

Definition 12.2.1

A subset S of a metric space X is called a *compact subset* of X if, and only if, the subspace S of X is compact.

Theorem 12.2.2

A subset S of a metric space X is compact if, and only if, every open cover for S in X has a finite subcover.

Proof

Suppose first that S is a compact subset of X and that $\mathcal{C}$ is an open cover for S in X. Then $\{U \cap S \mid U \in \mathcal{C}\}$ is an open cover for S in S, so, by definition, there is a finite subset $\mathcal{F}$ of $\mathcal{C}$ such that $\{U \cap S \mid U \in \mathcal{F}\}$ covers S. Then $\mathcal{F}$ is a finite subcover of $\mathcal{C}$ for S in X.

Conversely, suppose every open cover for S in X has a finite subcover. Suppose $\mathcal{G}$ is an open cover for S in S. Each member of $\mathcal{G}$ can be written $S \cap W$ for some open subset W of X. The set $\{W \subseteq X \mid W$ open in $X,\ S \cap W \in \mathcal{G}\}$ is an open cover for S in X and so has a finite subcover $\mathcal{F}$. Then $\{S \cap W \mid W \in \mathcal{F}\}$ is a finite subcover of $\mathcal{G}$ for S in S, so S is a compact subset of X. $\qquad\square$

Theorem 12.2.3

Suppose X is a compact metric space and S is a subset of X. Then S is compact if, and only if, S is closed in X.

Proof

S inherits total boundedness from X (7.9.1), so, by the spatial criterion of 12.1.3, S is compact if, and only if, S is complete, which occurs if, and only if, S is closed in X (10.3.2) because X is complete. $\qquad\square$

12.3 Compactness and Continuity

Much of the importance of compact metric spaces centres around the way in which continuous functions behave on them. We have seen that every continuous function on a compact metric space is automatically uniformly continuous and that every continuous image of a compact set is compact. We show here that every injective continuous map on a compact set has continuous inverse.

Theorem 12.3.1

Suppose X and Y are metric spaces, X is compact and $f: X \to Y$ is continuous. Then f is uniformly continuous on X and the subspace $f(X)$ of Y is compact.

Proof

This is a restatement of 9.1.5 and part of 9.2.1. It is, however, worth giving a direct proof of the second assertion using open covers. Suppose $\mathcal{C}$ is an open cover for $f(X)$. Then $\left\{ f^{-1}(U) \mid U \in \mathcal{C} \right\}$ is an open cover for X because f is continuous. Since X is compact, there exists a finite subset $\mathcal{F}$ of $\mathcal{C}$ such that $\left\{ f^{-1}(U) \mid U \in \mathcal{F} \right\}$ covers X. Then $\mathcal{F}$ covers $f(X)$. So $f(X)$ satisfies the open cover criterion for compactness and is thus compact. $\square$

If X and Y are metric spaces, the notation $\mathcal{C}(X,Y)$ is used for the space of all continuous bounded functions from X into Y (8.9.1). Now 12.3.1 tells us that, if X is compact, then all continuous functions from X into Y have compact range and are therefore automatically bounded. So, when X is compact, $\mathcal{C}(X,Y)$ comprises all continuous functions from X into Y; moreover, they are all uniformly continuous.

The familiar *Extreme Value Theorem*, which states that every real continuous function defined on $[0,1]$ attains its bounds, can be seen as a simple corollary of the fact that continuous images of compact sets are always compact. Furthermore, it is easy to believe from pictures that such a function, if it is injective, has continuous inverse from its range onto $[0,1]$; this, too, is a special application of a general theorem (12.3.2).

Theorem 12.3.2 (Inverse Function Theorem)

Suppose X and Y are metric spaces and X is compact. Suppose $f: X \to Y$ is injective and continuous. Then $f^{-1}: \operatorname{ran}(f) \to X$ is uniformly continuous.

Proof

Suppose S is a closed subset of X. Then S is compact, by 12.2.3, and its continuous image $f(S)$ is compact, by 12.3.1, and so is closed in $f(X)$ (12.1.3). But $f(S) = (f^{-1})^{-1}(S)$. Since S is an arbitrary closed subset of X, f^{-1} is continuous by 8.3.1. That f^{-1} is uniformly continuous follows from 12.3.1 because the domain of f^{-1} is $f(X)$, which is compact (12.3.1). $\square$

12.4   Unions and Intersections of Compact Subsets

As we might expect from studying closed sets and complete sets, compactness is preserved under finite unions and under arbitrary intersections. And every nest of non-empty compact sets has non-empty intersection.

Theorem 12.4.1

Suppose X is a metric space and $\mathcal{C}$ is a non-empty collection of compact subsets of X. Then:

(i) $\bigcap \mathcal{C}$ is a compact subset of X.

(ii) If $\mathcal{C}$ is finite, then $\bigcup \mathcal{C}$ is a compact subset of X.

Proof

Since every member of $\mathcal{C}$ is closed in X, so is $\bigcap \mathcal{C}$ (4.3.2); being a closed subset of any individual member of $\mathcal{C}$, it is compact (12.2.3). For (ii), if $\mathcal{C}$ is finite, then $\bigcup \mathcal{C}$ is complete (10.4.1) and totally bounded (7.9.2) and so is compact by 12.1.3.   □

Example 12.4.2

Every finite union of closed bounded intervals is a compact subset of $\mathbb{R}$. $\mathbb{R}$ itself, which is not compact, can be expressed as the countable union $\bigcup \{[n, n+1] \mid n \in \mathbb{Z}\}$ of closed bounded intervals, so finiteness of $\mathcal{C}$ cannot be dropped in 12.4.1.

Theorem 12.4.3

Suppose X is a metric space and $\mathcal{C}$ is a non-empty collection of compact subsets of X. Suppose $\mathcal{C}$ has the finite intersection property (12.1.2). Then $\bigcap \mathcal{C} \neq \varnothing$. In particular, every nest of non-empty compact subsets of X has non-empty intersection.

Proof

Let $A \in \mathcal{C}$. Then $\{A \cap C \mid C \in \mathcal{C}\}$ is a collection of non-empty closed subsets of the compact metric space A and it has the finite intersection property. So $\bigcap \{A \cap C \mid C \in \mathcal{C}\}$ is non-empty (12.1.3), whence $\bigcap \mathcal{C}$ is non-empty.   □

12.5 Compactness of Products

A finite product of compact metric spaces is compact if the product is endowed with a conserving metric: it is complete by 10.5.1 and totally bounded by 7.10.2. We know, however, that neither completeness nor total boundedness, nor indeed the nearest-point property (7.13.4), need be preserved when the product is endowed with an arbitrary product metric. Despite these deficiencies, compactness is preserved under every product metric.

Lemma 12.5.1

Suppose A and B are compact metric spaces and endow $A \times B$ with a product metric. Then $A \times B$ is compact.

Proof

Suppose $((a_n, b_n))$ is an arbitrary sequence in $A \times B$. Since A is compact, (a_n) has a subsequence (a_{m_n}) that converges in A. Then, since B is compact, (b_{m_n}) has a subsequence $(b_{p_{m_n}})$ that converges in B, and, by 6.7.1, $(a_{p_{m_n}})$ converges in A. So, by 6.5.1, $((a_{p_{m_n}}, b_{p_{m_n}}))$ is a subsequence of $((a_n, b_n))$ that converges in $A \times B$. Then $A \times B$ satisfies the convergence criterion for compactness and is therefore compact.   □

Theorem 12.5.2

Suppose $n \in \mathbb{N}$ and, for each $i \in \mathbb{N}_n$, (X_i, τ_i) is a non-empty metric space. Endow the product $P = \prod_{i=1}^n X_i$ with any product metric. Then P is compact if, and only if, X_i is compact for all $i \in \mathbb{N}_n$.

Proof

If all the coordinate spaces are compact, the compactness of P follows from a finite number of applications of 12.5.1. If P is compact, then each of the coordinate spaces, being the image of P under the appropriate projection map, is compact by 12.3.1 because the projections are all continuous (8.8.1).   □

12.6 Compactness and Nearest Points

We know now that the nearest-point property lies between completeness and compactness: every compact metric space has the nearest-point property and every metric space with the nearest-point property is complete. Furthermore, the concept of compactness gives us some more characterizations of the nearest-point property.

Theorem 12.6.1

Suppose X is a metric space. The following are equivalent:

(i) X has the nearest-point property.

(ii)   every closed bounded subset of X is compact.

(iii) every closed ball of X is compact.

Proof

If $X = \varnothing$, the statements are trivially true, so we suppose otherwise.

Suppose that X has the nearest-point property and that S is a closed bounded subset of X. Being closed, S has the nearest-point property (7.12.1); being also bounded, S is compact (9.1.4). So (i) implies (ii). That (ii) implies (iii) is clear because every closed ball is closed and bounded.

Last, suppose that X satisfies (iii). Suppose (x_n) is any bounded sequence in X. Then there exists a closed ball B of X that includes $\{x_n \mid n \in \mathbb{N}\}$. But B is compact by hypothesis, so (x_n) has a subsequence that converges in B and therefore in X. As (x_n) is an arbitrary bounded sequence in X, X satisfies the convergence criterion of 7.11.1 and so has the nearest-point property. $\square$

12.7   Local Compactness

The real line is not compact, but it does have the nearest-point property—its closed bounded subsets are compact—which means that the theory of compactness is applicable in a local sense to $\mathbb{R}$. Possessing a somewhat weaker property than the nearest-point property entitles a metric space to be called *locally compact*.

Theorem 12.7.1

Suppose X is a metric space and $x \in X$. The following statements are equivalent:

(i)   There exists $r \in \mathbb{R}^+$ such that $\flat[x\,;s]$ is compact for all $s \in (0\,,r]$.

(ii)   There is a closed ball of X centred at x that is compact.

(iii) There is an open subset U of X such that $x \in U$ and $\overline{U}$ is compact.

(iv)   There exist an open subset U and a compact subset K of X such that $x \in U \subseteq K$.

Proof

If (i) is satisfied, then so is (ii). If (ii) is satisfied, let $r \in \mathbb{R}^+$ be such that $\flat[x\,;r]$ is compact and set $U = \flat[x\,;r)$; then U is open, $x \in U$ and $\mathrm{Cl}_X(U)$ is a closed subset of the compact set $\flat[x\,;r]$ (Q 4.6) and so is compact by 12.2.3. So (iii) holds.

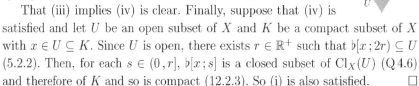

That (iii) implies (iv) is clear. Finally, suppose that (iv) is satisfied and let U be an open subset of X and K be a compact subset of X with $x \in U \subseteq K$. Since U is open, there exists $r \in \mathbb{R}^+$ such that $\flat[x\,;2r) \subseteq U$ (5.2.2). Then, for each $s \in (0\,,r]$, $\flat[x\,;s]$ is a closed subset of $\mathrm{Cl}_X(U)$ (Q 4.6) and therefore of K and so is compact (12.2.3). So (i) is also satisfied. □

Definition 12.7.2

A metric space X is said to be *locally compact* if, and only if, for each $x \in X$, there exist an open subset U of X and a compact subset K of X such that $x \in U \subseteq K$.

Because of 12.7.1, a metric space X is locally compact if, and only if, every $x \in X$ satisfies any one of the criteria listed there.

Example 12.7.3

Every discrete metric space X is locally compact since every singleton subset is both open and compact. If X is infinite and has the discrete metric, then all closed balls of radius less than 1, being singleton sets, are compact; all other balls, being equal to X, are not compact.

Example 12.7.4

Each compact metric space is both open in itself and compact and so is necessarily locally compact. If a metric space X has the nearest-point property, then, by 12.6.1, every closed ball of X is compact, so that X is certainly locally compact. The direct converse is not true because locally compact spaces need not be complete (Q 12.11), but even complete locally compact metric spaces need not have the nearest-point property. In fact, every infinite set with the discrete metric is complete (10.2.3) and locally compact (12.7.3) but does not have the nearest-point property (10.2.5). So nearest points are not guaranteed to exist in complete locally compact spaces.

Question 12.7.5

Which subspaces of locally compact spaces are locally compact? Not all are:

the subspace $\mathbb{Q}$ of $\mathbb{R}$ is not locally compact; indeed, there are no non-empty open subsets of $\mathbb{Q}$ with compact closure in $\mathbb{Q}$. Closed subsets of locally compact spaces are locally compact; interestingly, so are open subsets (Q 12.11).

Theorem 12.7.6

Suppose (X, d) is a locally compact metric space and $\tilde{X}$ is a completion of X. Then the designated isometric copy of X in $\tilde{X}$ is open in $\tilde{X}$.

Proof

Because designated copies of X in different completions are isometric with respect to an isometry between the completions themselves (10.12.5), we may assume that $\tilde{X}$ is the set $\delta(X) \cup \mathrm{vp}(X)$ of 10.12.2, where $\delta(X)$ is the isometric copy of X and the metric is $s : (u, v) \mapsto \sup\{|u(x) - v(x)| \mid x \in X\}$. Suppose $z \in X$. Since X is locally compact, there exists $r \in \mathbb{R}^+$ such that $\flat_X[z\,;r]$ is compact. We claim that $\flat_{\tilde{X}}[\delta_z\,;r] \subseteq \delta(X)$. Suppose $u \in \flat_{\tilde{X}}[\delta_z\,;r]$. Then $u(z) = u(z) - \delta_z(z) \le s(u, \delta_z) < r$. Note that, for all $x \in X \backslash \flat_X[z\,;r]$, we have $u(x) \ge d(x, z) - u(z) > r - u(z)$, so that $\inf u(X \backslash \flat_X[z\,;r]) \ne 0$. But $\inf u(X) = 0$ by definition, so $\inf u(\flat_X[z\,;r]) = 0$. The ball $\flat_X[z\,;r]$ is compact, so that u attains its minimum value on it (12.1.3), forcing $0 \in \mathrm{ran}(u)$. So $u \in \delta(X)$, as claimed. Since z is arbitrary in X, it follows that $\delta(X)$ is open in the completion $\tilde{X}$. $\qquad\square$

Example 12.7.7

Every locally compact metric space is a Baire space. The standard proof is very similar to that given for complete metric spaces in 10.11.4 (Q 12.12). Alternatively, it can be proved using the fact we have just proved in 12.7.6 that every locally compact metric space is isometric to an open subset of its completion because every open subset of a complete metric space, as we shall see in 13.6.3, is a Baire space.

12.8 Compact Subsets of Function Spaces

A common feature of compact subsets is that they are closed and bounded; in metric spaces that have the nearest-point property, that alone is sufficient to identify them (12.6.1). But even the space $\mathcal{C}([0, 1])$, being infinite-dimensional, does not have the nearest-point property (see 12.10.2); in such function spaces, the standard criterion for compactness is given by the rather difficult *Arzelà–Ascoli Theorem*.

Theorem 12.8.1

Suppose X is a non-empty set, (Y, e) is a metric space and $S \subseteq B(X, Y)$. For each $x \in X$, let $\hat{x}$ denote the function $f \mapsto f(x)$ defined on S and suppose $\hat{X} = \{\hat{x} \mid x \in X\} \subseteq B(S, Y)$ and $\hat{X}$ is totally bounded. Then S is totally bounded in $B(X, Y)$ if, and only if, $\hat{x}(S)$ is totally bounded in Y for all $x \in X$.

Proof

If S is totally bounded, then $\hat{x}(S)$ is totally bounded (9.2.1) for each $x \in X$ because the maps $\hat{x}$ are all uniformly continuous (9.4.6). For the converse, we let $\epsilon \in \mathbb{R}^+$ and suppose that, for every $x \in X$, $\hat{x}(S)$ is totally bounded in Y. Since $\hat{X}$ is totally bounded, there exists $k \in \mathbb{N}$ and a finite subset $\{z_i \mid i \in \mathbb{N}_k\}$ of X such that $\hat{X} \subseteq \bigcup \{\flat_{B(S,Y)}[\hat{z}_i; \epsilon/3] \mid i \in \mathbb{N}_k\}$. Suppose (g_n) is an arbitrary sequence in S. By hypothesis, $\hat{z}_1(S)$ is totally bounded in Y, so that the sequence $(g_n(z_1))$ has a Cauchy subsequence, say $(g_{t_n}(z_1))$; similarly, the sequence $(g_{t_n}(z_2))$ has a Cauchy subsequence, say $(g_{v_n}(z_2))$, and we note that $(g_{v_n}(z_1))$, being a subsequence of $(g_{t_n}(z_1))$, is also Cauchy. Repeating this argument for each member of the finite set $\{\hat{z}_i \mid i \in \mathbb{N}_k\}$, we get at last a subsequence (g_{m_n}) of (g_n) such that $(g_{m_n}(z_i))$ is Cauchy in Y for each $i \in \mathbb{N}_k$. Because these sequences are Cauchy and $\mathbb{N}_k$ is finite, there exists $j \in \mathbb{N}$ such that, for all $i \in \mathbb{N}_k$, we have $g_{m_n}(z_i) \in \flat_Y[g_{m_j}(z_i); \epsilon/3]$ for all $n \in \mathbb{N}$ with $n \geq j$. But the z_i were picked so that, for each $x \in X$, there exists $i \in \mathbb{N}_k$ such that $e(f(x), f(z_i)) < \epsilon/3$ for all $f \in S$. Therefore $g_{m_n}(x) \in \flat_Y[g_{m_j}(x); \epsilon]$ for all $n \in \mathbb{N}$ with $n \geq j$ and all $x \in X$, whence $g_{m_n} \in \flat_S[g_{m_j}; \epsilon)$ for all $n \in \mathbb{N}$ with $n \geq j$. Since ϵ is arbitrary in $\mathbb{R}^+$, (g_{m_n}) is Cauchy. Since (g_n) is an arbitrary sequence in S, this establishes that S is totally bounded (7.8.2). $\square$

Corollary 12.8.2 (Arzelà–Ascoli Theorem)

Suppose (X, d) is a compact metric space and (Y, e) is a metric space with the nearest-point property. Suppose S is a closed bounded subset of $\mathcal{C}(X, Y)$. For each $x \in X$, define $\hat{x} \colon S \to Y$ by $\hat{x}(f) = f(x)$ for each $f \in S$ and set $\hat{X} = \{\hat{x} \mid x \in X\}$. Then:

(i) $\hat{X} \subseteq B(S, Y)$; in other words, the functions $\hat{x}$ are all bounded.

(ii) S is compact if, and only if, the mapping $x \mapsto \hat{x}$ from X to $B(S, Y)$ is continuous.

Proof

For each $x \in X$, the map $\hat{x}$ is bounded because S is bounded (Q 9.10), so that $\hat{X} \subseteq B(S, Y)$, proving (i).

For the forward implication in (ii), we suppose that S is compact, $z \in X$ and $\epsilon \in \mathbb{R}^+$. Because S is totally bounded (12.1.3), there exist $k \in \mathbb{N}$ and a subset $\{g_i \mid i \in \mathbb{N}_k\}$ of $\mathcal{C}(X,Y)$ such that $S \subseteq \bigcup\{\flat_{e(X,Y)}[g_i ; \epsilon/3] \mid i \in \mathbb{N}_k\}$. Since the g_i are continuous at z and there is only a finite number of them, there exists $\delta \in \mathbb{R}^+$ such that, for all $x \in X$ with $d(z,x) < \delta$, we have $e(g_i(z), g_i(x)) < \epsilon/3$ for all $i \in \mathbb{N}_k$, and it follows that $e(f(z), f(x)) < \epsilon$ for all $f \in S$ because $S \subseteq \bigcup\{\flat_{e(X,Y)}[g_i ; \epsilon/3] \mid i \in \mathbb{N}_k\}$. This then gives $s(\hat{z}, \hat{x}) \leq \epsilon$, where s denotes the supremum metric on $B(S,Y)$. So the map $x \mapsto \hat{x}$ is continuous at z and, because z is arbitrary in X, it is a continuous map.[2]

For the backward implication in (ii), suppose that $x \mapsto \hat{x}$ is continuous. Then its range $\hat{X}$ is, like its domain X, compact (12.3.1) and therefore totally bounded. Moreover, for each $x \in X$, $\hat{x}(S)$, being bounded in Y by (i), is totally bounded because Y has the nearest-point property (7.11.1). It follows from 12.8.1 that S is totally bounded. Also, $\mathcal{C}(X,Y)$ is complete because Y is complete (10.8.3), so that S, being closed in $\mathcal{C}(X,Y)$, is also complete (10.3.2). S, being complete and totally bounded, is compact (10.3.2). $\square$

12.9   Paths of Minimum Length

Even the most general paths admit a well-defined concept of length. This length need not be finite. But even when two points are the endpoints of a path of finite length, there may be no path of minimum length that joins them (12.9.4). Inspired by [8], we now apply the theory we have developed to show that the existence of a path of finite length in a metric space with the nearest-point property implies the existence of a path of minimum length with the same endpoints (12.9.8).

Definition 12.9.1

Suppose (X,d) is a metric space and $f \colon [0,1] \to X$ is a path in X. For each $t \in [0,1]$, let $\mathcal{P}_t$ denote the set of all tuples $a \in \bigcup\{\mathbb{R}^n \mid n \in \mathbb{N}\backslash\{1\}\}$ for which $a_1 = 0$, $a_{\nu(a)} = t$ and $a_i \leq a_{i+1}$ for all $i \in \mathbb{N}_{\nu(a)-1}$, where $\nu(a)$ is the unique $n \in \mathbb{N}$ such that $a \in \mathbb{R}^n$. We define the *length* $\mathrm{lth}_t(f)$ of $f|_{[0,t]}$ to be $\sup\left\{\sum_{i=1}^{\nu(a)-1} d(f(a_i), f(a_{i+1})) \ \middle|\ a \in \mathcal{P}_t\right\}$ and the length $\mathrm{lth}(f)$ of f to be $\mathrm{lth}_1(f)$.

---

[2] What we have described as continuity of the point evaluation mapping $x \mapsto \hat{x}$, where the maps $\hat{x}$ are determined by a particular set S of functions, is more usually styled *equicontinuity* of the members of S. Suppose X and Y are metric spaces, $z \in X$ and S is a set of functions from X to Y. We say that the members of S are *equicontinuous* at z if, and only if, for every $\epsilon \in \mathbb{R}^+$, there exists $\delta \in \mathbb{R}^+$ such that, for all $x \in \flat[z ; \delta]$, we have $f(x) \in \flat[f(z) ; \epsilon]$ for all $f \in S$.

Example 12.9.2

The length of a path need not be finite. Let $f(x) = x\cos(\pi/x)$ for all $x \in (0,1]$ and $f(0) = 0$. It is easily verified that f is a path in $\mathbb{R}$. For each $k \in \mathbb{N}$, we have $f(1/k) = (-1)^k/k$, so that $|f(1/k) - f(1/(k+1))| > 1/k$. Since the harmonic series $\sum_{k=1}^{\infty} 1/k$ tends to ∞, it follows that the length of f is infinite.

Example 12.9.3

In a pathwise connected metric space, there may be points that are not connected by a path of finite length. $\Gamma = \{(0,0)\} \cup \{(x, x\cos(\pi/x)) \mid x \in (0,1]\}$ is a pathwise connected subset of $\mathbb{R}^2$, but there is no path in Γ of finite length that connects $(0,0)$ to any other point of Γ (12.9.2).

Example 12.9.4

When two points of a metric space can be connected by a path of finite length, there may be no path of minimum length that joins them. In the subspace $X = \{z \in \mathbb{C} \mid |z| > 1\}$ of $\mathbb{C}$, for example, the real numbers -2 and 2 are the endpoints of many paths; the infimum of the set of lengths of those paths is $\pi/3 + 2\sqrt{3}$, but there is no path in X of that length joining -2 to 2.

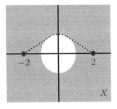

Theorem 12.9.5

Suppose (X, d) is a metric space and $f : [0,1] \to X$ is a path in X. Suppose $r, s \in [0,1]$ with $r < s$. Then $\mathrm{lth}_r(f) + d(f(r), f(s)) \le \mathrm{lth}_s(f)$.

Proof

We use the notation of 12.9.1. For each $a \in \mathcal{P}_r$, we have, by definition, $\left(\sum_{i=1}^{\nu(a)-1} d(f(a_i), f(a_{i+1})) \right) + d(f(r), f(s)) \le \mathrm{lth}_s(f)$, and the theorem follows by taking the supremum over $a \in \mathcal{P}_r$. $\square$

Theorem 12.9.6

Suppose X is a metric space and $f : [0,1] \to X$ is a path of finite length $L \in \mathbb{R}$. Then the function $x \mapsto \mathrm{lth}_x(f)$ is continuous on $[0,1]$ and its range is $[0, L]$.

Proof

Let $\epsilon \in \mathbb{R}^+$. f, being continuous on the compact set $[0,1]$, is uniformly continuous, so there exists $\delta \in \mathbb{R}^+$ such that, for all $r, s \in [0,1]$ with $|r - s| < \delta$, we have $d(f(r), f(s)) < \epsilon/2$. Using the notation of 12.9.1, let $a \in \mathcal{P}_1$ be such that $a_{i+1} - a_i < \delta$ for all $i \in \mathbb{N}_{\nu(a)-1}$ and $\sum_{i=1}^{\nu(a)-1} d(f(a_i), f(a_{i+1})) > L - \epsilon/2$. Suppose $t \in [0,1]$. Let $\gamma = \text{dist}(t, \{a_i \mid i \in \mathbb{N}_{\nu(a)}, a_i \neq t\})$. Then $0 < \gamma < \delta$. Suppose that $s \in [0,1]$ with $0 < |s - t| < \gamma$; note that $s \notin \{a_i \mid i \in \mathbb{N}_{\nu(a)}\}$ and let $k \in \mathbb{N}_{\nu(a)-1}$ be such that $a_k < s < a_{k+1}$. The definition of γ yields also $a_k \leq t \leq a_{k+1}$. So both $\text{lth}_s(f)$ and $\text{lth}_t(f)$ lie in the interval

$$\left[\sum_{i=1}^{k-1} d(f(a_i), f(a_{i+1})), \ L - \sum_{i=k+1}^{\nu(a)-1} d(f(a_i), f(a_{i+1})) \right],$$

where the former summation is interpreted to be 0 if $k = 1$ and the latter is interpreted to be 0 if $k = \nu(a) - 1$. It follows that

$$|\text{lth}_t(f) - \text{lth}_s(f)| \leq L + d(f(a_k), f(a_{k+1})) - \sum_{i=1}^{\nu(a)-1} d(f(a_i), f(a_{i+1})) < \epsilon.$$

Since ϵ and t are arbitrary, the function $x \mapsto \text{lth}_x(f)$ is continuous on $[0,1]$. Last, since this continuous function is non-decreasing (12.9.5) and takes the value 0 at 0 and L at 1, it follows from 11.3.3 that its range is $[0, L]$.   □

Theorem 12.9.7

Suppose (X, d) is a metric space and $f: [0,1] \to X$ is a path in X with non-zero finite length L. Then there exists a path $g: [0,1] \to X$ from $f(0)$ to $f(1)$ that has the same image as f and satisfies $\text{lth}_t(g) = tL$ for all $t \in [0,1]$. In particular g is Lipschitz with Lipschitz constant L.

Proof

For each $x \in [0,1]$, let $\alpha(x) = \text{lth}_x(f)/L$. Then α is a non-decreasing continuous function from $[0,1]$ onto $[0,1]$ (12.9.5, 12.9.6). Note that, if $r, s \in [0,1]$ and $\alpha(r) = \alpha(s)$, then $\text{lth}_r(f) = \text{lth}_s(f)$ and, by 12.9.5, $f(r) = f(s)$. So a function g is well defined on $[0,1]$ by the equation $g \circ \alpha = f$; also $g(0) = g(\alpha(0)) = f(0)$, $g(1) = g(\alpha(1)) = f(1)$ and, by 12.9.6, $g([0,1]) = g(\alpha([0,1])) = f([0,1])$. Moreover, g is Lipschitz with Lipschitz constant L because, for each $r, s \in [0,1]$, there exist $x, y \in [0,1]$ such that $r = \alpha(x)$ and $s = \alpha(y)$ and then, using 12.9.5,

$$d(g(r), g(s)) = d(f(x), f(y)) \leq |\text{lth}_x(f) - \text{lth}_y(f)| = L|\alpha(x) - \alpha(y)| = L|r - s|.$$

In particular, g is continuous and is therefore a path in X, and it follows from Q 12.15 that $\text{lth}_t(g) \leq tL$ for all $t \in [0,1]$. The reverse inequality is shown as follows. Suppose $t \in [0,1]$ and let $s \in \alpha^{-1}(\{t\})$. Then $\text{lth}_s(f) = tL$. For $a \in \mathcal{P}_t$, let $b \in \mathcal{P}_s$ be such that $b_i \in \alpha^{-1}(\{a_i\})$. Then $\nu(a) = \nu(b)$ and $g(a_i) = f(b_i)$ for each $i \in \mathbb{N}_{\nu(a)}$, so that

$$\sum_{i=1}^{\nu(a)-1} d(g(a_i), g(a_{i+1})) = \sum_{i=1}^{\nu(b)-1} d(f(b_i), f(b_{i+1})) \leq \mathrm{lth}_s(f) = tL$$

and it follows, because a is arbitrary in $\mathcal{P}_t$, that $\mathrm{lth}_t(g) \leq tL$. □

Theorem 12.9.8

Suppose (X, d) is a metric space with the nearest-point property and $a, b \in X$ with $a \neq b$. Suppose there is a path of finite length in X from a to b and let m be the infimum of the lengths of all paths from a to b. Then there is a path $g \colon [0, 1] \to X$ from a to b that satisfies $\mathrm{lth}_t(g) = tm$ for all $t \in [0, 1]$. In particular, g is Lipschitz, has length m and is injective.

Proof

For each $k \in \mathbb{R}^+$, let C_k be the set of all paths in X from a to b that are Lipschitz with Lipschitz constant not exceeding $k + m$. Each such path has length not exceeding $k + m$ by Q 12.15 and $C_k \neq \varnothing$ by 12.9.7. It follows immediately from Q 9.13 and 6.6.3 that C_k is a closed subset of $\mathcal{C}([0, 1], X)$. Moreover, C_k is bounded in $\mathcal{C}([0, 1], X)$ because, for $f \in C_k$ and $t \in [0, 1]$, we have $d(a, f(t)) \leq \mathrm{lth}(f)$, which yields $f \in \flat_{\mathcal{C}([0,1],X)}[a'; k + m]$, where a' is the constant function with value a on $[0, 1]$. For each $r \in [0, 1]$, define $\hat{r} \colon C_k \to X$ by $\hat{r}(f) = f(r)$. By 12.8.2, $\hat{r} \in B(C_k, X)$ and since, for all $u, v \in [0, 1]$ and all $f \in C_k$,

$$d(\hat{u}(f), \hat{v}(f)) = d(f(u), f(v)) \leq (k + m)|u - v|,$$

the function $r \mapsto \hat{r}$ from $[0, 1]$ to $B(C_k, X)$ is continuous. By 12.8.2, C_k is compact. So, by 12.4.3, $\bigcap\{C_k \mid k \in \mathbb{R}^+\}$ is not empty. Each path in this intersection has Lipschitz constant m and therefore length m (Q 12.15) and, by 12.9.7, each such path has an associated path g that satisfies the equation $\mathrm{lth}_t(g) = tm$ for all $t \in [0, 1]$. That such a path is necessarily injective is left as an exercise (Q 12.16). □

12.10 Finite-Dimensional Normed Linear Spaces

We have remarked that a norm on a linear space is entirely determined by the concomitant open unit ball (5.4.7). What is much more surprising is that the structure of the unit ball captures entirely one of the most basic algebraic properties of the metric space, namely its finite- or infinite-dimensionality.

Consider the two-dimensional linear subspace $\{a \in \mathbb{R}^3 \mid a_3 = 0\}$ of $\mathbb{R}^3$. There is a vector of length 1 of distance 1 from it, namely $(0, 0, 1)$. Similarly, such a vector can be found for any two-dimensional linear subspace S of $\mathbb{R}^3$. It is still believable, and provable using Q 7.23, that if S is any finite-dimensional subspace of a normed linear space X of larger dimension, then there is a vector of length 1 in X of distance 1 from S (Q 12.18). The corresponding statement for infinite-

dimensional closed subspaces is little less believable, but is it true? Alas, it is not. There are infinite-dimensional closed subspaces S of normed linear spaces X such that every vector of X of length 1 is of distance less than 1 from S. That is why the approximation given by the Riesz Lemma (12.10.1) is important. Using it, we show that the nearest-point property and local compactness are equivalent in a normed linear space and that the spaces that have these properties are precisely those that are finite-dimensional (12.10.2).

Lemma 12.10.1 (Riesz Lemma)

Suppose $(X, \|\cdot\|)$ is a normed linear space and S is a linear subspace of X that is not dense in X. Let $\epsilon \in (0, 1)$. Then there exists a vector $x \in X$ of unit norm such that $\operatorname{dist}(x, S) \geq 1 - \epsilon$.

Proof

Let $z \in X \backslash \overline{S}$. Since $\overline{S}$ is closed, $\operatorname{dist}(z, S) > 0$. Then there exists $w \in S$ such that $0 < \|w - z\| < \operatorname{dist}(z, S) / (1 - \epsilon)$. Let $x = (w - z)/\|w - z\|$. Then $\|x\| = 1$ and, for $s \in S$, we have $x - s = ((w - \|w - z\| s) - z)/\|w - z\|$. So, since $w - \|w - z\| s \in S$, we have $\|x - s\| \geq \operatorname{dist}(z, S) / \|w - z\| > 1 - \epsilon$.   □

Theorem 12.10.2

Suppose X is a normed linear space. The following statements are equivalent:

(i) X is finite-dimensional.

(ii) X has the nearest-point property.

(iii) X is locally compact.

(iv) The closed unit ball of X is compact.

Proof

If X is finite-dimensional, then X has the nearest-point property by Q 7.23. This in turn implies that X is locally compact (12.7.4). If X is locally compact, then there exists $r \in \mathbb{R}^+$ such that $\flat[0\,;r]$ is compact (12.7.1). As $\flat[0\,;1]$ is the image of $\flat[0\,;r]$ under the continuous map $x \mapsto x/r$, it, too, is compact by 12.3.1.

Last, we suppose that $\flat[0\,;1]$ is compact. The open cover for $\flat[0\,;1]$ consisting of all open balls of radius $1/2$ in X has a finite subcover. Let $n \in \mathbb{N}$ and $\{x_i \mid i \in \mathbb{N}_n\}$ be such that $\flat[0\,;1] \subseteq \bigcup\{\flat[x_i\,;1/2) \mid i \in \mathbb{N}_n\}$. No element of $\flat[0\,;1]$ is of distance greater than $1/2$ from the finite-dimensional linear subspace S of X generated by $\{x_i \mid i \in \mathbb{N}_n\}$, so S is dense in X by 12.10.1. But S, being finite-dimensional, is closed in X (10.2.4). So $S = X$ and X is finite-dimensional. $\square$

12.11   A Host of Norms

At last we shall keep the promise made in 1.1.10 to provide a general result from which the triangle inequality for the Euclidean metric on $\mathbb{R}^n$ follows. In fact, we shall define an infinity of related norms on $\mathbb{R}^n$, all of which bear a resemblance to the Euclidean norm. Several proofs of 12.11.3 are available; the proof given here employs in 12.11.1 a nice application of the Intermediate Value Theorem (11.3.3) and some elementary calculus.

Theorem 12.11.1

Let $n \in \mathbb{N}$ and $x, y \in \mathbb{R}^n$ with $x \neq y$. Let $p \in (1\,,\infty)$. The function f defined on $[0\,,1]$ to be

$$t \mapsto \sum_{i=1}^{n} |tx_i + (1-t)y_i|^p$$

attains its maximum value at either 0 or 1 and nowhere in between.

Proof

Since f is continuous and $[0\,,1]$ is compact, f certainly attains its maximum value (12.1.3). Suppose w is arbitrary in $(0\,,1)$. We must show that f does not attain this maximum value at w. Let $\mathbb{I} = \{i \in \mathbb{N}_n \mid x_i \neq y_i\}$ and $\mathbb{J} = \{i \in \mathbb{N}_n \mid wx_i + (1-w)y_i \neq 0\}$. Note that $\mathbb{I} \neq \varnothing$ because $x \neq y$. Moreover, if $\mathbb{I} \cap \mathbb{J} = \varnothing$, then it is easy to check that f does not attain its maximum value at w (Q 12.24), so we suppose $\mathbb{I} \cap \mathbb{J} \neq \varnothing$.

Let $S = \{0,1\} \cup \{y_i/(y_i - x_i) \mid i \in \mathbb{I}\}$. Let u be the largest member of S in $[0\,,w)$ and v be the smallest member of S in $(w\,,1]$; u and v are well-defined because S is finite and contains 0 and 1. Define g on $[u\,,v]$ by setting

$$g(t) = \sum_{i \in \mathbb{J}} |tx_i + (1-t)y_i|^p$$

for each $t \in [u, v]$. We shall show that g attains its maximum value at either u or v and does not attain it at w.

For each $i \in \mathbb{J}$, consider the function $t \mapsto tx_i + (1-t)y_i$. If this function had value 0 for some $t \in (u, v)$, then either $x_i = y_i$, in which case $x_i = y_i = 0$, or $t = y_i/(y_i - x_i) \in S$ and therefore $t = w$, by definition of u and v. Either outcome would contradict the fact that $i \in \mathbb{J}$. Therefore, for each $i \in \mathbb{J}$, the continuous real function $t \mapsto tx_i + (1-t)y_i$ does not take the value 0 anywhere in (u, v). It follows from the Intermediate Value Theorem (11.3.3) that this function is either positive throughout (u, v) or negative throughout (u, v). Set $\alpha_i = 1$ in the former case and $\alpha_i = -1$ in the latter. This constancy of sign ensures that, although the absolute-value function fails to be differentiable at 0, the function g is nonetheless doubly differentiable throughout (u, v). Specifically, for $t \in (u, v)$, we have

$$g(t) = \sum_{i \in \mathbb{J}} (\alpha_i(tx_i + (1-t)y_i))^p,$$

$$g'(t) = \sum_{i \in \mathbb{J}} p(\alpha_i(tx_i + (1-t)y_i))^{(p-1)}\alpha_i(x_i - y_i), \quad \text{and}$$

$$g''(t) = \sum_{i \in \mathbb{J}} p(p-1)|tx_i + (1-t)y_i|^{(p-2)}(x_i - y_i)^2.$$

Because $p > 1$ and $\mathbb{I} \cap \mathbb{J} \neq \varnothing$, it follows that $g''(t) > 0$ for all $t \in (u, v)$. We invoke elementary calculus to infer that, since g is continuous on $[u, v]$, the maximum value of g on $[u, v]$ is attained at either u or v and is not attained in (u, v). In particular, $g(w) < \max\{g(u), g(v)\}$. But $g(w) = f(w)$, $g(u) \leq f(u)$ and $g(v) \leq f(v)$, so that f does not attain its maximum value at w. Since w is arbitrary in $[0, 1]$, f must attain its maximum at either 0 or 1. $\square$

Definition 12.11.2

Suppose $n \in \mathbb{N}$ and $p \in \mathbb{R}^+$. For each $x \in \mathbb{R}^n$, we define $\|x\|_p = (\sum_{i=1}^n |x_i|^p)^{1/p}$.

The notation $\|\cdot\|_p$ is intended to convey the impression that the function that it stands for is a norm. But is it always a norm? We know already that $\|\cdot\|_2$ is a norm, at least when it is defined on $\mathbb{R}^2$ or $\mathbb{R}^3$ (1.7.4). We know also that $\|\cdot\|_1$ is a norm on $\mathbb{R}^2$ (1.7.5). We claim that $\|\cdot\|_p$ is a norm when defined on $\mathbb{R}^n$ for any $n \in \mathbb{N}$ and $p \in [1, \infty)$. This claim is justified next in 12.11.3. As we see in the diagram, $\|\cdot\|_p$ is never a norm when $p \in (0, 1)$ and $n > 1$ (Q 12.20).

$\{z \in \mathbb{R}^2 \mid \|z\|_{1/2} < 1\}$ is not convex.

Theorem 12.11.3 (Minkowski's Theorem)

Suppose $n \in \mathbb{N}$ and $p \in [1, \infty) \cup \{\infty\}$. Then $\|\cdot\|_p$ is a norm on $\mathbb{R}^n$.

Proof

The only norm property that we need to discuss is the triangle inequality since the other norm properties are clearly satisfied by $\|\cdot\|_p$ for all p. The triangle inequalities for $\|\cdot\|_\infty$ and $\|\cdot\|_1$ follow easily from the triangle inequality for the absolute-value function (Q 1.24), so we suppose that $p \in (1, \infty)$.

Suppose $x, y \in \mathbb{R}^n$. If either $x = 0$ or $y = 0$, the triangle inequality is trivial. Otherwise, $\left\| x/\|x\|_p \right\|_p = 1 = \left\| y/\|y\|_p \right\|_p$, so $\left\| tx/\|x\|_p + (1-t)y/\|y\|_p \right\|_p \leq 1$ for all $t \in [0,1]$, by 12.11.1; in particular, setting $t = \|x\|_p /(\|x\|_p + \|y\|_p)$, we have $\|x + y\|_p \leq \|x\|_p + \|y\|_p$. $\qquad\square$

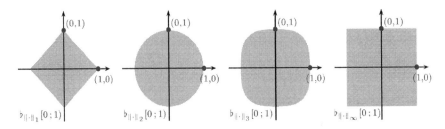

Example 12.11.4

Not just for $p = 1$ and $p = 2$ (7.4.5) but for each $p \in [1, \infty)$, there are, amongst the real bounded sequences, those sequences $a = (a_n)$ for which the real series $\sum_{n=1}^\infty |a_n|^p$ converges. This set is denoted by $\ell_p(\mathbb{R})$, or simply by ℓ_p. It is a subset of ℓ_∞ but is usually endowed with a different norm, defined for $a \in \ell_p$ by $\|a\|_p = (\sum_{n=1}^\infty |a_n|^p)^{1/p}$. That ℓ_p is algebraically closed under addition (B.20.5), and hence is a linear space, and that $\|\cdot\|_p$ is a norm on ℓ_p are both consequences of Minkowski's Theorem for $\mathbb{R}^n$ (12.11.3) by a limiting argument (Q 12.25). These ℓ_p spaces are good illustrations of infinite-dimensional normed linear spaces. We noted in 10.8.6, without proof, that ℓ_1 and ℓ_2 are complete; the same is true for all ℓ_p because, for each $p \in (1, \infty)$, ℓ_p can be identified as an isometric copy of the dual space $\ell^*_{p/(p-1)}$ (see [6]), which is automatically complete (10.8.6). There are similarly defined complex complete normed linear spaces $\ell_p(\mathbb{C})$ for $p \in [1, \infty)$ and $p = \infty$.

Summary

In this chapter, we have given some criteria for compactness, we have outlined the relationship between continuity and compactness, we have augmented our solution to the nearest-point problem and we have introduced local compactness. We have examined compactness in function spaces and found some paths of minimum length. We have shown that, amongst normed linear spaces, being finite-dimensional, having the nearest-point property, having compact closed balls and being locally compact are all equivalent properties. And we have introduced a lot of new norms on sequence spaces.

EXERCISES

Q 12.1 Suppose X is a metric space and A and B are disjoint subsets of X. Suppose A is closed and B is compact. Show that $\operatorname{dist}(A,B) > 0$.

†Q 12.2 Suppose X is a metric space and (x_n) is a convergent sequence in X with limit z. Show that the subset $\{z\} \cup \{x_n \mid n \in \mathbb{N}\}$ of X is compact.

Q 12.3 State, giving reasons, which of the following subsets of $\mathbb{R}^2$ are compact:

(i) $\{(x,y) : 2x^2 + y^2 = 1\}$.

(ii) $\{(x,y) : xy < 1\}$.

(iii) $\{(x,y) : e^x = \cos y\}$.

(iv) $\{(x,y) : 0 \le x \le 1, 0 \le y \le 1\}$.

†Q 12.4 Suppose $f : \mathbb{R} \to \mathbb{R}$ is continuous and I is a closed bounded interval of $\mathbb{R}$. Show that $f(I)$ is also a closed bounded interval of $\mathbb{R}$.

Q 12.5 Find an example of a complete metric space in which all sufficiently small closed balls are compact but large ones are not.

Q 12.6 Find a function between metric spaces that maps every compact connected subset of its domain onto a compact connected subset of its codomain but is not continuous.

Q 12.7 Suppose X is a totally bounded metric space and $\tilde{X}$ is a completion of X. Is $\tilde{X}$ compact?

Q 12.8 Suppose X is a metric space and $\tilde{X}$ is a completion of X. Show that $\tilde{X}$ has the nearest-point property if, and only if, every bounded subset of X is totally bounded.

†Q 12.9 Show that every compact metric space is separable (Q 4.26).

Q 12.10 Suppose X is a metric space that can be expressed as a countable union of compact subspaces. Show that X is separable.

†Q 12.11 Suppose X is a locally compact metric space and S is a subspace of X. Show that if S is either open or closed in X, then S is locally compact.

†Q 12.12 Prove that every locally compact metric space is a Baire metric space.

Q 12.13 Find a Baire metric space that is not locally compact.

Q 12.14 Why is it that closed balls of a locally compact metric space need not be compact but must be so if the metric is determined by a norm?

†Q 12.15 Suppose (X, d) is a metric space and $f \colon [0, 1] \to X$ is a path in X that is Lipschitz with Lipschitz constant $k \in \mathbb{R}^+$. Show that, for each $t \in [0, 1]$, $\operatorname{lth}_t(f) \leq kt$.

†Q 12.16 Show that the path g produced in 12.9.8 is injective.

Q 12.17 Consider $\mathcal{C}([0, 1])$, the linear space of continuous (and automatically bounded) scalar functions defined on $[0, 1]$. Show that the closed unit ball of $\mathcal{C}([0, 1])$, despite being closed and bounded, is not compact. Deduce that the linear space $\mathcal{C}([0, 1])$ is infinite-dimensional.

†Q 12.18 Suppose X is a normed linear space and S is a finite-dimensional proper linear subspace of X. Show that there exists a vector of length 1 of X that is of distance 1 from S.

Q 12.19 Suppose $p \in [1, \infty)$ and (M_1, d_1) and (M_2, d_2) are metric spaces. Define $e(x, y) = ([d_1(x_1, y_1)]^p + [d_2(x_2, y_2)]^p)^{\frac{1}{p}}$ for each $x = (x_1, x_2)$ and $y = (y_1, y_2) \in M_1 \times M_2$. Show that e is a metric on $M_1 \times M_2$.

†Q 12.20 Suppose $n \in \mathbb{N} \backslash \{1\}$ and $p \in (0, 1)$. Is $\|\cdot\|_p$ a norm on $\mathbb{R}^n$?

Q 12.21 Suppose $n \in \mathbb{N}$ and $r, s \in \mathbb{R}$ with $r, s \geq 1$. Show that, for $a \in \mathbb{R}^n \backslash \{0\}$, we have $n^{-1/s} \leq \|a\|_r / \|a\|_s \leq n^{1/r}$.

Q 12.22 Suppose $n \in \mathbb{N}$ and $a \in \mathbb{R}^n$. Show that $\lim_{p \to \infty} \|a\|_p = \|a\|_\infty$.

Q 12.23 Give an example of a sequence of real numbers that converges to 0 but is not in any $\ell_p(\mathbb{R})$ for $1 \leq p < \infty$.

†Q 12.24 With reference to 12.11.1, verify that f does not attain its maximum value at w if $\mathbb{I} \cap \mathbb{J} = \varnothing$.

†Q 12.25 Let $p \in [1, \infty)$. With reference to 12.11.4, use 12.11.3 to show that $\ell_p(\mathbb{R})$ is a linear space and that $\|\cdot\|_p$ is a norm on $\ell_p(\mathbb{R})$.

Q 12.26 Give an example of a sequence that is in $\ell_p(\mathbb{R})$ for all $p \in (1, \infty)$ and is not in $\ell_1(\mathbb{R})$. Use this sequence to show that the inclusion $\ell_r(\mathbb{R}) \subset \ell_s(\mathbb{R})$ is strict whenever $1 \leq r < s < \infty$.

13
Equivalence

My theory stands as firm as a rock;
every arrow directed against it will quickly
return to the archer. How do I know this?
Because I have studied it from all sides for many years;
because I have followed its roots, so to speak,
to the first infallible cause of all created things. *Georg Cantor, 1845–1918*

We began this book by discussing a variety of metrics and have seen that there may be several different natural metrics on a given set. In many situations, particular importance is attached to features of a metric space other than the metric itself—to its open subsets, its convergent subsequences, its nearest points and, above all, its continuous functions. Sometimes old metrics can be swapped for new ones that may make calculations easier but do not change these essential features. Now is the time to catalogue the properties that different changes of metric preserve.

13.1 Topological Equivalence of Metrics

A metric d on a set X is *stronger* than a metric e on the same set if, and only if, the topology generated by d includes that generated by e. This condition can be expressed in different ways (13.1.1).

Theorem 13.1.1 (Criteria for Comparability of Metrics)

Suppose X is a set and d and e are metrics on X. Then the following statements are equivalent:

(i)   (OPEN BALL CRITERION) Every open ball of (X, e) includes an open ball of (X, d) with the same centre.

(ii) (OPEN SET CRITERION) Every open subset of (X, e) is open in (X, d).

(iii) (CLOSED SET CRITERION) Every closed subset of (X, e) is closed in (X, d).

(iv) (IDENTITY FUNCTION CRITERION) The identity function from (X, d) to (X, e) is continuous.

(v) (CONVERGENCE CRITERION) Every sequence that converges in (X, d) converges in (X, e) to the same limit.

(vi) (DOMAIN CONTINUITY CRITERION) Every function from X into a metric space that is continuous with respect to e is continuous with respect to d, the codomain being assumed to have its metric unaltered.

(vii) (CODOMAIN CONTINUITY CRITERION) Every function from a metric space into X that is continuous with respect to d is continuous with respect to e, the domain being assumed to have its metric unaltered.

Proof

Suppose (i) holds and U is open in (X, e). Let $\mathcal{C}$ be the collection of all open balls of (X, d) that are included in U. Certainly, $\bigcup \mathcal{C} \subseteq U$. For each $x \in U$, there exists an open ball of (X, e) centred at x and included in U (5.2.2). By hypothesis, every such ball includes an open ball of (X, d) centred at x; in particular, $x \in \bigcup \mathcal{C}$. Since x is arbitrary in U, this gives $U \subseteq \bigcup \mathcal{C}$ and hence $U = \bigcup \mathcal{C}$. This set is open in (X, d) by 5.2.2. So (i) implies (ii). That (ii) and (iii) are equivalent is an immediate consequence of 4.1.4. That (ii) and (iv) are equivalent is a consequence of 8.3.1.

Suppose now that (x_n) is a sequence that converges in (X, d) with limit z. If the identity function from (X, d) to (X, e) is continuous, then (x_n) converges to z in (X, e) also because continuous functions preserve convergence (8.3.1). So (iv) implies (v).

Suppose that (v) holds, that Z is a metric space and that $f \colon X \to Z$ is continuous with respect to e. Suppose (x_n) is a sequence in X that converges in (X, d) to some $z \in X$. By hypothesis, (x_n) converges to z in (X, e) and, as f is continuous with respect to e, $(f(x_n))$ converges to $f(z)$ in Z. Since z is arbitrary in X, f is continuous with respect to d (8.3.1). So (v) implies (vi).

(vii) also follows from (v). Suppose that (v) holds, that W is a metric space and that $f \colon W \to X$ is continuous with respect to d. Suppose $v \in W$ and (w_n) is a sequence in W that converges to v in W. Since f is continuous with respect to d, $(f(w_n))$ converges to $f(v)$ in (X, d); by hypothesis, it also converges to $f(v)$ in (X, e). Since v is arbitrary in W, this implies that f is continuous with respect to e (8.3.1).

Finally, it is a special case of either (vi) or (vii) that the identity function from (X, d) to (X, e) is continuous. This in turn implies that every open ball $\flat_e[x\,; r]$ of (X, e) is an open subset of (X, d) and therefore includes some ball $\flat_d[x\,; s]$ with the same centre (5.2.2). So each of (vi) and (vii) implies (i), and equivalence of all seven conditions is proven.   □

Definition 13.1.2

Suppose X is a set and d and e are metrics on X. We say

- that d is *topologically stronger* than e and that e is *topologically weaker* than d if, and only if, every open subset of (X, e) is open in (X, d);
- that d and e are *topologically equivalent* if, and only if, d is both weaker and stronger than e; and
- that d and e are *not comparable* if, and only if, d is neither topologically stronger nor topologically weaker than e.

In the light of the preceding definition, we can now use 13.1.1 to write down a number of equivalent formulations of topological equivalence of metrics.

Corollary 13.1.3

Suppose d and e are metrics on a set X. Then the following statements are equivalent:

(i) d and e are topologically equivalent metrics.

(ii)   The topologies of (X, d) and (X, e) are identical.

(iii)   The collection of closed subsets of (X, d) is the same as that of (X, e).

(iv)   The identity functions from (X, d) to (X, e) and from (X, e) to (X, d) are both continuous.

(v)   Every convergent sequence of (X, d) is convergent in (X, e) with the same limit, and vice versa.

(vi)   Every function from X into a metric space is continuous with respect to d if, and only if, it is continuous with respect to e, the metric on the codomain being assumed to be unaltered.

(vii) Every function from a metric space into X is continuous with respect to d if, and only if, it is continuous with respect to e, where the metric on the domain is assumed to be unaltered.

Theorem 13.1.4

Suppose X is a set. Then topological equivalence is an equivalence relation (B.16.1) on the collection of all metrics on X.

Proof

This follows immediately from the fact that metrics on X are topologically equivalent if, and only if, they generate the same topology (13.1.3).   □

Example 13.1.5

It is easy to show that the metrics μ_1, μ_2 and μ_∞ (1.6.1) defined on $\mathbb{R}^2$ are all topologically equivalent; our diagram illustrates how each ball for any one of these metrics includes balls for the others with the same centre. We shall show presently (13.5.1) that these metrics display an even stronger type of equivalence.

Example 13.1.6

Suppose (X, d) is a metric space. All scalar multiples of d are topologically equivalent to d. So is the metric $(a, b) \mapsto d(a, b)/(1 + d(a, b))$ (Q 4.11). The metric on $\mathbb{R}^+$ given by $(a, b) \mapsto \left| a^{-1} - b^{-1} \right|$ is topologically equivalent to the Euclidean metric (5.3.2).

Example 13.1.7

Suppose (X, e) is a discrete metric space (4.3.7). Then e is topologically equivalent to the discrete metric on X.

Question 13.1.8

We have derived new metrics from old ones using injective functions on metric spaces (Q 1.12). Is there any simple property of the function that ensures that the new metric is comparable with the original metric? Suppose (X, d) and (Y, m) are metric spaces and $f \colon X \to Y$ is injective. The epsilon–delta criterion for continuity of f at a is that, for each $\epsilon \in \mathbb{R}^+$, there exists $\delta \in \mathbb{R}^+$ such that $d(a, b) < \delta \Rightarrow m(f(a), f(b)) < \epsilon$. Denote by e the metric on X given by $(a, b) \mapsto m(f(a), f(b))$ (Q 1.12) to see that this is precisely the condition for continuity of the identity function $\iota_{d,e}$ from (X, d) to (X, e) at a. So continuity of f from (X, d) to (Y, m) is a necessary and sufficient condition for e to be topologically weaker than d. Similarly, continuity of f^{-1} from $(f(X), m)$ to (X, d) is a necessary and sufficient condition for e to be topologically stronger than d (Q 13.3). For example, the metric $(a, b) \mapsto \left| e^b - e^a \right|$ determined on $\mathbb{R}$ by the exponential function is topologically weaker than the Euclidean metric $|\cdot|$ because the exponential function from $(\mathbb{R}, |\cdot|)$ to $(\mathbb{R}, |\cdot|)$ is continuous. Moreover, since its inverse—the logarithmic function from $(\mathbb{R}^+, |\cdot|)$ to $(\mathbb{R}, |\cdot|)$—is continuous, this metric is also topologically stronger than the Euclidean metric and is therefore topologically equivalent to it.

Example 13.1.9

Suppose (X, d) is a metric space and suppose $f \colon X \to \mathbb{R}$ is a continuous function.

Let e be the metric on X given by $(a,b) \mapsto d(a,b) + |f(a) - f(b)|$ (Q 1.16). Certainly $d \leq e$, so that the identity function from (X,e) to (X,d) is Lipschitz and therefore continuous. To show that the identity function $\iota_{d,e}$ from (X,d) to (X,e) is continuous, we proceed as follows. Suppose that $x \in X$ and $\epsilon \in \mathbb{R}^+$. Because f is continuous at x, there exists $\delta \in (0, \epsilon/2)$ such that, for all $b \in X$, $d(x,b) < \delta \Rightarrow |f(x) - f(b)| < \epsilon/2$. Then $d(x,b) < \delta \Rightarrow e(x,b) < \epsilon$, so that $\iota_{d,e}$ is continuous at x. Since x is arbitrary in X, $\iota_{d,e}$ is continuous. It follows that e and d are topologically equivalent metrics (13.1.3).

Question 13.1.10

Topologically equivalent metrics produce the same convergent sequences. Do they produce the same Cauchy sequences? Alas, they need not. Consider $\mathbb{R}^+$ with its usual metric $|\cdot|$ and the metric d given by $d(a,b) = |a^{-1} - b^{-1}|$ for all $a, b \in \mathbb{R}^+$. We saw in 5.3.2 that these two metrics are topologically equivalent. The sequence $(1/n)$ is Cauchy in $(\mathbb{R}^+, |\cdot|)$ but not Cauchy in $(\mathbb{R}^+, d)$, and the sequence (n) is Cauchy in $(\mathbb{R}^+, d)$ and not in $(\mathbb{R}^+, |\cdot|)$.

Question 13.1.11

Topologically equivalent metrics produce the same open sets and closed sets. They also produce the same compact subsets (12.3.1) and the same locally compact subsets (Q 13.7). They produce the same connected subsets (11.3.1). Do they produce the same dense sets? Indeed they do. Suppose X is a set and d and e are topologically equivalent metrics on X. Suppose S is dense in (X,d) and V is open in (X,e). Then V is open in (X,d) because d and e produce the same topology, so that $S \cap V \neq \varnothing$ and, because V is an arbitrary open subset of (X,e), S is dense in (X,e) by 4.2.1. The converse is proved the same way, so that d and e produce the same dense subsets of X.

Example 13.1.12

Equivalent metrics need not produce the same bounded sets. Consider again $\mathbb{R}^+$ with its usual metric $|\cdot|$ and the topologically equivalent metric d given by $d(a,b) = |a^{-1} - b^{-1}|$ for all $a, b \in \mathbb{R}^+$ (5.3.2). The interval $(1/2, \infty)$ is unbounded in $(\mathbb{R}^+, |\cdot|)$, but in $(\mathbb{R}^+, d)$ it is the ball $\flat_d[1\,;1]$ and has diameter 2.

Example 13.1.13

Topological equivalence does not preserve completeness because it does not preserve Cauchy sequences. This paves the way for many unlikely examples of metrics that are topologically equivalent to complete metrics. In 10.3.4, we saw how to make an open subset of a complete metric space complete by altering its metric. The function f used in 10.3.4 is continuous, so it follows from 13.1.9 that the new metric is topologically equivalent to the original metric. We can go further. If $\mathcal{U}$ is a countable collection of open subsets of a complete metric space, then $\bigcap \mathcal{U}$ can be endowed with a topologically equivalent metric that makes $\bigcap \mathcal{U}$ complete (Q 13.2). In particular, $\mathbb{R}\backslash\mathbb{Q} = \bigcap\{\mathbb{R}\backslash\{q\} \mid q \in \mathbb{Q}\}$ has a metric topologically equivalent to the Euclidean metric with respect to which it is complete: such a metric is $(a,b) \mapsto |a - b| + \sum_{n=1}^{\infty} 2^{-n}\big|(a - q_n)^{-1} - (b - q_n)^{-1}\big|$, where (q_n) is an enumeration of $\mathbb{Q}$.

Summary 13.1.14

Suppose X is a set and d and e are equivalent metrics on X. Then (X,d) and (X,e) admit the same

- open subsets;
- closed subsets;
- dense subsets;
- compact subsets;
- locally compact subsets;
- connected subsets;
- convergent sequences and limits;
- continuous functions with X as domain; and
- continuous functions with X as codomain.

13.2 Uniform Equivalence of Metrics

When we are manipulating convergence or continuity, we can substitute a given metric by a topologically equivalent one if such substitution makes our work easier. There are, however, some possible difficulties. In particular, Cauchy sequences may cease to be Cauchy, and totally bounded sets may cease to be bounded. Uniform equivalence of metrics makes these difficulties disappear.

Definition 13.2.1

Suppose X is a set and d and e are metrics on X. We say that d is *uniformly stronger* than e and that e is *uniformly weaker* than d if, and only if, the identity function from (X,d) to (X,e) is uniformly continuous. We say that d and e are *uniformly equivalent* if, and only if, each is uniformly stronger than the other.

Theorem 13.2.2

Suppose X is a set and d and e are metrics on X.

(i)   If d is uniformly stronger than e, then d is topologically stronger than e.

(ii) If d and e are uniformly equivalent, then they are topologically equivalent.

Proof

If d is uniformly stronger than e, then the identity function from (X, d) to (X, e) is uniformly continuous. It is therefore continuous (9.1.2) and d is topologically stronger than e (13.1.1). This proves (i); (ii) follows easily. $\square$

Theorem 13.2.3

Suppose X is a set. Then uniform equivalence is an equivalence relation on the collection of metrics on X.

Proof

Suppose d, e and m are metrics on X, d is uniformly stronger than e and e is uniformly stronger than m. The identity maps $\iota_{d,e}\colon (X, d) \to (X, e)$ and $\iota_{e,m}\colon (X, e) \to (X, m)$ are uniformly continuous, so their composition, which is $\iota_{d,m}\colon (X, d) \to (X, m)$, is also uniformly continuous by 9.6.1, which means that d is uniformly stronger than m. Similarly, if d is uniformly weaker than e and e is uniformly weaker than m, then d is also uniformly weaker than m. So uniform equivalence is transitive; that it is reflexive and symmetric is clear from the definition. $\square$

Example 13.2.4

Reasoning similar to that used in 13.1.8 shows that, if (X, d) and (Y, m) are metric spaces and $f\colon X \to Y$ is injective, then d is uniformly stronger than the metric e on X given by $(a, b) \mapsto m(f(a), f(b))$ if, and only if, f is uniformly continuous; similarly, uniform continuity of f^{-1} from $(f(X), m)$ to (X, d) is a necessary and sufficient condition for d to be uniformly weaker than e (Q 13.4).

The exponential function is not uniformly continuous on $\mathbb{R}$. Therefore the metric $(a, b) \mapsto \left| e^b - e^a \right|$ defined on $\mathbb{R}$, although it is topologically equivalent to the Euclidean metric, is not uniformly equivalent to it. However, the restriction of the exponential function to $[0, 1]$ is uniformly continuous and the logarithmic function is uniformly continuous on the range $[1, e]$ of this restriction. So the metric $(a, b) \mapsto \left| e^b - e^a \right|$ on $[0, 1]$ is uniformly equivalent to the Euclidean metric on $[0, 1]$ (Q 13.4).

Question 13.2.5

Metrics are topologically equivalent if, and only if, they produce the same open sets; they need not produce the same open balls (13.1.5). If they do produce the same open balls they are topologically equivalent, but is their equivalence uniform? It need not be. The metric on $\mathbb{R}$ determined by the function $f\colon \mathbb{R} \to \mathbb{R}$, for which $f(x) = x$ when $x \in \mathbb{R}^-$ and $f(x) = x^2$ when $x \in \mathbb{R}^{\oplus}$ (Q1.15), gives the same open balls as the Euclidean metric (Q5.12), but, since the function f is not uniformly continuous, the two metrics are not uniformly equivalent.

Question 13.2.6

Does uniform equivalence of metrics preserve boundedness? Let d be the usual metric on $\mathbb{N}$ and let e be the discrete metric. The identity map from $(\mathbb{N}, d)$ to $(\mathbb{N}, e)$ is a Lipschitz map and its inverse is uniformly continuous by 9.1.10. So these metrics are uniformly equivalent, but $(\mathbb{N}, e)$ is a bounded space and $(\mathbb{N}, d)$ is not. This example also serves to show that the nearest-point property is not necessarily preserved by uniform equivalence: the ball $\flat[1\,;1]$ of $(\mathbb{N}, e)$ is bounded but not totally bounded, so that $(\mathbb{N}, e)$ does not have the nearest-point property, whereas $(\mathbb{N}, d)$, being a closed subset of $\mathbb{R}$ with the Euclidean metric, certainly does.

Question 13.2.7

Uniform equivalence preserves the same features as topological equivalence because it satisfies a stronger condition. It also preserves Cauchy sequences and totally bounded sets by 9.2.1. Because it preserves both convergent sequences and Cauchy sequences, it preserves completeness (but see also 10.6.1). And, because every composition of a uniformly continuous function with a uniformly continuous identity function yields a uniformly continuous function by 9.6.1, uniform equivalence also preserves uniform continuity.

Summary 13.2.8

Suppose X is a set and d and e are uniformly equivalent metrics on X. Then (X, d) and (X, e) admit the same

- open subsets;
- closed subsets;
- dense subsets;
- compact subsets;
- locally compact subsets;
- connected subsets;

- convergent sequences and limits;
- Cauchy sequences;
- totally bounded subsets;
- complete subsets;
- continuous functions with X as domain;
- continuous functions with X as codomain;
- uniformly continuous functions with X as domain; and
- uniformly continuous functions with X as codomain.

13.3 Lipschitz Equivalence of Metrics

Lipschitz equivalence is a very much more powerful property than uniform equivalence. It is the form of equivalence most common in linear spaces.

Definition 13.3.1

Suppose X is a set and d and e are metrics on X. We say that d is *Lipschitz stronger* than e and that e is *Lipschitz weaker* than d if, and only if, the identity function from (X, d) to (X, e) is a Lipschitz function. We say that d and e are *Lipschitz equivalent* if, and only if, each of d and e is Lipschitz stronger than the other.

Theorem 13.3.2

Suppose X is a set and d and e are metrics on X.

(i) If d is Lipschitz stronger than e, then d is uniformly stronger than e.

(ii) If d and e are Lipschitz equivalent, then they are uniformly equivalent.

Proof

Suppose that d is Lipschitz stronger than e, and let $k \in \mathbb{R}^+$ be such that $e(a, b) \le kd(a, b)$ for all $a, b \in X$. Let $\epsilon \in \mathbb{R}^+$. Set $\delta = \epsilon/k$. Then, for all $a, b \in X$ with $d(a, b) < \delta$, we have $e(a, b) \le k\delta = \epsilon$. So d is uniformly stronger than e. This proves (i); (ii) follows easily. $\square$

Theorem 13.3.3

Suppose X is a set. Lipschitz equivalence is an equivalence relation on the collection of all metrics on X.

Proof

Suppose that d, e and m are metrics on X and that d is Lipschitz stronger than e and e is Lipschitz stronger than m. Then the identity maps $\iota_{d,e} \colon (X,d) \to (X,e)$ and $\iota_{e,m} \colon (X,e) \to (X,m)$ are Lipschitz functions, so that their composition $\iota_{d,m} \colon (X,d) \to (X,m)$ is also a Lipschitz function by 9.6.1, which means that d is Lipschitz stronger than m. Similarly, if d is Lipschitz weaker than e and e is Lipschitz weaker than m, then d is Lipschitz weaker than m. So Lipschitz equivalence is transitive; it is certainly reflexive and symmetric.   □

Example 13.3.4

When testing for Lipschitz equivalence of a given metric with a metric determined by an injective function, we use a result (Q 13.6) similar to those already given for the other types of equivalence (13.1.8 and Q 13.4). Consider, for example, the function $f : x \mapsto \sqrt{1 - x^2}$ defined on $[0,1]$ (9.5.2). This function is uniformly continuous but is not Lipschitz on $[0,1]$: specifically, if $k \in \mathbb{R}^+$ and $x \in \big((k^2 - 1)/(k^2 + 1),1\big)$, then $(1+x) > k^2(1-x)$, whence $\sqrt{1 - x^2} > k(1-x)$ or, in other words, $|f(x) - f(1)| > k|x - 1|$. Moreover, f is equal to its own inverse. So the metric on $[0,1]$ given by $(a,b) \mapsto \left|\sqrt{1 - a^2} - \sqrt{1 - b^2}\right|$, although it is uniformly equivalent to the Euclidean metric, is neither Lipschitz stronger nor Lipschitz weaker than the Euclidean metric.

Example 13.3.5

We have used two metrics on $\mathcal{C}([0,1])$, the usual supremum metric s and the integral metric of 7.7.5. How do they compare? For all $f, g \in \mathcal{C}([0,1])$, we have

$$\int_0^1 |f(x) - g(x)|\, dx \le \sup\{|f(x) - g(x)| \mid x \in [0,1]\} = s(f,g),$$

so that s is Lipschitz stronger than the integral metric. Is there any sort of equivalence between the two metrics? Is the supremum metric weaker in any sense than the integral metric? To answer this, consider the sequence (f_n) in $\mathcal{C}([0,1])$, where, for each $n \in \mathbb{N}$, f_n is given by $f_n(x) = (1 - nx)$ for $x \in [0,1/n]$ and $f_n(x) = 0$ otherwise. Clearly $\int_0^1 |f_n(x)|\, dx = 1/2n$, whereas $\sup\{f_n(x) \mid x \in [0,1]\} = 1$. With respect to the integral metric, $f_n \to 0$, but the distance, with respect to s, from each f_n to the zero function is 1. So, by the convergence criterion (13.1.1), the integral metric is not even topologically stronger than the supremum metric.

$\int_0^1 |f_7(x)|\, dx = 1/14.$

Theorem 13.3.6

Suppose X is a set and d and e are metrics on X. Suppose that d is Lipschitz stronger than e. Then every bounded subset of (X, d) is bounded in (X, e).

Proof

Let $k \in \mathbb{R}^+$ be such that $e(a, b) \leq kd(a, b)$ for all $a, b \in X$. Suppose S is a subset of X that is bounded in (X, d). Then, for each $a, b \in S$, we have $e(a, b) \leq kd(a, b) \leq k \operatorname{diam}_d(S) < \infty$, so that S is bounded in (X, e). $\square$

It follows from 13.3.6 that Lipschitz equivalence, unlike uniform equivalence, preserves boundedness. Since it also preserves convergence of sequences, it preserves the nearest-point property as well (7.11.1).

Summary 13.3.7

Suppose X is a set and d and e are Lipschitz equivalent metrics on X. Then (X, d) and (X, e) admit the same

- open subsets;
- closed subsets;
- dense subsets;
- bounded subsets;
- compact subsets;
- locally compact subsets;
- connected subsets;
- convergent sequences and limits;
- Cauchy sequences;
- totally bounded subsets;
- complete subsets;
- subsets with the nearest-point property;
- continuous functions with X as domain;
- continuous functions with X as codomain;
- uniformly continuous functions with X as domain;
- uniformly continuous functions with X as codomain;
- Lipschitz functions with X as domain; and
- Lipschitz functions with X as codomain.

13.4 The Truth about Conserving Metrics

All conserving metrics on a product are equivalent by 4.5.1. It is natural to ask whether or not they are necessarily Lipschitz equivalent because there is an obvious similarity between the definition of a conserving metric and that of Lipschitz equivalence. They are, of course (13.4.1).

Theorem 13.4.1

Suppose $n \in \mathbb{N}$ and, for each $i \in \mathbb{N}_n$, (X_i, τ_i) is a metric space. All conserving metrics on $\prod_{i=1}^{n} X_i$ are Lipschitz equivalent. In particular, each is Lipschitz equivalent to the Euclidean product metric μ_2 on $\prod_{i=1}^{n} X_i$ (1.6.1).

Proof

Suppose d is a conserving metric on $\prod_{i=1}^{n} X_i$. Then d and μ_2 are Lipschitz stronger than μ_∞ and Lipschitz weaker than μ_1. Moreover, we have $\mu_\infty(a,b) \leq \mu_2(a,b) \leq \mu_1(a,b) \leq n\mu_\infty(a,b)$ for all $a, b \in \prod_{i=1}^{n} X_i$, so that μ_1 and μ_∞ are Lipschitz equivalent. The proof is completed by 13.3.3. $\square$

Question 13.4.2

The cat is out of the bag. Why bother with conserving metrics at all? The reader who cares to go back over all our theorems that involve conserving metrics will discover that, in every case—except, for obvious reasons, in 1.6.4— the theorem has an extension to all metrics that belong to the same Lipschitz equivalence class as μ_1, μ_2 and μ_∞. This applies to 4.5.1, 6.10.1, 7.2.1, 7.10.2, 7.10.3, 7.10.4, 7.13.1, 7.13.2, 9.1.7, 9.8.1, 9.8.2, 10.5.1 and 10.5.2. We coined the term *conserving metric* and confined ourselves to such metrics in order to make our proofs a little cleaner and thereby smooth the reader's path to understanding. By injecting Lipschitz constants into the proofs, theorems for the wider collection of metrics can be obtained easily. But the story does not end there. The astute reader will notice that, although arbitrary product metrics are not appropriate in any of the theorems listed above, some of them can be extended by using uniform equivalence. Which ones?

13.5 Equivalence of Norms

What was meant by saying in our introduction to Lipschitz equivalence (13.3) that it is the form of equivalence most commonly encountered when we are dealing with linear spaces? Precisely that in a linear space, norms that are topologically equivalent—by which we mean that the metrics they determine are topologically equivalent—are Lipschitz equivalent; generally, we say simply that they are equivalent.

Theorem 13.5.1

Suppose X is a linear space. Two norms on X are topologically equivalent if, and only if, they are Lipschitz equivalent.

Proof

If the norms are Lipschitz equivalent, then they are uniformly equivalent and so topologically equivalent. For the converse, if they are topologically equivalent, then the two identity maps are continuous. Since the identity maps are linear, they are Lipschitz functions by 9.4.7.   □

Example 13.5.2

Suppose X is a linear space and $\|\cdot\|_A$ and $\|\cdot\|_B$ are norms on X, both of which make X into a complete space. If $\|\cdot\|_A$ is topologically weaker than $\|\cdot\|_B$, then $\|\cdot\|_A$ and $\|\cdot\|_B$ are equivalent. This follows from 13.1.3 and 10.11.8, though the latter is unproven in our exposition.

Example 13.5.3

In Q 10.1, we used 6.8.6 to show that $\mathcal{C}([0,1])$ with the integral metric is not complete. An alternative, but indirect, way of doing this is to observe that both the integral metric and the supremum metric are determined by norms, that the supremum metric makes the space complete (10.8.3) and that the two metrics are comparable but not equivalent (13.3.5); it follows from the unproven 13.5.2 that the integral metric does not produce a complete metric space. The proof of Theorem 13.5.4 below could similarly be shortened by appealing to 13.5.2.

Theorem 13.5.4

Suppose X is a finite-dimensional linear space. All norms on X are equivalent.

Proof

Let S be a basis for X. Each vector v of X is uniquely represented as a sum $\sum_{s \in S} \lambda_{v,s} s$ for some scalars $\lambda_{v,s}$ (1.7.10), and it is easily verified that the map $v \mapsto \sum_{s \in S} |\lambda_{v,s}|$ is a norm on X. We label this norm $\|\cdot\|_S$. Suppose $\|\cdot\|$ is any norm on X. The set $\{\|s\| \mid s \in S\}$ is finite and so has a maximum element; let m be this maximum. Then, by the triangle inequality for $\|\cdot\|$, for each $v \in X$,

$$\|v\| = \left\| \sum_{s \in S} \lambda_{v,s} s \right\| \le m \sum_{s \in S} |\lambda_{v,s}| = m\|v\|_S \,.$$

So $\|\cdot\|_S$ is Lipschitz stronger than $\|\cdot\|$ and the function $\|\cdot\| \colon (X, \|\cdot\|_S) \to \mathbb{R}^{\oplus}$ is Lipschitz and therefore continuous. The set $C = \{x \in X \mid \|x\|_S = 1\}$ is clearly closed in the closed unit ball of $(X, \|\cdot\|_S)$ and is therefore compact (12.10.2, 12.2.3). So, by 12.3.1, the set $\{\|x\| \mid x \in C\}$ is a compact subset of $\mathbb{R}^{\oplus}$. It does not contain 0, so there exists $k \in \mathbb{R}^+$ such that $k \le \|x\|$ for all $x \in C$. It follows from the properties of norms that $\|v\|_S \le k^{-1}\|v\|$ for all $v \in X$. Therefore $\|\cdot\|_S$ is weaker than $\|\cdot\|$, so these norms are equivalent. Since $\|\cdot\|$ is an arbitrary norm on X, it follows from 13.3.3 that all norms on X are equivalent. $\square$

13.6 Equivalent Metric Spaces

A concept closely related to equivalence of metrics, but not to be confused with it, is equivalence of metric spaces. Here we are concerned not with different metrics on a single set but with metrics on sets that are identified with each other by one-to-one correspondence. A bijective map that preserves the metric—and therefore every other intrinsic property of a metric space—we know as an isometry (1.4.1); it is the strongest form of equivalence between metric spaces. Our discussion on equivalent metrics prompts us to discuss some weaker forms of equivalence between metric spaces than that determined by an isometry.

Definition 13.6.1

Suppose (X, d) and (Y, e) are metric spaces.

- X and Y are said to be *homeomorphic* or *topologically equivalent* if, and only if, there exists a bijective function $f \colon X \to Y$ that is continuous and has continuous inverse; such a function is called a *homeomorphism*.
- X and Y are said to be *uniformly equivalent* if, and only if, there exists a bijective function $f \colon X \to Y$—called a *uniform equivalence*—that is uniformly continuous and has uniformly continuous inverse.
- X and Y are said to be *Lipschitz equivalent* if, and only if, there exists a bijective function $f \colon X \to Y$—called a *Lipschitz equivalence*—that is a Lipschitz function and has a Lipschitz function for its inverse.

In Definition 13.6.1, X and Y may be equal. In that case, equivalence of the metrics d and e, in any of the three forms, certainly implies the corresponding type of equivalence of the spaces. The converse is not true. Equivalence of (X, d) and (X, e) does not necessarily imply equivalence of the metrics d and e (13.6.6); for the metrics to be equivalent, in any of the three senses, the identity function must satisfy the properties ascribed to f in Definition 13.6.1.

Theorem 13.6.2

Suppose (X, d) and (Y, e) are metric spaces.

(i)   If X and Y are homeomorphic, then any homeomorphism between them identifies open sets with open sets, closed sets with closed sets, dense sets with dense sets, compact subsets with compact subsets, locally compact subsets with locally compact subsets, connected sets with connected sets, convergent sequences and their limits with convergent sequences and their limits, continuous functions defined on X with continuous functions defined on Y and continuous functions into X with continuous functions into Y.

(ii) If X and Y are uniformly equivalent, any uniform equivalence between them identifies also Cauchy sequences with Cauchy sequences, totally bounded sets with totally bounded sets, complete subspaces with complete subspaces, uniformly continuous functions defined on X with uniformly continuous functions defined on Y and uniformly continuous functions into X with uniformly continuous functions into Y.

(iii) If X and Y are Lipschitz equivalent, any Lipschitz equivalence between them identifies also bounded sets with bounded sets, sets exhibiting the nearest-point property with sets exhibiting the nearest-point property, Lipschitz functions defined on X with Lipschitz functions defined on Y and Lipschitz functions into X with Lipschitz functions into Y.

Proof

A bijective map $f\colon X \to Y$ identifies X and Y as sets. So we are merely looking at the various types of equivalence of the metrics d and e when Y is identified with X through the map $x \mapsto f(x)$. The proposition therefore follows from 13.1.3, 13.1.11, 9.2.1, 13.3.6, 12.3.1, 12.3.2, 10.6.1 and the results concerning compositions of functions of similar continuity type.   □

Example 13.6.3

Every homeomorphic copy of a Baire metric space is a Baire metric space because every homeomorphism identifies open subsets of the domain with open subsets of the range and dense subsets of the domain with dense subsets of the range (13.6.2). In particular, every open subset of a complete metric space is a Baire space because it is homeomorphic to a complete metric space (13.1.13). It is in fact true that a metric space is homeomorphic to a complete metric space if, and only if, it can be expressed as the intersection of a countable number of dense open subsets of its own completion.[1]

Example 13.6.4

$\mathbb{Q}$ is not homeomorphic to a complete metric space; if it were, then it would be a Baire space and the set $\bigcap\{\mathbb{Q}\setminus\{q\} \mid q \in \mathbb{Q}\}$ would be dense in $\mathbb{Q}$. It is plain to see that it is, in fact, empty.

Question 13.6.5

The celebrated and very important Schröder–Bernstein Theorem tells us that if A and B are sets and if there exists an injective function from each into the other, then there is a bijective function between A and B (B.17.2). This prompts us to ask whether or not two metric spaces are necessarily homeomorphic if there exist continuous bijective functions from each to the other. But the answer to this question is *no*. Consider the subsets X and Y of $\mathbb{R}$ given by $X = \bigcup\{(3n, 3n+1) \mid n \in \mathbb{N} \cup \{0\}\} \cup \{3n+2 \mid n \in \mathbb{N} \cup \{0\}\}$ and $Y = \{1\} \cup X\setminus\{2\}$; endow X and Y with the usual Euclidean metric. Consider the functions $f: X \to Y$ and $g: Y \to X$ given by $f(2) = 1$ and $f(x) = x$ for all $x \in X\setminus\{2\}$ and

$$g(y) = \begin{cases} y/2, & \text{if } y \in (0, 1]; \\ (y-2)/2, & \text{if } y \in (3, 4); \\ y - 3, & \text{if } y \in Y\setminus(0, 4). \end{cases}$$

Both these functions are bijective and continuous; this is more or less obvious from the picture and is easy to check. Now suppose there is a homeomorphism $h: X \to Y$. Let $z \in X$ be such that $h(z) = 1$. The singleton sets $\{3n+2\}$, for $n \in \mathbb{N} \cup \{0\}$, are all open in X, so their images in Y are also singleton open sets. In particular, $z \neq 3n + 2$ for any $n \in \mathbb{N} \cup \{0\}$, so $z \in (3m, 3m+1)$

---

[1] This is discussed in [3].

for some $m \in \mathbb{N} \cup \{0\}$. Consider the interval $(3m, z]$. It is connected (11.3.2). Because h is continuous, its image $S = h((3m, z])$ is also connected (11.3.1) and is therefore also an interval (11.3.2), and, because h is injective, S is not a degenerate interval. Note also that $S \subseteq Y$. Therefore there exists $r \in (0, 1)$ such that $(r, 1] \subseteq S$. The same argument applies to the set $S' = h([z, 3m + 1))$, yielding some $r' \in (0, 1)$ such that $(r', 1] \subseteq S'$. Then $S \cap S' \neq \{1\}$, contradicting injectivity of h. So X and Y are not homeomorphic metric spaces.

Example 13.6.6

Suppose d and e are metrics on a set X. If d and e are topologically equivalent, then certainly (X, d) and (X, e) are homeomorphic, because the identity function $\iota_{d,e}: (X, d) \to (X, e)$ is continuous and has continuous inverse $\iota_{e,d}: (X, e) \to (X, d)$. The converse is not true. Here is an extraordinary example of Lipschitz equivalent copies of $\mathbb{R}^{\oplus}$ with metrics that are not even topologically equivalent. Define metrics d and e on $\mathbb{R}^{\oplus}$ by $d(a, b) = |a - b| + ||a| - |b||$ and $e(a, b) = |a - b| + ||2a| - |2b||/2$, where, for each $x \in \mathbb{R}$, $|x|$ denotes the integer part of x (B.6.9). The identity map $\iota_{d,e}$ is not continuous because the open subset $[1/2, 1)$ of $(\mathbb{R}^{\oplus}, e)$ is not open in $(\mathbb{R}^{\oplus}, d)$, so d and e are not topologically equivalent metrics. But $e(a, b) = d(2a, 2b)/2$ for all $a, b \in \mathbb{R}^{\oplus}$; the function $x \mapsto x/2$ from $(\mathbb{R}^{\oplus}, d)$ to $(\mathbb{R}^{\oplus}, e)$ is Lipschitz with Lipschitz constant $1/2$ and its inverse is Lipschitz with Lipschitz constant 2. So the spaces are Lipschitz equivalent.

Summary

In this chapter, we have examined three different types of equivalence of metrics. Topological equivalence is important because it preserves all those properties of a metric space that depend only on the topology; uniform equivalence is important because it is the usual form of equivalence in compact metric spaces; and Lipschitz equivalence is important because equivalence of norms is always Lipschitz. Last, we have considered briefly corresponding, but broader, notions of equivalence between metric spaces.

EXERCISES

Q 13.1   Suppose X and Y are non-empty sets and m_1 and m_2 are metrics on X and e_1 and e_2 are metrics on Y. Suppose m_2 is stronger than m_1 and e_2 is weaker than e_1. Suppose $f: X \to Y$ is continuous with respect to m_1 and e_1. Show that f is also continuous with respect to m_2 and e_2.

†Q 13.2   Suppose (X, d) is a complete metric space and $\mathcal{U}$ is a countable collection of open subsets of X. Show that $\bigcap \mathcal{U}$ can be endowed with a metric topologically equivalent to d with respect to which $\bigcap \mathcal{U}$ is complete.

†Q 13.3   Suppose (X, d) and (Y, m) are metric spaces and $f: X \to Y$ is an injective function. Show that the metric e given by $(a, b) \mapsto m(f(a), f(b))$ on X is topologically stronger than d if, and only if, $f^{-1}: f(X) \to X$ is continuous.

†Q 13.4   Suppose (X, d) and (Y, m) are metric spaces and $f: X \to Y$ is an injective function. Show that the metric e given by $(a, b) \mapsto m(f(a), f(b))$ on X is uniformly weaker than d if, and only if, f is uniformly continuous. Show also that e is uniformly stronger than d if, and only if, f^{-1} from $(f(X), m)$ to (X, d) is uniformly continuous.

Q 13.5   Consider the interval $[1, \infty)$ with its usual metric $|\cdot|$ and the metric d given by $(a, b) \mapsto \left| a^{-1} - b^{-1} \right|$ for all $a, b \in [1, \infty)$. Show that d is Lipschitz weaker than $|\cdot|$ but that the metrics are not Lipschitz equivalent.

†Q 13.6   Suppose (X, d) and (Y, m) are metric spaces and $f: X \to Y$ is an injective function. Show that the metric e given by $(a, b) \mapsto m(f(a), f(b))$ on X is Lipschitz weaker than d if, and only if, f is a Lipschitz function and that e is Lipschitz stronger than d if, and only if, f^{-1} from $(f(X), m)$ to (X, d) is a Lipschitz function.

†Q 13.7   Suppose d and e are topologically equivalent metrics on a set X and that a subset S of X is locally compact with respect to d. Show that S is locally compact with respect to e.

Q 13.8   Which of the intervals $(0, 1)$, $[0, 1)$ and $[0, 1]$ is homeomorphic to $\mathbb{R}$?

Q 13.9   Show that $\mathbb{R}$ is not homeomorphic to $\mathbb{R}^2$ and that neither space is homeomorphic to $\mathbb{R}^3$. Can this statement be generalized?

†Q 13.10   Is every locally compact metric space homeomorphic to a complete metric space?

Q 13.11   Is any of the intervals $(0, 1)$, $[0, 1)$ and $[0, 1]$ uniformly equivalent to $\mathbb{R}$?

Q 13.12   Suppose (X, d) is a metric space in which every pair of points can be connected by a path of finite length. For each $a, b \in X$, set $e(a, b)$ to be the infimum of the set of lengths of paths in X from a to b. Show that e is a metric on X and that e is Lipschitz stronger than d.

Appendix A
Language and Logic

Ambiguity of language is philosophy's
main source of problems. That is why
it is of the utmost importance to examine
attentively the very words we use.   Giuseppe Peano, 1858–1932

When students of mathematics are introduced to logical arguments, they are generally expected to get used to the language and methods of such arguments as they hear and see them used. There are a few things that should, at some stage of that process, be stated explicitly. This appendix seeks to clarify certain points that the student who embarks on a rigorous course in mathematics, such as the present one, should know. In particular, it gives some standard tricks that are used in proofs and are not always mentioned in presentations.

A.1 Theorems and Proofs

Mathematics uses reasoned argument to reach conclusions; the conclusions we call *theorems*, and the arguments we call *proofs*. The theorem–proof approach to the writing of mathematics is followed throughout this book. It provides greater clarity than any other method of presenting mathematics, and the student who gets used to it will be well rewarded in understanding.

It is true, of course, that reasoned argument must have assumptions to work on and that conclusions are dependent on the truth of those assumptions. Mathematicians have, however, distilled this process to a fine art. Our assumptions—the axioms of set theory—are few and utterly basic. It would take us too far afield to discuss the axioms here. It suffices to say that from the axioms follow all the fundamental results about sets, numbers and functions that, at this level, we take for granted; results needed in this book are listed in Appendix B.

A.2 Truth of Compound Statements

Suppose P and Q are statements. They can be combined together to form the compound statements

- P and Q;
- P or Q;
- $P \Rightarrow Q$ (if P then Q); and
- $P \Leftrightarrow Q$ (P and Q are equivalent).

The truth or falsehood of these compound statements depends only on the truth or falsehood of the individual statements P and Q. The logical position is as follows. Suppose P and Q are propositions that are either true or false. The first statement, 'P and Q', is true when P and Q are both true; otherwise it is false. The second statement, 'P or Q', is false when P and Q are both false; otherwise it is true. The third statement, '$P \Rightarrow Q$', is false when P is true and Q is false; otherwise it is true. The fourth statement is true when P and Q are either both true or both false; otherwise it is false.

The two compound statements that the student of logic or mathematics needs to be most careful about are the second and third. In our discipline, '*or*' is always inclusive; 'P or Q' always means 'P or Q or both', and implications such as the following, that might be the cause of dispute, are true:

- If $0 = 1$, then $2 = 2$.
- If all integers are even, then all integers are odd.

A.3 If, and Only If

There are some phrases that mathematicians trot out frequently, as if everyone has a clear understanding of what they mean. The most common of such phrases is '*if, and only if*'. This phrase always joins two *statements* together, is often rolled off the tongue as if it were a single word, and is sometimes horribly abbreviated to '*iff*'. Here is an attempt to explain why it means that the statements are equivalent.

Suppose P and Q are statements. There are many ways of expressing the logical implication '$P \Rightarrow Q$' in English, depending on the context; here are some of them:

- P implies Q;
- if P then Q;
- P only if Q;
- P is sufficient for Q;
- Q if P; and

- Q is necessary for P.

The statement 'P *if, and only if,* Q' is the conjunction of the statements 'P *if* Q' and 'P *only if* Q'; the first of these statements is the same as '$Q \Rightarrow P$' and the second is the same as '$P \Rightarrow Q$'. It follows that 'P *if, and only if,* Q' is the same as '$P \Leftrightarrow Q$' and can be written variously as

- $P \Rightarrow Q$ and $Q \Rightarrow P$;
- P is necessary and sufficient for Q;
- Q is necessary and sufficient for P;
- P and Q are equivalent; or
- Q if, and only if, P.

The phrase '*if, and only if*' is often seen also in definitions. In giving a formal definition of a *subset*, for example, we might say that a set B is a subset of a set A if, and only if, every member of B is a member of A. The point here is that we want to indicate not only that B is to be called a subset of A if every member of B is a member of A but also, to close off the issue, that in no other circumstances is B to be called a subset of A. This closing off may seem to be a little pedantic, but it does help when strict formal definitions are to leave us in no doubt whatsoever about what is being described. Mathematics does, after all, aspire to a level of accuracy and certainty undreamed of in other disciplines.

A.4   Transitivity of Implication

We often use the *transitivity of implication* in proofs, particularly when three or more statements are to be proved equivalent. Suppose we want to show that statements P, Q and R are equivalent; in other words, that $P \Leftrightarrow Q$, $Q \Leftrightarrow R$ and $R \Leftrightarrow P$. There are six implications requiring proof: $P \Rightarrow Q$, $Q \Rightarrow P$, $Q \Rightarrow R$, $R \Rightarrow Q$, $R \Rightarrow P$ and $P \Rightarrow R$. It is, however, a matter of pure logic that implication is transitive—that if $A \Rightarrow B$ and $B \Rightarrow C$ both hold, then so does $A \Rightarrow C$. So our problem can be solved by proving a mere three implications, $P \Rightarrow Q$, $Q \Rightarrow R$ and $R \Rightarrow P$, transitivity doing the rest. In practice, of course, we try to pick the most efficient way of achieving the equivalence, which may or may not involve proving more than the minimum number of implications.

A.5   Proof by Counterexample

We often refute a statement by citing a *counterexample*. We have a proposition $P(x)$ that refers to a variable x that is allowed to vary throughout some set A.

Suppose that, after many failed attempts to prove the proposition, we begin to doubt the universal truth of $P(x)$ on A and set out to prove it false. In order to succeed, we need only find one value of x in A for which the statement $P(x)$ does not hold. This is a matter of pure logic, but saying that without qualification does not do justice to the beauty of a method that enables us to solve the problem of the universal truth of $P(x)$ *without knowing very much at all about the values x or whether or not most of them satisfy $P(x)$.* Consider the example in which $P(x)$ says '*x is green*' and x is allowed to vary through the set of all cars in Ireland. I know that it is not true that *all cars in Ireland are green* simply because my own car is black; I do not need to travel down every laneway in the land in order to be certain that this universal statement is false.

A.6 Vacuous Truth

Some statements are true because they assert nothing; for this reason, they may seem less convincing than other statements of truth. Consider a statement $P(x)$ that refers to a variable x that takes values in some set A. Suppose we discover that the set A is empty. We then say without more ado that the statement $P(x)$ is *vacuously true* for all x in A. The statement is true—no less true than any other true statement—simply because it claims nothing. For example, if there is irrefutable evidence that there are no inhabitants of Mars, then we can confidently say that the statement *'All blue Martians wear silk hats'* is true; a lawyer might read other things into it, but pure logic dictates that it is true simply because there are no blue Martians. The reader who is not convinced might like to try refuting the statement by finding a counterexample; that is, by producing a blue Martian who does not wear a silk hat. The abstract logical position is this: the statement *'for all x in A, $P(x)$ holds'* is the negation of the statement *'there exists some x in A such that $P(x)$ does not hold'*, and when A is empty, the latter statement is certainly false, so that the former, its negation, is true.

A.7 Proof by Contradiction

A proposition P is to be proved. We may start by assuming that P is false; in other words, that its negation $\neg P$ is true. We present a reasoned argument that produces a contradiction. This may be in the shape of something we already know to be false, like $1 = 0$, or in the argument's yielding two opposing statements Q and $\neg Q$ that cannot both be true, or it may be that we arrive at

P itself, which of course contradicts our assumption $\neg P$. We then conclude that our assumption $\neg P$ was false to start with and that P is therefore true. Such a conclusion is justified by pure logic: if $\neg P$ implies something that is false, then P has to be true. This is proof by contradiction or reductio ad absurdum.

A.8 Proof by Contraposition

We sometimes use *contraposition* in order to prove an implication. Recall that knowing the truth of an implication $P \Rightarrow Q$ does not tell us that either P or Q is true; the useful information such a truth carries is that if P is true, then so is Q. We wish to prove that $P \Rightarrow Q$. Instead, we prove the *contrapositive* $\neg Q \Rightarrow \neg P$, where $\neg Q$ and $\neg P$ are the negations of Q and P, respectively. We then conclude that $P \Rightarrow Q$. Such a conclusion is justified by pure logic because the implication $P \Rightarrow Q$ is equivalent to its contrapositive $\neg Q \Rightarrow \neg P$.

Asked to prove an implication $P \Rightarrow Q$, we often use a hybrid of contraposition and contradiction: we assume both P and $\neg Q$ and arrive at a contradiction; from such a contradiction, we can conclude that $P \Rightarrow Q$ because this implication is logically equivalent to the negation of the conjunction (P and $\neg Q$).

A.9 Proof by Induction

The Principle of Mathematical Induction can be stated in several ways, but the following formulation covers all that is needed. Suppose $P(n)$ is a statement about integers n and we wish to prove that $P(n)$ is true for every integer n with $n \geq m$, where m is some given integer. The Principle of Mathematical Induction tells us that it is sufficient to show two things: first, that $P(m)$ is

true; second, that, for each integer k with $k \geq m$, the truth of $P(a)$ for all integers a such that $m \leq a \leq k$ implies the truth of $P(k + 1)$. This is not a matter of pure logic; it is based on an understanding of the nature of integers that has been built into the modern axiomatic presentation of numbers as mathematical

© Eoghan Ó Searcóid, 2005

objects. It is, nonetheless, a method of proof just as sound and irrefutable as any other method of proof in mathematics. It is necessary to stress this because the word *induction* is sometimes used in other disciplines to describe a much more relaxed process of inference, a process that does not stand up to the rigorous requirements of mathematical proof. The word *induction* is never used in such a way in mathematics.

A.10 Existence

In logic and in mathematics, existence is a much broader concept than it is in the computational sciences. The statement *'there exists x such that $P(x)$ holds'*, or in symbols *'$(\exists x)(P(x))$'*, is an abbreviation of *'it is not the case that, for all x, $P(x)$ does not hold'*, or in symbols *'$\neg((\forall x)(\neg P(x)))$'*. When we assert that there exists x satisfying $P(x)$, we are not inferring that we have identified such an x; we are not inferring that we know a method by which such an x can be identified; and we are not inferring even that there is a method for identifying such an x. In fact, it sometimes happens that, while asserting existence of x, we know, and can prove without any contradiction being implied, that there is no method by which any such x can be identified. An assertion of existence is thus considerably weaker than identification.

We need not be unduly worried about the distinction between existence and identification. But it does raise one immediate question. Once we know, or have assumed, that $(\exists x)(P(x))$, is it valid to *'let z be such that $P(z)$ holds'*? The former statement is an assertion of existence, while the latter looks a bit like an identification (but is not because z, though fixed, is not determined). We assure the reader that it is logically valid to make the leap from the one to the other in a mathematical argument; the reasons for this are not trivial.[1]

A.11 Let and Suppose

The words *let* and *suppose* are not synonymous. We may *suppose* what we like, irrespective of truth or falsehood, and leave it to logic to unravel the consequences, but we can *let* only after existence has been established or assumed. We use phrases such as *'let z satisfy $P(z)$'* only when it is already known, or has been assumed in the context, that there is some object that satisfies $P(x)$. We say *'suppose z satisfies $P(z)$'* whenever we will. We may, for example, suppose there is a colony of blue martians alive and well and resident in Conamara, and then let z be one of the group. But we cannot without duplicity *'let z be a blue martian'* while agreeing with all the world that martians are green.

---

[1] The reader who is interested in knowing them is advised to follow a course in mathematical logic or to read the section on *Rule C* in [9].

Appendix B
Sets

Reductio ad absurdum, which Euclid loved so much,
is one of a mathematician's finest weapons.
It is a far finer gambit than any chess play:
a chess player may offer the sacrifice
of a pawn or even a piece, but a
mathematician offers the game.   G. H. Hardy, 1877–1947

The terms *set* and *collection* are used synonymously throughout the book. Every set is uniquely determined by the *members*—also called *elements* or *points*—that belong to it. In this appendix, we give some basic facts about a variety of sets, including number sets, relations and functions, and we explain the set-theoretic terms and notation that are used in the book. We state some basic theorems, mostly without proof. We discuss briefly, but with some care, the different ways in which sequences may appear in mathematical proofs. Last, we define the algebraic structures that are used in the book.

B.1 Notation for Sets

We write $a \in S$ to indicate that a is a member of the set S; we say that a *belongs* to S and that S *contains* a. We write $a \notin S$ to indicate that a is not a member of S. There is exactly one set with no members; it is called the *empty set* and is denoted by $\varnothing$. Any set with just one element is called a *singleton set*. For the sake of simplicity, we shall assume for the time being that we know the difference between a *finite set* and an *infinite set* (see B.17.3).

Except in special cases, we use uppercase italic letters, such as A, B and C, to denote sets, and lowercase letters, such as a, b and c, for their members. It is important, however, to be aware of the fact that a set may have members that are themselves sets.[1] When we are dealing with a set whose members are sets which themselves have members that are of interest to us, we shall usually

---

[1] In the most widely known axiomatic theory of sets, *all* members of sets are themselves sets.

adopt a hierarchy of notation: the overarching set may be denoted by a script letter, say $\mathcal{A}$; its members, themselves sets, may be denoted by uppercase italic letters, so that we may have for instance $\mathcal{A} = \{A, B, D, G\}$. Then the members of the members of $\mathcal{A}$ may be denoted by lowercase letters. The reader will see immediately that if $x \in B$ and $B \in \mathcal{A}$, then it does not necessarily follow—and, indeed, will usually not be true—that x is in $\mathcal{A}$.

B.2 Subsets and Supersets

Definition B.2.1

Suppose A and B are sets. We say

- that A and B are equal, written $A = B$, if, and only if, A and B have exactly the same members;
- that A is a *subset* of B and that B is a *superset* of A, written $A \subseteq B$, if, and only if, every member of A is also a member of B; and
- that A is a *proper subset* of B and that B is a *proper superset* of A, written $A \subset B$, if, and only if, every member of A is a member of B and there is a member of B that is not in A.

Evidently, $A \subset B$ if, and only if, $A \subseteq B$ and $A \neq B$. When $A \subseteq B$, we say that A is *included* in B and that B *includes* A. Since it is vacuously true that every member of the empty set is a member of every set, it is also true that $\varnothing \subseteq A$ for every set A.

Definition B.2.2

Suppose A is a set. The collection of all subsets of A is a set; it is called the *power set* of A and is denoted by $\mathcal{P}(A)$.

The name *power set* reflects the easily verified fact that, if A is a finite set with n members, then $\mathcal{P}(A)$ has 2^n members. Note that $\mathcal{P}(\varnothing)$ is not empty; it has the empty set for its only member, reflecting the fact that $2^0 = 1$.

Definition B.2.3

A non-empty collection $\mathcal{N}$ of sets is called a *nest* if, and only if, for each $A, B \in \mathcal{N}$, either $A \subseteq B$ or $B \subseteq A$.

B.3 Universal Set

In any particular discussion, it is often the case that all the sets being considered are subsets of a single set, sometimes termed a *universal set*.

Definition B.3.1

Suppose A is a subset of some universal set X. We define the *complement* of A in X to be the set $\{x \in X \mid x \notin A\}$. We sometimes denote this complement by A^c. When we use this notation, it should be clear from the context which superset of A is being regarded as the universal set.

Theorem B.3.2

Suppose A is a subset of some universal set X. Then $(A^c)^c = A$.

Theorem B.3.3

Suppose A and B are subsets of a universal set X and $A \subseteq B$. Then $B^c \subseteq A^c$.

It is not obligatory to work within the confines of a universal set. We can widen a discussion by considering any objects of set theory; there is an inexhaustible supply of sets. In fact, B.3.4 below is an elementary consequence of the axioms of set theory. Given any set S in a mathematical argument, B.3.4 allows us to *let z be such that $z \notin S$* without even invoking A.10.

Theorem B.3.4

Suppose S is a set. Then there exists x such that $x \notin S$, and such an x can be identified.

B.4 Number Sets

Some of the most important sets that can be constructed using the axioms of set theory are the following number sets:

- the set $\mathbb{N} = \{1, 2, 3, \ldots\}$ of *natural numbers*;
- the set $\mathbb{Z} = \{\ldots, -2, -1, 0, 1, 2, \ldots\}$ of *integers*;
- the set $\mathbb{Q} = \{p/q \mid p \in \mathbb{Z}, q \in \mathbb{N}\}$ of *rational numbers*;
- the set $\mathbb{R}$ of *real numbers*; and
- the set $\mathbb{R} \backslash \mathbb{Q}$ of *irrational numbers*.

We assume that all these sets have been constructed with the inclusions $\mathbb{N} \subseteq \mathbb{Z} \subseteq \mathbb{Q} \subseteq \mathbb{R}$ and that it has been established that all the inclusions are proper. We assume that the usual algebraic operations have been defined on these sets and that the set $\mathbb{C} = \{a + ib \mid a, b \in \mathbb{R}\}$ of *complex numbers*, where $i^2 = -1$, has been constructed as a proper superset of $\mathbb{R}$.

Theorem B.4.1

Suppose $z = a + ib \in \mathbb{C}$, where $a, b \in \mathbb{R}$. Then the real numbers a and b are uniquely determined; we shall call a the *real part* of z and denote it by $\Re z$, and we shall call b the *imaginary part* of z and denote it by $\Im z$.

B.5   Ordered Pairs and Relations

Set theory is equipped with objects called *ordered pairs*. Members a and b of sets always determine an ordered pair (a, b), and it is not the same as the ordered pair (b, a) except in the special case when $a = b$. Ordered pairs, in turn, give rise to Cartesian products and other relations.

Definition B.5.1

If A and B are sets, then the set $\{(a, b) \mid a \in A, b \in B\}$ is denoted by $A \times B$ and is called the *Cartesian product* of A and B.

If either A or B is empty, then $A \times B$ and $B \times A$ are also clearly empty; if A and B are both non-empty, it is easily verified that it is not possible for $A \times B$ to be the same as $B \times A$ except in the special case when $A = B$.

Definition B.5.2

Suppose ρ is a set.

- ρ is called a *relation* if, and only if, all its members are ordered pairs.
- If ρ is a relation, then $\{a \mid \text{there exists } b \text{ such that } (a, b) \in \rho\}$ is called the *domain* of ρ and may be denoted by $\mathrm{dom}(\rho)$.
- If ρ is a relation, then $\{b \mid \text{there exists } a \text{ such that } (a, b) \in \rho\}$ is called the *range* of ρ and may be denoted by $\mathrm{ran}(\rho)$.

It is fundamental to set theory that domains and ranges of relations themselves are always sets. If ρ is a relation and S is any set that includes both the domain and the range of ρ, then ρ may be described as a *relation on S*. There are many relations, such as *is equal to* or *is less than*, that are familiar to all users of real numbers. It is therefore customary to use suggestive notation such as $<$, $\simeq$ or $\equiv$ to denote given relations and to adopt notation such as $a < b$ as an equivalent of the ugly alternative $(a, b) \in \,<$.

B.6   Totally Ordered Sets

Definition B.6.1

Suppose $<$ is a relation on a set X. The relation $<$ is said

- to be *transitive* if, and only if, for each $a, b, c \in X$, if $a < b$ and $b < c$, then $a < c$;

- to obey the *law of trichotomy* if, and only if, for each $a, b \in X$, exactly one of $a < b$, $a = b$ and $b < a$ is true; and

- to be a *total ordering* on X if, and only if, it is transitive and obeys the law of trichotomy.

Example B.6.2

It is easy to show that $\subset$ is a total ordering on any nest (B.2.3).

Definition B.6.3

Suppose X is a set, $s \in X$, A is a non-empty subset of X and $<$ is a total ordering on X. Define $\leq$ and $\geq$ in the obvious way. Then s is called

- a *lower bound* for A in X if, and only if, $s \leq a$ for all $a \in A$;

- an *upper bound* for A in X if, and only if, $a \leq s$ for all $a \in A$;

- the *minimum* or *smallest* or *least* member of A, denoted by $\min A$, if, and only if, $s \in A$ and s is a lower bound for A (the properties of the relation $<$ ensure that there can be at most one such s); and

- the *maximum* or *largest* or *greatest* member of A, denoted by $\max A$, if, and only if, $s \in A$ and s is an upper bound for A (the properties of the relation $<$ ensure that there can be at most one such s).

Theorem B.6.4

Every non-empty finite subset of a totally ordered set has a maximum and a minimum element.

Many subsets of totally ordered sets have no maximum or minimum element. This prompts us to look for substitutes in *suprema* and *infima* (B.6.5). If a set does have a maximum element, then that maximum is, of course, its supremum, and if it has a minimum element, that minimum is its infimum.

Definition B.6.5

Suppose A is a non-empty subset of a totally ordered set X.

- If the set of upper bounds of A in X has a minimum element, it is called the *least upper bound* or *supremum* of A in X and is denoted by $\sup A$ or, if there is need for greater clarity, by $\sup_X A$.
- If the set of lower bounds of A in X has a maximum element, it is called the *greatest lower bound* or *infimum* of A in X and is denoted by $\inf A$ or by $\inf_X A$.

Theorem B.6.6

Suppose X is a set with a totally ordering $<$. Suppose $z \in X$ and $A \subseteq X$.

(i)   If $\sup A$ exists, then $\sup A \leq z$ if, and only if, $a \leq z$ for all $a \in A$.

(ii) If $\inf A$ exists, then $z \leq \inf A$ if, and only if, $z \leq a$ for all $a \in A$.

In B.6.7 below, we give three refinements of total ordering. The first and second are characteristic of the ordering of the natural numbers, and the third is characteristic of the ordering of the real numbers. Indeed, it is intrinsic to the construction of the real numbers from the axioms of set theory, and vitally important for the theory of metric spaces, that the ordering of $\mathbb{R}$ is complete.

Definition B.6.7

Suppose X is a set equipped with a total ordering $<$. X is said to be

- *well ordered* by $<$ if, and only if, every non-empty subset of X has a minimum element;
- *enumeratively ordered* by $<$ if, and only if, each non-empty subset of X with an upper bound in X has a maximum element and each non-empty subset of X with an lower bound in X has a minimum element; and
- *completely ordered* by $<$ if, and only if, each non-empty subset of X that has an upper bound in X has a least upper bound in X.

Theorem B.6.8

$\mathbb{R}$ is completely ordered by its usual ordering $<$.

Definition B.6.9

For each $x \in \mathbb{R}$, we define the *integer part* of x, denoted by $\lfloor x \rfloor$, to be the largest integer that does not exceed x; in other words, $\sup\{n \in \mathbb{Z} \mid n \leq x\}$.

Definition B.6.10

For every $z = a + ib \in \mathbb{C}$, where $a, b \in \mathbb{R}$, we define the *modulus* $|z|$ of z to be $\sqrt{a^2 + b^2}$. When z is real, this is often called the *absolute value* of z; if $z \geq 0$, then $|z| = z$, and if $z < 0$, then $|z| = -z$.

Two fundamental properties that emerge in the construction of $\mathbb{R}$ from the axioms of set theory are stated without proof below in B.6.11 and B.6.12.

Theorem B.6.11

Suppose a and b are real numbers and $a < b$. Then there exist a rational number r and an irrational number s for which $a < r < b$ and $a < s < b$. It follows, using induction, that the number of such r and such s is infinite.

Theorem B.6.12

Suppose $r \in \mathbb{R}$ and $r > 0$. Then there exists $m \in \mathbb{N}$ such that $0 < 1/m < r < m$. It follows, in particular, that $\inf_{\mathbb{R}}\{r \in \mathbb{R} \mid r > 0\} = \inf\{1/n \mid n \in \mathbb{N}\} = 0$.

B.7 Extended Real Numbers

It is convenient to work with an extension of the real number system that has both an upper and a lower bound. Specifically, we use the familiar symbols ∞ and $-\infty$, spoken of as *infinity* and *minus infinity*, to stand for distinct mathematical objects that are not real numbers—they can occur as members of sets within set theory, but we are not going to give a justification of that here. We append these two objects to the set $\mathbb{R}$ to form a larger set $\tilde{\mathbb{R}}$. The ordering of $\mathbb{R}$ is then extended to $\tilde{\mathbb{R}}$ and the operations of addition and multiplication are extended in so far as that is possible.

Definition B.7.1

We define the set of *extended real numbers* to be the set $\tilde{\mathbb{R}} = \mathbb{R} \cup \{-\infty, \infty\}$ and the set of *extended natural numbers* to be the set $\tilde{\mathbb{N}} = \mathbb{N} \cup \{\infty\}$. We extend the ordering of $\mathbb{R}$ by saying that $-\infty < s$ for all $s \in \mathbb{R} \cup \{\infty\}$ and $s < \infty$ for all $s \in \mathbb{R} \cup \{-\infty\}$. Addition is partially extended to $\tilde{\mathbb{R}}$ by specifying that $s + \infty = \infty = \infty + s$ for all $s \in \mathbb{R} \cup \{\infty\}$ and $s + (-\infty) = -\infty = -\infty + s$ for all $s \in \mathbb{R} \cup \{-\infty\}$. Multiplication is partially extended to $\tilde{\mathbb{R}}$ by specifying that $s\infty = \infty = \infty s$ and $s(-\infty) = -\infty = -\infty s$ for all $s \in \left\{ x \in \tilde{\mathbb{R}} \mid x > 0 \right\}$ and $s\infty = -\infty = \infty s$ and $s(-\infty) = \infty = -\infty s$ for all $s \in \left\{ x \in \tilde{\mathbb{R}} \mid x < 0 \right\}$. We also define $|\infty| = |-\infty| = \infty$, but we do not define any sum of ∞ and $-\infty$ or product of 0 with either ∞ or $-\infty$.

Definition B.7.2

By convention, we define

- $\inf \varnothing = \infty$ and
- $\sup \varnothing = -\infty$.

Notice that the order completeness of $\mathbb{R}$ (B.6.8) implies that every non-empty subset S of $\mathbb{R}$ that has a lower bound in $\mathbb{R}$ has a greatest lower bound in $\mathbb{R}$, namely $- \sup\{-x \mid x \in S\}$. Notice also that ∞ is an upper bound in $\tilde{\mathbb{R}}$ for every non-empty set of extended real numbers and that $-\infty$ is a lower bound in $\tilde{\mathbb{R}}$ for every such set. These facts, together with the order completeness of $\mathbb{R}$ and the conventions of B.7.2, enable us to make below in B.7.3 a very succinct statement about the extended real numbers.

Theorem B.7.3

Every subset of $\tilde{\mathbb{R}}$ has both a supremum and an infimum in $\tilde{\mathbb{R}}$.

As a result of B.7.3, we habitually use $\inf S$ and $\sup S$ to denote the infimum and the supremum in $\tilde{\mathbb{R}}$ of any subset S of $\mathbb{R}$ or of $\tilde{\mathbb{R}}$. At the same time, we must, however, alert the reader to the singularly odd fact that $\sup \varnothing < \inf \varnothing$, whereas the expected $\inf A \leq \sup A$ holds for every non-empty subset A of $\tilde{\mathbb{R}}$.

Theorem B.7.4

Suppose A and B are non-empty subsets of $\tilde{\mathbb{R}}$ and $A \subseteq B$. Then we have the inequalities $\inf B \leq \inf A \leq \sup A \leq \sup B$.

B.8 Ordered Subsets of the Real Numbers

We use notation that is not absolutely standard for some of the distinguished subsets that arise from the ordering of $\mathbb{R}$.

Definition B.8.1

We define

- the set $\mathbb{N}_k = \{n \in \mathbb{N} \mid 1 \leq n \leq k\}$ for each $k \in \mathbb{N}$;
- the set $\mathbb{R}^+ = \{r \in \mathbb{R} \mid r > 0\}$ of *positive real numbers*;
- the set $\mathbb{R}^- = \{r \in \mathbb{R} \mid r < 0\}$ of *negative real numbers*;
- the set $\mathbb{R}^\oplus = \{r \in \mathbb{R} \mid r \geq 0\}$ of *non-negative real numbers*; and
- the set $\mathbb{R}^\ominus = \{r \in \mathbb{R} \mid r \leq 0\}$ of *non-positive real numbers*.

Definition B.8.2

A non-empty subset I of $\mathbb{R}$ is called an *interval* if, and only if, for every $x \in \mathbb{R}$, the implication $\inf I < x < \sup I \Rightarrow x \in I$ is true.

There are ten types of interval; we list them here with $a, b \in \mathbb{R}$ and $a < b$. Intervals of the type $[a, a]$ are said to be *degenerate*.

$$
\begin{aligned}
[a, a] &= \{a\}; \\
(a, b) &= \{x \in \mathbb{R} \mid a < x < b\}; \\
[a, b) &= \{x \in \mathbb{R} \mid a \leq x < b\}; \\
(a, b] &= \{x \in \mathbb{R} \mid a < x \leq b\}; \\
[a, b] &= \{x \in \mathbb{R} \mid a \leq x \leq b\}; \\
(a, \infty) &= \{x \in \mathbb{R} \mid a < x\}; \\
(-\infty, b) &= \{x \in \mathbb{R} \mid x < b\}; \\
[a, \infty) &= \{x \in \mathbb{R} \mid a \leq x\}; \\
(-\infty, b] &= \{x \in \mathbb{R} \mid x \leq b\}; \\
(-\infty, \infty) &= \mathbb{R}.
\end{aligned}
$$

B.9 Ordered Tuples

In addition to ordered pairs, set theory has *ordered triples* (a, b, c), ordered quadruples (a, b, c, d) and so on; in general, there are *ordered n-tuples* with n entries $(a_1, \ldots, a_n)$, where n is any natural number. In this book, we generally use a single letter with integer subscripts to indicate the entries in an ordered tuple: we write a for the ordered n-tuple $(a_1, \ldots, a_n)$; each entry a_i is called the ith *coordinate* of a. The idea of a Cartesian product is then extended to more than two sets (B.9.1), the members of the product being ordered tuples.

Definition B.9.1

Suppose $n \in \mathbb{N}$ and $(A_1, \ldots, A_n)$ is an ordered n-tuple of sets. We define the *Cartesian product*, or simply the *product*, $\prod_{i=1}^{n} A_i$ of these sets to be the collection of all ordered n-tuples $a = (a_1, \ldots, a_n)$ with $a_i \in A_i$ for each $i \in \mathbb{N}_n$. The sets A_i are called the *coordinate sets* of the product. If the coordinate sets are all equal to some set S, then it is usual to denote the product $\prod_{i=1}^{n} A_i$ by S^n. If $n = 1$, the product may be described as a *trivial product*.

Note, in particular, that a product of a positive number of sets is empty if, and only if, one or more of the coordinate sets is empty.

Example B.9.2

The principal example of a Cartesian product is the set of points in a plane. Each point is designated by an ordered pair (a_1, a_2) of coordinates, where a_1 and a_2 are real numbers, the first indicating the so-called x-coordinate and the second the so-called y-coordinate

A point a of $\mathbb{R}^2$.

of the point $a = (a_1, a_2)$. Clearly the plane is the set $\{(a_1, a_2) \mid a_1, a_2 \in \mathbb{R}\}$. This is the Cartesian product $\mathbb{R} \times \mathbb{R}$, otherwise denoted by $\mathbb{R}^2$ and known as *two-dimensional real space*. Similarly, *three-dimensional real space* consists of all ordered triples of real numbers and is thus the product $\mathbb{R}^3$. The set $\mathbb{R}^n$ of ordered n-tuples of real numbers is called n-*dimensional real space*.

B.10 Union, Intersection and Difference

Definition B.10.1

Suppose A and B are sets. We define

- the *union* $A \cup B$ to be the set $\{x \mid x \in A \text{ or } x \in B\}$ that has in it all members of A together with all members of B;

- the *intersection* $A \cap B$ to be the set $\{x \mid x \in A \text{ and } x \in B\}$ that has in it all those members that are common to both A and B; and

- the *set difference* $A \backslash B$ to be the set $\{x \in A \mid x \notin B\}$ that has in it all members of A that are not members of B.

The reader may like to note that, where there is an understood universal set, the set difference $A \backslash B$ is the same as $A \cap B^c$.

It is evident from their definitions that both union and intersection are *commutative*: $A \cup B = B \cup A$ and $A \cap B = B \cap A$ for all sets A and B. It is equally evident that $A \backslash B$ is not the same as $B \backslash A$ except when $A = B$.

It is easy to show that union *distributes* over intersection and vice versa; that is, that $A \cup (B \cap C) = (A \cup B) \cap (A \cup C)$ and $A \cap (B \cup C) = (A \cap B) \cup (A \cap C)$ for all sets A, B and C. It is also easy to show that set difference behaves in a regular non-distributive manner in that $A \backslash (B \cap C) = (A \backslash B) \cup (A \backslash C)$ and $A \backslash (B \cup C) = (A \backslash B) \cap (A \backslash C)$ for all sets A, B and C.

The concepts of union and intersection are extended to more than two sets in the obvious way. Both union and intersection are associative operations; that is to say that $(A \cup B) \cup C = A \cup (B \cup C)$ and $(A \cap B) \cap C = A \cap (B \cap C)$ for all sets A, B and C. We shall therefore dispense with the unnecessary parentheses and write $A \cup B \cup C$ for the former and $A \cap B \cap C$ for the latter. Difference of sets is not associative; in fact, $A \backslash (B \backslash C) = ((A \backslash B) \backslash C) \cup (A \cap C)$, and the reader may like to check that $A \backslash (B \backslash C) = (A \backslash B) \backslash C$ if, and only if, $A \cap C = \varnothing$.

Definition B.10.2

We say that sets A and B are *disjoint* if, and only if, $A \cap B = \varnothing$. Suppose $\mathcal{C}$ is a collection of sets. We say that the members of $\mathcal{C}$ are *mutually disjoint* if, and only if, for each $A, B \in \mathcal{C}$ with $A \neq B$, we have $A \cap B = \varnothing$.

B.11   Unions and Intersections of Arbitrary Collections

We extend union and intersection to an arbitrary non-empty collection $\mathcal{C}$ of sets. The union of all the sets that are members of $\mathcal{C}$ is that set whose members are the members of the members of $\mathcal{C}$, namely $\{x \mid x \in A \text{ for some } A \in \mathcal{C}\}$. It is denoted by $\bigcup \mathcal{C}$. The intersection of the sets that are members of $\mathcal{C}$ is $\{x \mid x \in A \text{ for every } A \in \mathcal{C}\}$. This intersection is denoted by $\bigcap \mathcal{C}$. Notation such as $\bigcup \mathcal{C}$ and $\bigcap \mathcal{C}$ is particularly useful when $\mathcal{C}$ is an infinite collection of sets or when the members of $\mathcal{C}$ are not easily specified. Of course, if $\mathcal{C}$ can be written simply as, for example, $\{A, B\}$, then we usually write $A \cup B$ for the union rather than $\bigcup \mathcal{C}$ or $\bigcup \{A, B\}$, and we write $A \cap B$ for the intersection rather than $\bigcap \mathcal{C}$ or $\bigcap \{A, B\}$. For a union of an unspecified number n of sets $A_1, \ldots, A_n$, however, we often prefer notation such as $\bigcup \{A_i \mid i \in \mathbb{N}_n\}$ to $A_1 \cup \ldots \cup A_n$. If $\mathcal{C}$ is a singleton, $\mathcal{C} = \{A\}$, then $\bigcup \mathcal{C} = A = \bigcap \mathcal{C}$, of course.

When $\mathcal{C}$ is the empty set, a special case arises. We define $\bigcup \varnothing$ to be $\varnothing$. On the other hand, we shall not assign any meaning to $\bigcap \varnothing$ as it could, with some justification, be defined in more than one way.

Definition B.11.1

Suppose X is a set and S is a subset of X. A collection $\mathcal{C}$ of subsets of X is called a *cover* for S in X, and we say that $\mathcal{C}$ *covers* S if, and only if, $S \subseteq \bigcup \mathcal{C}$. If $\mathcal{C}$ covers S and $\mathcal{A}$ is a subset of $\mathcal{C}$ that also covers S, then $\mathcal{A}$ is called a *subcover* of $\mathcal{C}$ for S in X.

The most important theorem relating union and intersection is that of De Morgan. It will be familiar to readers at least in its simplest case, where it says that, for subsets A and B of some universal set, we have $(A \cup B)^c = A^c \cap B^c$ and $(A \cap B)^c = A^c \cup B^c$. It can, however, be extended not merely to non-empty finite collections of sets but to all non-empty collections of sets, as stated below in B.11.2.

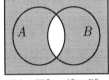

$(A \cap B)^c = A^c \cup B^c.$

Theorem B.11.2 (De Morgan's Theorem)

Suppose $\mathcal{C}$ is a non-empty collection of sets and Z is any given set. Then $Z \backslash \bigcap \mathcal{C} = \bigcup \{ Z \backslash A \mid A \in \mathcal{C} \}$ and $Z \backslash \bigcup \mathcal{C} = \bigcap \{ Z \backslash A \mid A \in \mathcal{C} \}$.

B.12 Functions

A *function* acts on a given set, assigning a single value to each member of the set; this need not be a number—it can be any member of any set. Formally, a function is a special type of relation; the set on which it acts is its domain, and the values assigned by it form its range (B.5.2). The domain and the range of a function may be the same, may overlap, or may have empty intersection.

Definition B.12.1

Suppose ρ is a relation. ρ is called a *function* if, and only if, for each $a \in \operatorname{dom}(\rho)$, there is exactly one b such that $(a, b) \in \rho$.

Notation B.12.2

Suppose f is a function. The value assigned by f to each $x \in \operatorname{dom}(f)$ is called the *image* of x under f or the *value* of f at x and is usually denoted by $f(x)$.

There are occasions when this notation is abbreviated; if f acts on ordered pairs, for example, we write $f(a, b)$ rather than $f((a, b))$.

Sometimes, when assigning a letter to a function, we indicate the action performed by that function; for example, $g : x \mapsto x^2 - 6$ says that the function that has value $x^2 - 6$ at each x of some domain—either specified or understood from the context—is being denoted by g. On occasion, it is not thought necessary to use a letter to denote a function. We might talk about the function $x \mapsto x^3$, meaning the function that has the value x^3 at each x in some domain, specified or understood. For many functions, even this notation is abbreviated. We talk of the function x^3 or the function $\sin x$ or the function $x + 2$ rather than of $x \mapsto x^3$, $x \mapsto \sin x$ or $x \mapsto x + 2$; here there is ambiguity, and the notation should not be used unless it is clear from the context that the expressions x^3, $\sin x$ and $x + 2$ refer to the appropriate functions rather than to specific values assigned by those functions at some particular point x of the domain.

Definition B.12.3

Suppose X is a set and $\mathcal{S}$ is a collection of functions defined on X. Then, for each $x \in X$, the function $f \mapsto f(x)$ defined on $\mathcal{S}$ is called the *point evaluation function* for x on $\mathcal{S}$. We shall denote it by $\hat{x}$, the set $\mathcal{S}$ being understood from the context. Note that, for each $x \in X$ and $f \in \mathcal{S}$, we have $\hat{x}(f) = f(x)$.

B.13 Restrictions, Extensions and Compositions

Definition B.13.1

Suppose f and g are functions. We say that f is a *restriction* of g and that g is an *extension* of f if, and only if, $\mathrm{dom}(f) \subseteq \mathrm{dom}(g)$ and $f(x) = g(x)$ for all $x \in \mathrm{dom}(f)$. If this is so and $A = \mathrm{dom}(f)$, we say that f is the restriction of g to A and may write f as $g|_A$.

Definition B.13.2

Suppose f and g are functions and $\mathrm{ran}(f) \subseteq \mathrm{dom}(g)$. The *composition* of g after f, denoted by $g \circ f$, is the function $x \mapsto g(f(x))$ defined on $\mathrm{dom}(f)$.

Even if $f \circ g$ and $g \circ f$ are both defined—when $\mathrm{ran}(g) \subseteq \mathrm{dom}(f)$ and $\mathrm{ran}(f) \subseteq \mathrm{dom}(g)$—they are not usually equal to each other. Composition of functions is associative: if f, g and h are functions with $\mathrm{ran}(f) \subseteq \mathrm{dom}(g)$ and

$\mathrm{ran}(g) \subseteq \mathrm{dom}(h)$, then $(h \circ g) \circ f = h \circ (g \circ f)$. If $\mathrm{ran}(f) \subseteq \mathrm{dom}(f)$, then the composite function $f \circ f$ is usually denoted by f^2 and the result of $(n-1)$ such compositions is denoted by f^n. We may write f^0 for the identity function $x \mapsto x$ defined on $\mathrm{dom}(f)$.

Any set can have a function defined on it; even the empty set admits the empty function, which does nothing. Many important functions are defined on Cartesian products. For example, the *binary operations* of addition and multiplication on the real numbers are the functions from $\mathbb{R} \times \mathbb{R}$ into $\mathbb{R}$ given by $(a,b) \mapsto a+b$ and $(a,b) \mapsto ab$, respectively.

Definition B.13.3

Suppose $n \in \mathbb{N}$ and $(A_1, \ldots, A_n)$ is an ordered n-tuple of sets. For each $i \in \mathbb{N}$ with $1 \le i \le n$, we define the *natural projection* π_i of $\prod_{i=1}^n A_i$ onto the *coordinate set* A_i by $\pi_i(a) = a_i$ for each $a \in \prod_{i=1}^n A_i$.

In general, the range of π_i in B.13.3 is A_i itself, but this fails to be the case when A_i is non-empty and some other A_k is empty—in such a case, the product is empty and π_i is the empty function, which has empty range.

B.14 Mappings

Strictly speaking, at least within axiomatic set theory, a function, being a relation, is a set of ordered pairs (B.12.1); in fact, the function f *is* the set $\{(a, f(a)) \mid a \in \mathrm{dom}(f)\}$. But mathematicians usually call this set the *graph* of f and think of a function as a *map* or *mapping* from its domain into another set called its *codomain*, using the terms function, map and mapping interchangeably. The only necessary condition on the codomain of a mapping is that it include the range. We write $f \colon X \to Y$ to indicate that f is a mapping with domain X and codomain Y, and we express this by saying that f is a mapping, map or function from X into Y. Sometimes the codomain Y specified in this notation is more important than the range of f. For example, we deal often with functions that have only real or only complex values; such functions are called *real functions* and *complex functions*, respectively.

Definition B.14.1

Suppose X and Y are sets and $f \colon X \to Y$. Then f is said to be

- *injective* or *one-to-one* if, and only if, for all $a, b \in X$, the implication $f(a) = f(b) \Rightarrow a = b$ is true;
- *surjective* or *onto* Y if, and only if, $\mathrm{ran}(f) = Y$; and
- *bijective* or a *one-to-one correspondence* if, and only if, f is both injective and surjective.

Definition B.14.2

Suppose X and Y are sets and $f\colon X \to Y$ is injective. The function from $\mathrm{ran}(f)$ to X that assigns to each $y \in \mathrm{ran}(f)$ the unique $x \in X$ such that $f(x) = y$ is called the *inverse* of f and is denoted by f^{-1}.

With reference to the foregoing definition, it is easy to show that f^{-1} is bijective, that its inverse $(f^{-1})^{-1}$ is $f\colon X \to \mathrm{ran}(f)$, that $f^{-1} \circ f = \iota_X$, the *identity function* $x \mapsto x$ on X, and that $f \circ f^{-1} = \iota_{\mathrm{ran}(f)}$, the identity function $y \mapsto y$ on $\mathrm{ran}(f)$.

Definition B.14.3

Suppose X and Y are sets, $A \subseteq X$, $B \subseteq Y$ and $f\colon X \to Y$. We define

- the *image* of A under f to be the set $\{f(x) \mid x \in A\}$, denoted by $f(A)$; and
- the *inverse image* of B by f to be $\{x \in X \mid f(x) \in B\}$, denoted by $f^{-1}(B)$.

Theorem B.14.4

Suppose X and Y are sets and $f\colon X \to Y$. Suppose $C \subseteq X$, $D \subseteq Y$, $\mathcal{U} \subseteq \mathcal{P}(X)$ and $\mathcal{V} \subseteq \mathcal{P}(Y)$. Then

(i) $C \subseteq f^{-1}(f(C))$ with equality if f is injective;

(ii) $f(f^{-1}(D)) \subseteq D$ with equality if f is surjective;

(iii) $f^{-1}(\bigcup \mathcal{V}) = \bigcup \{f^{-1}(B) \mid B \in \mathcal{V}\}$;

(iv) $f^{-1}(\bigcap \mathcal{V}) = \bigcap \{f^{-1}(B) \mid B \in \mathcal{V}\}$;

(v) $f(\bigcup \mathcal{U}) = \bigcup \{f(A) \mid A \in \mathcal{U}\}$; and

(vi) $f(\bigcap \mathcal{U}) \subseteq \bigcap \{f(A) \mid A \in \mathcal{U}\}$ with equality if f is injective.

Use of notation such as $f^{-1}(B)$ does not indicate that there is an inverse function f^{-1} for f; it should be regarded merely as useful notation. Of course, if f happens to be injective and therefore does have an inverse function f^{-1}, it is easily checked that the two meanings we have in this case attached to the notation $f^{-1}(B)$, namely the image $\{f^{-1}(y) \mid y \in B\}$ of B under f^{-1} and the inverse image $\{x \in X \mid f(x) \in B\}$ of B by f, are identical.

The reader should take special note of the first two items and of the last item in B.14.4, where the inclusions may well be proper.

Definition B.14.5

Suppose X and Y are sets, $A \subseteq X$, $z \in A$ and $f: X \to Y$, and suppose also that Y is equipped with a total ordering. We say that f *attains its maximum on A* at z if, and only if, $\sup_Y f(A)$ exists and equals $f(z)$. We say that f *attains its minimum on A* at z if, and only if, $\inf_Y f(A)$ exists and equals $f(z)$.

B.15   Chains

Definition B.15.1

A non-empty collection $\mathcal{C}$ of sets is said to be *chained* if, and only if, for each $A, B \in \mathcal{C}$, there exists an ordered n-tuple $(U_1, \ldots, U_n)$ of members of $\mathcal{C}$ with $A = U_1$ and $B = U_n$ such that $U_{i-1} \cap U_i \neq \varnothing$ for all $i \in \mathbb{N}_n \backslash \{1\}$. Such an n-tuple is called a *chain* from A to B in $\mathcal{C}$. Note that every member of a chained collection is necessarily non-empty.

Suppose $n \in \mathbb{N}$ and $(A_1, \ldots, A_n)$ is an ordered n-tuple of non-empty sets. Then the product $P = \prod_{i=1}^{n} A_i$ includes various special *copies* of each of the coordinate sets. These copies are chained and their union is the whole of the product. This is the substance of B.15.2.

Theorem B.15.2

Suppose $n \in \mathbb{N}$ and $(A_1, \ldots, A_n)$ is an ordered n-tuple of non-empty sets. Set $P = \prod_{i=1}^{n} A_i$. Then, for each $a \in P$ and $j \in \mathbb{N}_n$, let $A_{j,a}$ denote the special copy $\{x \in P \mid x_i = a_i$ for all $i \neq j\}$ of A_j. Let $\mathcal{C} = \{A_{j,a} \mid a \in P, j \in \mathbb{N}_n\}$. Then $\mathcal{C}$ is chained and $\bigcup \mathcal{C} = P$.

Proof

$P \neq \varnothing$ because the coordinate sets are all non-empty. It follows easily that each $A_{j,a}$ is in a natural one-to-one correspondence with A_j and that $\bigcup \mathcal{C} = P$. The claim that $\mathcal{C}$ is chained is justified as follows. Suppose $A_{j,a}$ and $A_{k,b}$ are arbitrary members of $\mathcal{C}$. For each $m \in \mathbb{Z}$ with $0 \leq m \leq n$, let $z(m)$ be that member of P that agrees with b in its first m coordinates and agrees with a in the rest. Then $z(0) = a$ and $z(n) = b$. In general, for $0 \leq m \leq n$ and $i \in \mathbb{N}_n$, we have $z(m)_i = b_i$ if $i \leq m$ and $z(m)_i = a_i$ otherwise. Now note the memberships $a = z(0) \in A_{j,a} \cap A_{1,z(0)}$, $z(m) \in A_{m,z(m-1)} \cap A_{m+1,z(m)}$ for every $m \in \mathbb{N}$ with $1 \leq m \leq n-1$, and $b = z(n) \in A_{n,z(n-1)} \cap A_{k,b}$, from which we deduce that $(A_{j,a}, A_{1,z(0)}, \ldots A_{n,z(n-1)}, A_{k,b})$ is a chain from $A_{j,a}$ to $A_{k,b}$. $\qquad\qquad\qquad\qquad\qquad\qquad\qquad\qquad\qquad\qquad\qquad\qquad\qquad \square$

B.16 Equivalence Relations

Definition B.16.1

Suppose X is a set and $\simeq$ is a relation on X. Then $\simeq$ is called an *equivalence relation* on X if, and only if,

- $\simeq$ is *reflexive*, in that $a \simeq a$ for every $a \in X$;
- $\simeq$ is *symmetric*, in that $a \simeq b \Rightarrow b \simeq a$ for all $a, b \in X$; and
- $\simeq$ is *transitive*, in that, for all $a, b, c \in X$, if $a \simeq b$ and $b \simeq c$, then $a \simeq c$.

If $\simeq$ is an equivalence relation on X and $a \in X$, then the set $\{b \in X \mid b \simeq a\}$ is called an *equivalence class* of X with respect to $\simeq$. The equivalence classes generated by $\simeq$ are mutually disjoint subsets of X and their union is X itself. We encapsulate these two facts into one word: we say that the equivalence classes of X with respect to $\simeq$ form a *partition* of X.

B.17 Cardinality and Countability

The *cardinality* of a finite set is the number of elements it contains. Two finite sets have the same cardinality if, and only if, there is a one-to-one correspondence between them; using this idea, we generalize the notion of having *the same number of elements* to all sets in B.17.1. The term *cardinality* is used here only for convenience; we do not assume that every set has a well defined cardinality, nor do we assume that every pair of sets can be compared using an injective function as in B.17.1.

Definition B.17.1

Sets X and Y are said to have the *same cardinality* if, and only if, there is a bijective function between X and Y. If there is an injective function from X to Y but no bijective one, we say that Y has *greater cardinality* than X.

Theorem B.17.2 (Schröder–Bernstein Theorem)

Suppose X and Y are sets and Y has greater cardinality than X. Then X does not have greater cardinality than Y.

Definition B.17.3

Suppose S is a set. S is said to be

- *finite* if, and only if, $\mathbb{N}$ has greater cardinality than S;
- *countable* if, and only if, S has the same cardinality as a subset of $\mathbb{N}$;
- *denumerable* or *countably infinite* if, and only if, S has the same cardinality as $\mathbb{N}$; and
- *uncountable* if, and only if, S is not countable.

Theorem B.17.4

(i)   Every countable union of countable sets is countable.

(ii) Every Cartesian product of a finite number of countable sets is countable.

It is demonstrated in many textbooks that $\mathbb{Q}$ is countable, that $\mathbb{R}$ is uncountable, that every non-degenerate interval is uncountable, that the collection of continuous functions defined on $[0, 1]$ is of a greater cardinality than $\mathbb{R}$, and that there are sets of greater and greater cardinality. All finite sets are trivially countable; in fact they are precisely those countable sets that do not have the same cardinality as $\mathbb{N}$ itself. A well-ordered set need not be countable and a total ordering on a countable set need not be a well ordering. Neither need a well ordering on a countable set be enumerative. But every enumeratively ordered set (B.6.7) must be countable.

B.18   Sequences

Definition B.18.1

Suppose X is a set. A function s into X is called a *sequence* in X if, and only if, its domain is an enumeratively ordered infinite set. If s is bijective, then X is necessarily countable and s is called an *enumeration* of X. The value of s at a member n of the domain, instead of being written $s(n)$, is usually written s_n and is called the nth *term* of the sequence; the sequence itself is often denoted by (s_n) rather than simply by s, or by $(s_n)_{n \in D}$ if it is thought necessary to indicate that some particular set D is its domain. Sometimes a sequence is defined on an infinite subset of $\mathbb{N}$ (with its inherited ordering) or on some other enumeratively ordered infinite set such as $\mathbb{N} \cup \{0\}$, but we shall assume that the domain is $\mathbb{N}$ with its standard ordering unless we state otherwise.

Definition B.18.2

Suppose $s = (s_n)$ is a sequence in a set X. Suppose $m = (m_n)$ is a sequence in $\mathbb{N}$ that is strictly increasing, by which we mean that $m_k < m_l$ for all $k, l \in \mathbb{N}$ with $k < l$. Then $s \circ m = (s_{m_n})_{n \in \mathbb{N}}$ is called a *subsequence* of s.

Definition B.18.3

Suppose $n \in \mathbb{N}$. A function defined on $\mathbb{N}_n$ is called a *finite sequence* of *length* n.

Definition B.18.4

Suppose (x_n) is a sequence of real or complex numbers. Then the sequence $(\sum_{i=1}^{n} x_i)_{n \in \mathbb{N}}$ is called a *series* and is denoted by $\sum_{n \in \mathbb{N}} x_n$ or by $\sum_{n=1}^{\infty} x_n$.

B.19   Infinite Selection

The student who has reached the level at which metrics are being discussed will already have noticed that there are some subtle variations in the way in which sequences are introduced into mathematical arguments. Here are five illustrative ways in which this might be done.

- For each $n \in \mathbb{N}$, let $a_n = n^2$.
- Set $b_0 = 1$. Then, for each $n \in \mathbb{N}$, let $b_n = nb_{n-1}$.
- Given a sequence (A_n) of non-empty sets, choose $x_n \in A_n$ for each $n \in \mathbb{N}$.
- Suppose that ρ is a non-empty relation on some given set S and suppose that each member of the range of ρ is also in the domain; that is to say, for each $(a, z) \in \rho$, there exists at least one $b \in S$ such that $(z, b) \in \rho$. Use this fact to choose a sequence (w_n) in S such that, for each $n \in \mathbb{N}$, $(w_n, w_{n+1}) \in \rho$.
- Given an infinite set C, choose a sequence (z_n) of distinct terms in C.

Let us look first at (a_n) and (b_n). There is a stark distinction between their definitions, and the reader might well ask whether the second has the same validity as the first. For the first sequence (a_n), we simultaneously allot values to a_n for all natural numbers n. The assignments for the second sequence (b_n), although effected in the twinkling of an eye and in one line of text, nonetheless make each term b_n depend on all the terms b_m with $m < n$. Such a definition is said to be *recursive* or *inductive*. Whichever it is called, and whether induction is mentioned or not, there is always an implicit appeal to the Principle of Induction in this type of definition; using induction, it can be proved—though we offer no proof here—that all such recursive assignments really do define sequences just as well and with equal validity as the straightforward definitions such as that of (a_n).

Let us focus now on a similarity between these two definitions. In both cases, there is an effective procedure for deciding what each term of the sequence actually is. To get the value of a_n, we simply multiply n by itself. To get the value of b_n, we have to work harder because we need to go through all the terms b_m with $m < n$ before we can evaluate b_n; nonetheless, this involves only a finite number of steps, and the process then terminates. Both a_n and b_n are truly and unambiguously defined, a_n directly and b_n by recursion.

When we turn to the sequences (x_n), (w_n) and (z_n), we notice that they are not well defined in the sense that (a_n) and (b_n) are. They are somewhat elusive. Each x_n has been *selected* from A_n rather than specified. No procedure is given for determining what each x_n might actually be, nor do we know that it is possible to identify any particular candidate for x_n. The same is true for the sequences (w_n) and (z_n). Are these really sequences or are they illusory objects of our fancy? Have (x_n), (w_n) and (z_n) just been pulled out of a hat? The answer to this last question, if we are honest, is that they have indeed been pulled out of a hat—out of a silk hat—with flair, panache and very sound judgement.

In our discussion on existence (A.10), we said that there is a logical justification for making the leap from '$(\exists x)(P(x))$' to '*let z be such that $P(z)$ holds*'. This justification can be extended to making a finite number of jumps but not to making an infinite number. So the knowledge that all the sets A_n are non-empty, while it permits us to '*let $x_n \in A_n$*' for all n in any finite subset of $\mathbb{N}$, does not permit us to make simultaneously the infinite number of choices necessary to get the sequence (x_n). What permits us to do this, in the absence of any further information about the sets A_n, is a weak form of an axiom independent of the axioms of set theory known as the *Axiom of Choice*. This weak form of the Axiom of Choice says precisely what is needed to give us the sequence (x_n) and is especially important in cases where it is not possible to *identify* any sequence with the required properties.

Axiom B.19.1 (Axiom of Countable Choice)

Suppose (A_n) is any sequence of non-empty sets. Then there exists a sequence (x_n) that has the property that $x_n \in A_n$ for each $n \in \mathbb{N}$.

It might appear that this axiom, together with induction, is sufficient to give us also the sequence (w_n). But it is not. Suppose ρ is a non-empty relation on a set S such that, for each $u \in \mathrm{ran}(\rho)$, there exists $v \in S$ with $(u, v) \in \rho$. If S is countable, it is not difficult to deduce from B.19.1 that there exists a function $f : \mathrm{dom}(\rho) \to \mathrm{dom}(\rho)$ such that $(a, f(a)) \in \rho$ for all $a \in \mathrm{dom}(\rho)$. Then we can pick any $w_1 \in \mathrm{dom}(\rho)$ and, for each $n \in \mathbb{N}$, recursively let $w_{n+1} = f(w_n)$. The second part of this procedure is effected by induction, but the first depends crucially on the countability of S—it is not sound when S is uncountable and we have no further information about S or ρ. Since sequences of this type appear in the book, we need to postulate a stronger axiom that is still weaker than the full Axiom of Choice. It is left as an exercise to show that B.19.2 implies B.19.1.

Axiom B.19.2 (Axiom of Dependent Choice)

Suppose ρ is a non-empty relation and $\mathrm{ran}(\rho) \subseteq \mathrm{dom}(\rho)$. Then there exists a sequence (a_n) such that, for every $n \in \mathbb{N}$, $(a_n, a_{n+1}) \in \rho$.

The sequence (z_n) fares better than (w_n); it can be chosen using the Axiom of Countable Choice. The reader may like to prove this as an exercise, but it is instructive (B.19.3) to choose (z_n) using the stronger Axiom of Dependent Choice.

Theorem B.19.3

Suppose C is an infinite set. Then there exists a sequence (z_n) in C all of whose terms are distinct.

Proof

Apply B.19.2 to the relation $\{(S, S \cup \{x\}) \mid S \subset C,\ S \text{ finite},\ x \in C \backslash S\}$ to get a sequence (A_n) of subsets of C each of which has exactly one element more than its predecessor. Then, for each $n \in \mathbb{N}$, we set z_n to be the unique member of $A_{n+1} \backslash A_n$. The terms of (z_n) are distinct for, if $k, m \in \mathbb{N}$ and $k < m$, then $z_k \in A_{k+1} \subseteq A_m$ and $z_m \notin A_m$. $\square$

Let it be clear that assignments of the third, fourth and fifth types, such as those that gave us the sequences (x_n), (w_n) and (z_n) at the beginning of this section, are not definitions. They all require infinite choice, and (w_n) requires infinite dependent choice. All can be validly used within mathematical arguments only because we have accepted weak forms of the Axiom of Choice. None of them actually defines a sequence with the given property; it is, despite the

concrete notation, no more in reality than an assertion of the existence of such
a sequence (and we have already seen that assertions of existence are inherently
weak). We do not focus on this point anywhere else in this book, but, for the
reader who is interested in knowing where such choices are being made, we
reserve the words *choose* and *choice*, using them only where there is an appeal
to a weak form of the Axiom of Choice. There are other words, such as *pick*
and *select*, or even *let*, that we use where the context demands some such word
and where there is no appeal to any Axiom of Choice. No stronger form of the
Axiom of Choice than the two given here is needed anywhere in this book.

B.20 Algebraic Structures

The algebraic structures that are considered in this book are linear spaces and
algebras. Both need an underlying field.

Definition B.20.1

Suppose X is a set equipped with binary operations from $X \times X$ to X, one
described as addition and denoted here by $+$ and one described as multiplication and denoted here by juxtaposition. X is called a *field* if, and only if,

- $a + b = b + a$ and $ab = ba$ for all $a, b \in X$;
- $(a + b) + c = a + (b + c)$ and $(ab)c = a(bc)$ for all $a, b, c \in X$;
- $a(b + c) = (ab) + (ac)$ for all $a, b, c \in X$;
- there exist distinct unique members $0, 1 \in X$, called respectively the *zero*
  and the *unity* of F, such that $0 + a = 1a = a$ for all $a \in X$; and
- for each $a \in X \backslash \{0\}$, there exist unique $-a, a^{-1} \in X$ such that $a + (-a) = 0$
  and $aa^{-1} = 1$.

Example B.20.2

$\mathbb{Q}, \mathbb{R}$ and $\mathbb{C}$ are all fields.

Definition B.20.3

Suppose F is a field and V is a non-empty set equipped with two binary operations, one from $V \times V$ to V, here called *addition* and denoted by $+$, and the
other from $F \times V$ to V, here called *scalar multiplication* and denoted by juxtaposition. Suppose further that there exists a unique member 0 of V that, for
each $v \in V$, satisfies $v + 0 = v$ and $v + (-1)v = 0$, where 1 is the unity of F,
and that, for all $u, v, w \in V$ and $\mu, \lambda \in F$, we have

- $u + v = v + u$;
- $u + (v + w) = (u + v) + w$;
- $1v = v$;
- $\lambda(u + v) = \lambda u + \lambda v$;
- $(\lambda + \mu)v = \lambda v + \mu v$;
- $\lambda(\mu v) = (\lambda \mu)v$.

Then V is called a *vector space over F* or a *linear space over F*. In this context, the members of V are called *vectors* and the members of F are called *scalars*. For each $v \in V$, we generally denote the vector $(-1)v$ by $-v$. For subsets A and B of V, we shall denote the set $\{a + b \mid a \in A, \, b \in B\}$ by $A + B$ or by $B + A$ and, for any scalar λ, the set $\{\lambda b \mid b \in B\}$ by λB; in the special case where A is a singleton set $\{a\}$, we may write $a + \lambda B$ instead of $\{a\} + \lambda B$. If $F = \mathbb{R}$, we shall call V a *real linear space*. If $F = \mathbb{C}$, we shall call V a *complex linear space*. In this book, it is to be understood that the term *linear space* always refers to a real or complex linear space.

Definition B.20.4

A linear space V over a field F, equipped with an extra binary operation from $V \times V$ to V, here called *multiplication* and denoted by juxtaposition, is called an *algebra* over the field F if, and only if, for every $u, v, w \in V$ and every $\alpha \in F$, we have

- $u(vw) = (uv)w$;
- $u(v + w) = uv + uw$ and $(v + w)u = vu + wu$; and
- $\alpha(uv) = (\alpha u)v = u(\alpha v)$.

An algebra over $\mathbb{R}$ is called a *real algebra*; an algebra over $\mathbb{C}$ is called a *complex algebra*.

Definition B.20.5

- A non-empty subset S of a linear space V is called a *linear subspace* of V if, and only if, it is *algebraically closed* under the operations of addition and scalar multiplication. This means that for each $a, b \in S$ and scalar λ, we have both $a + b \in S$ and $\lambda a \in S$. It is easy to check that S is then a linear space in its own right over the same field as V.

- A non-empty subset S of an algebra V is called a *subalgebra* of V if, and only if, it is algebraically closed under the operations of addition, multiplication and scalar multiplication. It is easy to check that S is then an algebra in its own right over the same field as V.

Example B.20.6

$\mathbb{R}$ itself is a real algebra; in this case, the vectors and scalars are all real numbers. $\mathbb{C}$ is a complex algebra but is also a real algebra. $\mathbb{C}$ and $\mathbb{R}$ are both algebras over $\mathbb{Q}$, and $\mathbb{Q}$ is an algebra over itself. For each $n \in \mathbb{N}$, $\mathbb{R}^n$ is a real linear space and $\mathbb{C}^n$ is a complex linear space. For each $m, n \in \mathbb{N}$, the set $\mathcal{M}_{m \times n}(\mathbb{R})$ of $m \times n$ matrices with real entries, endowed with the standard operations of addition and multiplication by scalars, is a real vector space. Similarly, the set $\mathcal{M}_{m \times n}(\mathbb{C})$ of $m \times n$ matrices with complex entries is a complex linear space. The set $\mathcal{M}_{n \times n}(\mathbb{R})$ is a real algebra and the set $\mathcal{M}_{n \times n}(\mathbb{C})$ is a complex algebra.

Example B.20.7

Suppose X is a non-empty set and V is a linear space. Then the collection $\mathcal{F}$ of functions from X into V is a linear space over the same field as V when addition and scalar multiplication are defined in $\mathcal{F}$ in the obvious way: for $f, g \in \mathcal{F}$ and scalar λ, we define $f + g$ and λf by the equations $(f + g)(x) = f(x) + g(x)$ and $(\lambda f)(x) = \lambda(f(x))$ for all $x \in X$. The various properties of a linear space are easily verified using the corresponding properties of V. It should be noted particularly that all this happens simply because the common codomain of the functions is a linear space; it has nothing to do with the nature of the domain X or with any algebraic structure that X may or may not have.

Example B.20.8

Suppose X is a non-empty set. When the linear space V of B.20.7 is $\mathbb{R}$ or $\mathbb{C}$, we have two important fundamental linear spaces: the collection of all real functions defined on X is a real linear space; and the collection of all complex functions defined on X is a complex linear space. Indeed, these are both algebras, the first real and the second complex, when multiplication of functions f and g is defined by the equations $fg(x) = f(x)g(x)$ for all $x \in X$.

Example B.20.9

Some special cases of linear spaces of the type described in B.20.8 spring to mind. The collection of all functions from $\mathbb{R}$ into $\mathbb{R}$ is a real linear space; the collection of all functions from $\mathbb{C}$ into $\mathbb{C}$ is a complex linear space. The collection of all real sequences, namely of all functions from $\mathbb{N}$ into $\mathbb{R}$, is a real linear space; similarly the collection of all complex sequences is a complex linear space.

Example B.20.10

For each $n \in \mathbb{N} \cup \{0\}$, the function x^n (that is, $x \mapsto x^n$) defined on $\mathbb{R}$ is known as a *power function*, and a function of the type $\sum_{i=0}^{\infty} \alpha_i x^i$, where the α_i are real numbers, all except a finite number of which are zero, and the x^i are the various power functions, is called a *polynomial function* from $\mathbb{R}$ to $\mathbb{R}$. The set poly$(\mathbb{R})$ of all polynomial functions from $\mathbb{R}$ to $\mathbb{R}$ is a subalgebra of the algebra of all real functions defined on $\mathbb{R}$.

Example B.20.11

For each polynomial function $p = \sum_{i=0}^{\infty} \alpha_i x^i$ in poly$(\mathbb{R})$, we define the *degree* $\deg p$ of p to be $-\infty$ if $p = 0$ and to be the maximum value of i for which $\alpha_i \neq 0$ otherwise. Then, for each $n \in \mathbb{N} \cup \{0\}$, the collection poly$_n(\mathbb{R})$ of all real polynomial functions with degree that does not exceed n is a linear subspace of poly$(\mathbb{R})$. It is not a subalgebra of poly$(\mathbb{R})$ because it is not algebraically closed under multiplication.

Example B.20.12

Suppose that X and Y are linear spaces over the same field. A mapping $f: X \to Y$ is called a *linear mapping* if, and only if, $f(a + b) = f(a) + f(b)$ and $f(\lambda a) = \lambda f(a)$ for all $a, b \in X$ and all scalars λ. For each f and g in the collection $\mathcal{L}(X, Y)$ of all linear maps from X to Y and each scalar λ, we define $f + g$ and λf by the equations $(f + g)(a) = f(a) + g(a)$ and $(\lambda f)(a) = \lambda f(a)$ for all $a \in X$. It is easy to check that $f + g$ and λf thus defined are linear maps and that, endowed with these operations, $\mathcal{L}(X, Y)$ is a linear space over the same field as X and Y. The linear space $\mathcal{L}(X, X)$ can be made into an algebra by using composition of functions as the multiplication.

B.21 Isomorphism

Isomorphic structures are structures that have the same form; the form being considered depends on the context. We give some examples.

Example B.21.1

The real-number system forms a completely ordered field (B.6.7). $\mathbb{Q}$ is an ordered field, but the ordering is not complete. $\mathbb{C}$ does not have a standard ordering. So, of the three fields we have encountered, just one is a completely ordered field. Actually we can go much further than this and say that the real-number system is effectively the only completely ordered field. By this we mean that every completely ordered field is merely a relabelling of $\mathbb{R}$ in that it cannot be distinguished from $\mathbb{R}$ except by using something other than the specified properties of a completely ordered field. Mathematicians use the phrase *up to isomorphism* to describe this extension of the concept of uniqueness. The word *isomorphic* means literally *of the same form*. An *isomorphism* in mathematics is thus a function that preserves whatever structure is under consideration; in the present context, it preserves the whole of the algebraic structure—the structure determined by the binary operations—and the order. Specifically, if F is a completely ordered field, then there exists a bijective function $f\colon F \to \mathbb{R}$, called a *field order isomorphism*, which preserves both the order (for all $a, b \in F$ with $a < b$, we have $f(a) < f(b)$), and the algebraic operations (for all $a, b \in F$, we have $f(a + b) = f(a) + f(b)$ and $f(ab) = f(a)f(b)$). We then say that F is an *order isomorphic copy* of $\mathbb{R}$.

Example B.21.2

Examples of isomorphic copies of $\mathbb{R}$ in common use are those subsets of $\mathbb{R}^2$ referred to as the x-axis and y-axis, namely the sets $\mathbb{R} \times \{0\} = \{(r, 0) \mid r \in \mathbb{R}\}$ and $\{0\} \times \mathbb{R} = \{(0, r) \mid r \in \mathbb{R}\}$. They can be ordered like $\mathbb{R}$ and are thought of as being identical to $\mathbb{R}$, though, strictly speaking, their members are not real numbers at all, but ordered pairs of real numbers. The reader may like to note that $\mathbb{R} \times \{0\}$ and $\{0\} \times \mathbb{R}$ are, in the notation of B.15.2, the copies $\mathbb{R}_{1,(0,0)}$ and $\mathbb{R}_{2,(0,0)}$ of $\mathbb{R}$ that are included in $\mathbb{R}^2$.

Example B.21.3

If $m, n \in \mathbb{N}$ and $m < n$, we regard $\mathbb{R}^m$ as a linear subspace of $\mathbb{R}^n$ by identifying it with an isomorphic copy that really is a subspace of $\mathbb{R}^n$.

Example B.21.4

The linear space $\mathrm{poly}(\mathbb{R})$ described in B.20.10 is a linear subspace of the linear space of all functions from $\mathbb{R}$ into $\mathbb{R}$ (B.20.9). It is, in fact, an isomorphic copy of the linear space of all real sequences; the mapping $(\alpha_n) \mapsto \sum_{i=0}^{\infty} \alpha_{i+1} x^i$ preserves in $\mathrm{poly}(\mathbb{R})$ all of the linear space properties of the space of sequences.

Example B.21.5

For each $n \in \mathbb{N} \cup \{0\}$, the collection $\mathrm{poly}_n(\mathbb{R})$ (B.20.11) of all real polynomial functions with degree that does not exceed n is a linear subspace of $\mathrm{poly}(\mathbb{R})$. In exactly the same way as $\mathrm{poly}(\mathbb{R})$ is an isomorphic copy of the linear space of real sequences, so $\mathrm{poly}_n(\mathbb{R})$ is an isomorphic copy of $\mathbb{R}^{n+1}$.

B.22   Finite-Dimensional Linear Spaces

There are various concepts of *dimension* used in mathematics. The simplest of them is the algebraic concept applicable to all linear spaces. We are not concerned in this book with dimension itself; we simply want to distinguish between finite-dimensional and infinite-dimensional linear spaces.

Definition B.22.1

Suppose V is a linear space over $\mathbb{R}$ or $\mathbb{C}$ and S is a subset of V.

- A vector v of V is called a *linear combination* of members of S if, and only if, there exists a finite subset $\{s_i \mid 1 \le i \le n\}$ of S such that $v = \sum_{i=1}^{n} \lambda_i s_i$ for some scalars $\lambda_1, \dots, \lambda_n$.

- S is called a *spanning set* for V if, and only if, every vector of V is a linear combination of members of S.

- S is called a *basis* for V if, and only if, S is a spanning set for V and no proper subset of S is a spanning set for V.

- V is said to be *finite -dimensional* if, and only if, it has a finite basis. It is shown in elementary courses on linear algebra that all such bases have the same number of elements; this number is called the *dimension* of V.

- If V has no finite basis, then it is said to be *infinite-dimensional*.

Example B.22.2

$\{(1,0,0), (0,1,0), (0,0,1)\}$ is the standard basis for the three-dimensional space $\mathbb{R}^3$. Similarly, for each $n \in \mathbb{N}$, the standard basis of the n-dimensional space $\mathbb{R}^n$ is the collection of those n-tuples consisting of $(n-1)$ zeroes and one 1.

Example B.22.3

For each $n \in \mathbb{N}$, the space $\mathrm{poly}_n(\mathbb{R})$, being isomorphic to $\mathbb{R}^{n+1}$, has dimension $n+1$; in fact, it is easy to verify that the set $\{x^i \mid i \in \{0\} \cup \mathbb{N}_n\}$ of power functions (one of them, x^0, being the constant function 1) is a basis for $\mathrm{poly}_n(\mathbb{R})$.

Example B.22.4

The linear space $\mathrm{poly}(\mathbb{R})$ has each of the spaces $\mathrm{poly}_n(\mathbb{R})$ as a linear subspace. One does not expect any linear space to have subspaces of greater dimension than itself, and it is easy enough to verify that this cannot happen. It follows that $\mathrm{poly}(\mathbb{R})$ is infinite-dimensional. It is not always possible to construct a basis for an infinite-dimensional space. (The existence of a basis may depend on the Axiom of Choice.) In the case of $\text{poly}(\mathbb{R})$, however, we can write down the standard basis, namely $\left\{ x^i \mid i \in \{0\} \cup \mathbb{N} \right\}$.

Example B.22.5

Every finite-dimensional linear space is isomorphic to $\mathbb{R}^n$ or $\mathbb{C}^n$, where n is the dimension of the space. In fact, if V is an n-dimensional linear space, then there is a basis $\{v_i \mid i \in \mathbb{N}_n\}$ of V and it is not difficult to establish that the map $(\alpha_1, \ldots, \alpha_n) \mapsto \sum_{i=1}^n \alpha_i v_i$ from $\mathbb{R}^n$ or $\mathbb{C}^n$ into V is bijective and preserves both addition and scalar multiplication. This effectively means that the study of finite-dimensional linear spaces, as linear spaces, can be confined to the study of $\mathbb{R}^n$ and $\mathbb{C}^n$ for all $n \in \mathbb{N} \cup \{0\}$; every property of finite-dimensional linear spaces is to be found in these spaces. There may, of course, be other properties that some finite-dimensional spaces have and others do not—$\text{poly}_n(\mathbb{R})$, for example, is a space of functions that can be evaluated at each point of the domain $\mathbb{R}$, whereas its isomorphic copy $\mathbb{R}^{n+1}$ is not.

Solutions

Chapter 1: Metrics

Q 1.4 When $n = 1$, μ_2 is simply $(a, b) \mapsto |b - a|$, the usual metric on $\mathbb{R}$. For all real numbers u, v, w, z, we have $(wv - uz)^2 \geq 0$, so that $2uvwz \leq w^2v^2 + u^2z^2$, from which we get $(uv + wz)^2 \leq (u^2 + w^2)(v^2 + z^2)$ and then $(uv + wz) \leq \sqrt{(u^2 + w^2)(v^2 + z^2)}$, yielding

$$(u + v)^2 + (w + z)^2 \leq \left(\sqrt{u^2 + w^2} + \sqrt{v^2 + z^2} \right)^2 .$$

Suppose inductively that $k \in \mathbb{N}$ and that the function μ_2 is a metric on $\mathbb{R}^k \times \mathbb{R}^k$. Let $a, b, c \in \mathbb{R}^{k+1}$. Then, by the inductive hypothesis,

$$\sum_{i=1}^{k+1}(b_i - a_i)^2 \leq \left(\sqrt{\sum_{i=1}^{k}(b_i - c_i)^2} + \sqrt{\sum_{i=1}^{k}(c_i - a_i)^2} \right)^2 + (b_{k+1} - a_{k+1})^2,$$

which, since certainly $b_{k+1} - a_{k+1} = (b_{k+1} - c_{k+1}) + (c_{k+1} - a_{k+1})$, does not exceed $\left(\sqrt{\sum_{i=1}^{k+1}(b_i - c_i)^2} + \sqrt{\sum_{i=1}^{k+1}(c_i - a_i)^2} \right)^2$, by the inequality previously displayed; taking square roots gives the triangle inequality for μ_2 on $\mathbb{R}^{k+1} \times \mathbb{R}^{k+1}$. The other metric properties are obvious. So the Principle of Induction ensures that the appropriate μ_2 is a metric on $\mathbb{R}^n$ for each $n \in \mathbb{N}$.

Q 1.7 One might add the difference in lengths of the words to the number of letters that occur in the same position in the words but are different. For example, $d(\text{kiss}, \text{curse}) = |4 - 5| + 3 = 4$ because the words have four and five letters, respectively, and they differ in their first, second and third letters. It is easy to show this is a metric. Moreover, $d(\text{complement}, \text{compliment}) = 1$. This method will not always produce good results because words like 'torture' and 'pleasure', despite their obvious similarity, are further apart than might be wished; in fact, $d(\text{torture}, \text{pleasure}) = |7 - 8| + 7 = 8$.

Q 1.12 Call the function e. It is clearly symmetric and non-negative; moreover, if $a, b \in Z$ and $e(a, b) = 0$, we have $d(f(a), f(b)) = 0$, which, since d is a metric on X, implies that $f(a) = f(b)$, which in turn implies that $a = b$ because f is injective. For the triangle inequality, we have $d(f(a), f(b)) \leq d(f(a), f(c)) + d(f(c), f(b))$ for all $a, b, c \in Z$ because d is a metric on X; but this is precisely $e(a, b) \leq e(b, c) + e(c, a)$.

Q 1.14 The inverse function is an injective function from $\mathbb{N}$ to $\mathbb{R}$, so d gives a metric on $\mathbb{N}$. Then, for $m, n \in \mathbb{N}$, $d(m, \infty) = m^{-1} \leq \left| m^{-1} - n^{-1} \right| + n^{-1} = d(m, n) + d(n, \infty)$ and $d(m, n) = \left| n^{-1} - m^{-1} \right| \leq m^{-1} + n^{-1} = d(m, \infty) + d(\infty, n)$.

Q 1.15 Define $f \colon \mathbb{R} \to \mathbb{R}$ by $f(x) = x$ if $x < 0$ and $f(x) = x^2$ if $x \geq 0$. Then f is injective and $d(a, b) = |f(a) - f(b)|$ for all $a, b \in \mathbb{R}$. So d is a metric by Q 1.12.

Q 1.16 Label the function e. Clearly e satisfies the positive and symmetric conditions for being a metric. For the triangle inequality, suppose $a, b, c \in X$. Then, using the triangle inequality for d and for the absolute-value function, we have

$$
\begin{aligned}
e(a, b) = d(a, b) + |f(a) - f(b)| \ &\leq \ d(a, c) + d(c, b) + |f(a) - f(c)| + |f(c) - f(b)| \\
&= \ e(a, c) + e(c, b).
\end{aligned}
$$

Q 1.17 $v(a) - v(b) = \delta_z(a) - \delta_z(b) \leq d(a, b) \leq \delta_z(a) + \delta_z(b) < v(a) + v(b)$ for all $a, b \in X$.

Q 1.19 d is symmetric and non-negative and, for $u, v \in Y$, if $d(u, v) = 0$, we must have u and v either both in X or both in $Y \backslash X$, so that $u = v$ because d is a metric on each of these sets. For the triangle inequality, suppose $u, v, w \in Y$; if all or none of them are in X, then certainly the inequality holds. Suppose $u, v \in X$ and $w \in Y \backslash X$. Then we have $d(u, v) \leq d(u, a) + d(a, v) \leq d(u, w) + d(w, v)$ and $d(u, w) = d(u, a) + 1 + d(b, w) = d(u, a) + d(v, w) - d(v, a) \leq d(u, v) + d(v, w)$. A similar argument holds if $u, v \in Y \backslash X$ and $w \in X$.

Q 1.20 Let $a = (1, 0) \in \mathbb{R}^2$; then $d(a, (0, 0)) = 2 > 1 = \mu_1(a, (0, 0))$.

Q 1.22 No. For the metric determined by the norm to be preserved under ϕ, we need $\|\phi(a) - \phi(b)\|_Y = \|a - b\|_X$ for all $a, b \in X$. We have $\|\phi(a - b)\|_Y = \|a - b\|_X$, so that, if $\phi(a - b) = \phi(a) - \phi(b)$ for all $a, b \in X$, then ϕ is an isometry.

Q 1.24 Suppose $a, b \in \mathbb{R}^n$. For each $i \in \mathbb{N}_n$, we have $(a + b)_i = a_i + b_i$, so that $|(a + b)_i| \leq |a_i| + |b_i|$, which yields $\sum_{i=1}^{n} |(a + b)_i| \leq \sum_{i=1}^{n} |a_i| + \sum_{i=1}^{n} |b_i|$ and $\max\{|(a + b)_i| \mid i \in \mathbb{N}_n\} \leq \max\{|a_i| + |b_i| \mid i \in \mathbb{N}_n\} \leq \|a\|_\infty + \|b\|_\infty$.

Chapter 2: Distance

Q 2.3 The set of upper bounds of I in X is J and the least member of J is 4, so $\sup I = 4$ and $\mathrm{dist}(\sup I, I) = 3$.

Q 2.5 Since $\mathrm{iso}(S) \subseteq S$, this follows from the second part of 2.6.4.

Q 2.8 By 2.6.5, $\mathrm{acc}(\bigcap \mathcal{C}) \subseteq \mathrm{acc}(S) \subseteq \mathrm{acc}(\bigcup \mathcal{C})$ for each $S \in \mathcal{C}$ and the stated inclusions follow immediately. The inclusions may be proper. If $\mathcal{C} = \{(0, r) \mid r \in \mathbb{R}^+\}$, then $\bigcap \mathcal{C} = \varnothing$, which has no accumulation point in $\mathbb{R}$, whereas $0 \in \mathrm{acc}_\mathbb{R}(S)$ for every $S \in \mathcal{C}$. On the other hand, if $\mathcal{C} = \{(r, \infty) \mid r \in \mathbb{R}^+\}$, then 0 is an accumulation point in $\mathbb{R}$ of $\bigcup \mathcal{C} = \mathbb{R}^+$ but $0 \notin \mathrm{acc}_\mathbb{R}(S)$ for any $S \in \mathcal{C}$.

Q 2.9 There is implication in one direction. If $a_i \in \mathrm{iso}(\pi_i(S))$ for all $i \in \mathbb{N}_n$, then $r = \min\{\mathrm{dist}(a_i, \pi_i(S) \backslash \{a_i\}) \mid i \in \mathbb{N}_n\} > 0$ and, for each $x \in S \backslash \{a\}$, we have $x_j \neq a_j$ for at least one $j \in \mathbb{N}_n$, so that $r \leq \tau_j(x_j, a_j) \leq d(x, a)$. Since x is arbitrary in $S \backslash \{a\}$, $r \leq \mathrm{dist}(a, S \backslash \{a\})$ and $a \in \mathrm{iso}(S)$.

The converse is true if $S = P$. Suppose that $j \in \mathbb{N}_n$ and $a_j \notin \mathrm{iso}(X_j)$. Let $r \in \mathbb{R}^+$. Then there exists $w \in X_j \backslash \{a_j\}$ such that $\tau_j(a_j, w) < r$. Let $x \in P$ be such that $x_i = a_i$ for all $i \in \mathbb{N}_n \backslash \{j\}$ and $x_j = w$. Then $d(x, a) \leq \sum_{i=1}^{n} \tau_i(a_i, x_i) < r$. So $\mathrm{dist}(a, P \backslash \{a\}) < r$. Since r is arbitrary in $\mathbb{R}^+$, it follows that $\mathrm{dist}(a, P \backslash \{a\}) = 0$ and that $a \notin \mathrm{iso}(P)$. If $S \neq P$, anything might happen. Consider, for example, the subset $S = (\mathbb{R}^+ \times \{1\}) \cup (\{1\} \times \mathbb{R}^+) \cup \{(0, 0)\}$ of $\mathbb{R}^2$. Here, $\pi_1(S) = \mathbb{R}^\oplus = \pi_2(S)$ and 0 is not isolated in either, but $(0, 0)$ is of distance 1 from the rest of S and so is isolated in S.

Q 2.10 As in Q 2.9, there is implication in one direction. Suppose that, for all $i \in \mathbb{N}_n$, $a_i \notin \mathrm{acc}(\pi_i(S))$. Then $r = \min\{\mathrm{dist}(a_i, \pi_i(S) \backslash \{a_i\}) \mid i \in \mathbb{N}_n\}$ is greater than zero. Moreover, for each $x \in S\backslash\{a\}$, we have $x_j \neq a_j$ for some $j \in \mathbb{N}_n$ and therefore $r \leq \tau_j(x_j, a_j) \leq d(x, a)$. Since x is arbitrary in $S\backslash\{a\}$, it follows that $\text{dist}(a, S\backslash\{a\}) \neq 0$ and $a \notin \text{acc}(S)$.

The converse does not hold in general. Consider the subset $(\mathbb{R}^+ \times \{1\}) \cup (\{1\} \times \mathbb{R}^+)$ of $\mathbb{R}^2$. We have $\pi_1(S) = \mathbb{R}^+ = \pi_2(S)$, and 0 is an accumulation point of both, whereas the distance from $(0,0)$ to S is 1. If, however, $S = P$, then the reverse implication does hold. Suppose $j \in \mathbb{N}_n$ and $a_j \in \text{acc}(X_j)$. Let $r \in \mathbb{R}^+$. Then there exists $w \in X_j\backslash\{a_j\}$ with $\tau_j(w, a_j) < r$. Let $x \in P$ be such that $x_i = a_i$ for all $i \in \mathbb{N}_n\backslash\{j\}$ and $x_j = w$. Then $x \neq a$ and $d(x, a) \leq \sum_{i=1}^n \tau_i(x_i, a_i) < r$, so that $\text{dist}(a, P\backslash\{a\}) < r$. Since r is arbitrary in $\mathbb{R}^+$, we deduce that $\text{dist}(a, P\backslash\{a\}) = 0$ and that $a \in \text{acc}(P)$.

Chapter 3: Boundary

Q 3.1 For $x \in (a, b)$, we have $\text{dist}(x, \mathbb{R}\backslash(a, b)) = \min\{|x - a|, |x - b|\}$, which is not zero. Similarly, for $x \in \mathbb{R}\backslash[a, b]$, we have $\text{dist}(x, (a, b)) = \min\{|x - a|, |x - b|\} \neq 0$. In neither case is x a boundary point of (a, b). However, $a = \inf(a, b)$ and $b = \sup(a, b)$, so that both are zero distance from (a, b) by 2.2.5 and, since neither is in (a, b), they are both boundary points of (a, b).

Q 3.4 An argument similar to that used in 3.3.4 will show that $\Gamma \subseteq \partial\Gamma$. Now suppose $y \in [-1, 1]$ and $s \in \mathbb{R}^+$. By B.6.12, there is $n \in \mathbb{N}$ such that $(1 - s\sin^{-1} y)/(2\pi s) < n$. Set $x = 1/(\sin^{-1} y + 2n\pi)$. Then $x < s$ and $\sin(1/x) = y$. Therefore $\text{dist}((0, y), \Gamma) < s$, so, since s is arbitrary in $\mathbb{R}^+$, we must have $\text{dist}((0, y), \Gamma) = 0$ and $(0, y) \in \partial\Gamma$. We now need to show that no other point of $\mathbb{R}^2\backslash\Gamma$ is in the boundary of Γ. Towards this, suppose $(b, c) \in \mathbb{R}^2\backslash(\Gamma \cup \{(0, y) \mid y \in [-1, 1]\})$. If $b \leq 0$, then it is easy to show that $(b, c) \notin \partial\Gamma$, so we suppose also that $b > 0$. Since $(b, c) \notin \Gamma$, we have $c \neq \sin(1/b)$. We let $t = |\sin(1/b) - c|$ and set $r = \min\{b/2, tb^2/(b^2 + 2)\}$. We claim that, for $x \in \mathbb{R}^+$, $(x, \sin(1/x))$ cannot be a distance less than r from (b, c). Our claim is justified by showing that, if $|x - b| < r$, then $|\sin(1/x) - c| \geq r$. So we let $x \in \mathbb{R}^+$ be such that $|x - b| < r$. Then $x > b - r \geq b/2$, so that $|1/x - 1/b| = |x - b|/xb \leq 2r/b^2$. It can be verified using elementary trigonometry that $|\sin\alpha - \sin\beta| \leq |\alpha - \beta|$ for all $\alpha, \beta \in \mathbb{R}$, so that we now get $|\sin(1/x) - \sin(1/b)| \leq 2r/b^2$. By the triangle inequality of 1.1.2, $|\sin(1/x) - c| \geq |\sin(1/b) - c| - |\sin(1/x) - \sin(1/b)|$; so that, using the fact that $r \leq tb^2/(b^2 + 2)$, we now get $|\sin(1/x) - c| \geq t - 2r/b^2 \geq r$ to complete our calculations.

Q 3.8 Consider the definition of the Cantor set given in 3.3.3. The interval $\left(\frac{1}{3}, \frac{2}{3}\right)$ consists of all those numbers in $[0, 1]$ whose ternary expansions must have 1 in the first place. These are deleted from $[0, 1]$ on the first bout of deletion to create I_1. The intervals $\left(\frac{1}{9}, \frac{2}{9}\right)$ and $\left(\frac{7}{9}, \frac{8}{9}\right)$ consist of all those numbers in I_1 whose ternary expansions must have 1 in the second place. The deletions continue and what are left finally in the Cantor set are those numbers in $[0, 1]$ that have a ternary expansion made up entirely of zeroes and twos. The reader who is dissatisfied with this heuristic account may like to make it mathematically precise.

Q 3.9 Suppose $a, b \in \mathcal{K}$ and $a \neq b$. Let $a = \sum_{n=1}^\infty x_n/3^n$ and $b = \sum_{n=1}^\infty y_n/3^n$, where the x_n and y_n are all zeroes or twos. Then $(a + b)/2 = \sum_{n=1}^\infty z_n/3^n$, where $z_n = (x_n + y_n)/2$ for each $n \in \mathbb{N}$. Note that this is a ternary expansion of $(a + b)/2$ because $z_n \in \{0, 1, 2\}$ for each $n \in \mathbb{N}$. Since $a \neq b$, there is a least $s \in \mathbb{N}$ such that $x_s \neq y_s$, so that one of these numbers is 0 and the other 2, making $z_s = 1$. Now $(a + b)/2 \in \mathcal{K}$ if, and only if, it has a ternary expansion in which no 1 occurs. Since $z_s = 1$, this occurs if, and only if, $z_n = 0$ (in which case $x_n = y_n = 0$) for all $n > s$ or $z_n = 2$ (in which case $x_n = y_n = 2$) for all $n > s$. In other words, $(a + b)/2 \in \mathcal{K}$ if, and only if, $(a + b)/2$ has a ternary expansion that terminates in either 1 or 2 and has in it otherwise only zeroes and twos. This is precisely the same as saying that $(a + b)/2$ is of the form $k/3^n$, where $k \in \mathbb{N}$ and k is not divisible by 3.

Q 3.11 Consider the subset $U = \{(1/n, n) \mid n \in \mathbb{N}\}$ of $\mathbb{R}^2$ with its Euclidean metric. $0 \in \partial_\mathbb{R} \pi_1(U)$ and $n \in \partial_\mathbb{R} \pi_2(U)$ for each $n \in \mathbb{N}$, whereas $(0, r) \notin \partial_{\mathbb{R}^2} U$ for any $r \in \mathbb{R}$.

Conversely, consider $S = (\mathbb{R}^\oplus \times \mathbb{R}^\ominus) \cup (\mathbb{R}^\ominus \times \mathbb{R}^\oplus)$. Then $(0, 0) \in S$ and, for each $r \in \mathbb{R}^+$, $(r, r) \notin S$, so that $(0, 0) \in \partial_{\mathbb{R}^2} S$. However, $\pi_1(S) = \mathbb{R} = \pi_2(S)$, and both have empty boundary in $\mathbb{R}$.

Q 3.15 No. $\partial \mathbb{Q} = \mathbb{R}$ but $\partial \overline{\mathbb{Q}} = \partial \mathbb{R} = \varnothing$.

Q 3.19 No. Look at $\mathbb{Q}$ as a subset of $\mathbb{R}$. $\mathbb{Q}^\circ = \varnothing$, so $(\mathbb{Q}^\circ)^c = \mathbb{R}$, whereas $\overline{\mathbb{Q}} = \mathbb{R}$ and $(\overline{\mathbb{Q}})^c = \varnothing$, which has empty closure.

Chapter 4: Open, Closed and Dense Subsets

Q 4.2 By 3.6.9, $(S^\circ)^c = \overline{S^c}$. It follows that $S^\circ = \varnothing$ if, and only if, $\overline{S^c} = X$.

Q 4.5 Suppose first that F is a closed subset of X. Then $X \backslash F$ is open in X by 4.1.4, so, by 4.4.1, $Z \backslash F = Z \cap (X \backslash F)$ is open in Z and its complement in Z, namely $F \cap Z$, is closed in Z by 4.1.4.

For the converse, we suppose S is a subset of Z that is closed in Z. Then $Z \backslash S$ is open in Z by 4.1.4, so that, by 4.4.1, there exists an open subset U of X such that $Z \backslash S = U \cap Z$. Then $S = Z \backslash (U \cap Z) = Z \cap (X \backslash U)$. Since $X \backslash U$ is closed in X by 4.1.4, this completes the proof.

Q 4.6 If the inclusion holds, then, since Z is closed in Z, it follows that Z is closed in X, as claimed. On the other hand, if Z is closed in X, then so is $F \cap Z$ for every closed subset F of X by 4.3.2, so that the stated inclusion holds by Q 4.5.

Q 4.7 The two conditions are satisfied if, and only if, Z is both open and closed in X by 4.4.2 and Q 4.6. This is the same as saying that Z includes its boundary in X while containing no point of that boundary, a circumstance clearly equivalent to having empty boundary.

Q 4.10 Since $\mathbb{R}^+$ is open in $\mathbb{R}$, the topology of $\mathbb{R}^+$ is included in that of $\mathbb{R}$ by 4.4.2. But closed subsets of $\mathbb{R}^+$ need not be closed in $\mathbb{R}$: the interval $(0, 1]$, for example, is closed in $\mathbb{R}^+$ but not in $\mathbb{R}$—indeed, $\mathbb{R}^+$ itself is closed in $\mathbb{R}^+$ but not closed in $\mathbb{R}$. So the fact that every open subset of $\mathbb{R}^+$ is open in $\mathbb{R}$ does not imply that every closed subset of $\mathbb{R}^+$ is closed in $\mathbb{R}$.

Q 4.12 The function $f : \mathbb{R} \to \mathbb{R}$ defined by $f(x) = x$ if $x \in \mathbb{R}^-$ and $f(x) = 2x$ otherwise is injective. Since $d(a, b) = |f(a) - f(b)|$ for all $a, b \in \mathbb{R}$, d is a metric on $\mathbb{R}$ by Q 1.12. Moreover, if $S \subseteq \mathbb{R}$ and $z \in \mathbb{R}$, then $\mathrm{dist}_{|\cdot|}(z, S) \leq \mathrm{dist}_d(z, S) \leq 2\mathrm{dist}_{|\cdot|}(z, S)$, so that $\mathrm{dist}_d(z, S) = 0 \Leftrightarrow \mathrm{dist}_{|\cdot|}(z, S) = 0$. The same is true if S is replaced by S^c, so $z \in \partial_d S \Leftrightarrow z \in \partial_{|\cdot|} S$. Since z is arbitrary in $\mathbb{R}$, we then get $\partial_d S = \partial_{|\cdot|} S$. So S is open with respect to d if, and only if, S is open with respect to the Euclidean metric.

Q 4.13 The two topologies are $\{V \cap S \mid V \text{ open in } X\}$ and $\{U \cap S \mid U \text{ open in } Z\}$, respectively. But, since Z is a subspace of X, U is open in Z if, and only if, $U = V \cap X$, where V is open in X.

Q 4.14 If $z \in \bigcap \mathcal{F}$ and $a \in X \backslash \{z\}$, then $d(a, z) > 0$ and there exists $F \in \mathcal{F}$ with $\mathrm{diam}(F) < d(a, z)$. Then, since $z \in F$, we have $a \notin F$ and therefore $a \notin \bigcap \mathcal{F}$.

Q 4.15 $\mathbb{Q}$.

Q 4.19 Suppose S is a subset of X. Since $S \subseteq \overline{S}$, 3.7.1 gives $S^\circ \subseteq (\overline{S})^\circ$. Also 3.6.9 gives $(S^\circ)^c = \overline{S^c}$. So $(\overline{S})^\circ = \varnothing \Rightarrow S^\circ = \varnothing \Leftrightarrow (S^\circ)^c = X \Leftrightarrow \overline{S^c} = X$. Therefore, if S is nowhere dense in X, S^c is dense in X. Moreover, if S is closed in X, then the implication $(\overline{S})^\circ = \varnothing \Rightarrow S^\circ = \varnothing$ above is reversible, showing that, for closed sets, being dense and having nowhere dense complement are the same thing.

Q 4.22 Suppose X is a metric space and A is a dense subset of X. Then $\overline{A} = X$. If A is also nowhere dense, then $\varnothing = (\overline{A})^\circ = X^\circ = X$.

Q 4.23 Suppose first that, for each $i \in \mathbb{N}_n$, D_i is dense in X_i. Let V be an arbitrary non-empty open subset of P. Then, for each $i \in \mathbb{N}_n$, there exists a non-empty open subset U_i of X_i such that $\prod_{i=1}^n U_i \subseteq V$. Because, for each $i \in \mathbb{N}_n$, D_i is dense in X_i, $U_i \cap D_i \neq \varnothing$ (4.2.1). Then $\prod_{i=1}^n (U_i \cap D_i)$ is a non-empty subset of $V \cap \prod_{i=1}^n D_i$ and, since V is arbitrary, it follows that $\prod_{i=1}^n D_i$ is dense in P.

For the converse, suppose $j \in \mathbb{N}_n$ and D_j is not dense in X_j. Then there exists a non-empty open subset U_j of X_j such that $D_j \cap U_j = \varnothing$ (4.2.1). For each $k \in \mathbb{N}_n \backslash \{j\}$, set $U_k = X_k$. Then $(\prod_{i=1}^n D_i) \cap (\prod_{i=1}^n U_i) = \varnothing$, so that, since $\prod_{i=1}^n U_i$ is non-empty and open in P, $\prod_{i=1}^n D_i$ is not dense in P.

Q 4.25 Indeed it can. Consider the subset $(\mathbb{R}^+ \times \mathbb{Q}) \cup (\mathbb{R}^- \times (\mathbb{R} \backslash \mathbb{Q}))$ of $\mathbb{R}^2$ with the usual Euclidean metric.

Q 4.26 That $\mathbb{Q}^n$ is dense in $\mathbb{R}^n$ follows immediately from Q 4.23. Since $\mathbb{Q}^n$ is a finite product of countable sets, it is countable (B.17.4). So $\mathbb{R}^n$ is separable.

Q 4.27 For each $A \in \mathcal{C}$, choose a countable dense subset of A (B.19.1) and let D be the union of the chosen sets. Then D is dense in $\bigcup \mathcal{C}$ because each $x \in \bigcup \mathcal{C}$ is of distance 0 from a subset of D and therefore from D itself by 2.3.1. D is countable by B.17.4.

Chapter 5: Balls

Q 5.2 By 5.2.2, there exists $t \in \mathbb{R}$ such that $\flat[u\,;t] \subseteq U$. Let $s = \min\{t/2, r/2\}$. Then $\flat[u\,;s] \subseteq \flat[u\,;t] \subseteq U$, as required.

Q 5.4 If $z \in \mathrm{iso}(X)$, then $\mathrm{dist}(z, X \backslash \{z\}) > 0$; for $r \in \mathbb{R}^+$ with $r \le \mathrm{dist}(z, X \backslash \{z\})$, we have $\flat[z\,;r] = \{z\}$. Conversely, if $r \in \mathbb{R}^+$ is such that $\flat[z\,;r] = \{z\}$, then $d(x,z) \ge r$ for all $x \in X \backslash \{z\}$, so that $z \in \mathrm{iso}(X)$.

Q 5.9 For each $x \in E$, define $\delta(x) = \mathrm{dist}(x, F)$. Similarly, for each $x \in F$, define $\delta(x) = \mathrm{dist}(x, E)$. Then $\delta(x) > 0$ for each $x \in E \cup F$ because each of E and F is closed and they are disjoint. Let $U = \bigcup \{\flat[x\,;\delta(x)/2] \mid x \in E\}$ and $V = \bigcup \{\flat[x\,;\delta(x)/2] \mid x \in F\}$. Then U and V are open by 5.2.2, and $E \subseteq U$ and $F \subseteq V$. Moreover, U and V are disjoint, for, if we had $z \in U \cap V$, then there would be $a \in E$ and $b \in F$ with $z \in \flat[a\,;\delta(a)/2] \cap \flat[b\,;\delta(b)/2]$, from which we should get $d(a,b) \le d(a,z) + d(z,b) < \delta(a)/2 + \delta(b)/2 \le \max\{\delta(a), \delta(b)\}$, yielding either $\mathrm{dist}(a, F) < \delta(a)$ or $\mathrm{dist}(b, E) < \delta(b)$, both contradictions of the definition of δ.

This does not necessarily imply that $\mathrm{dist}(E, F) = 0$. The subsets $\{(x, e^x) \mid x \in \mathbb{R}\}$ and $\{(x, 0) \mid x \in \mathbb{R}\}$ of $\mathbb{R}^2$ are disjoint and closed in $\mathbb{R}^2$ with the usual metric, but their distance apart is 0.

Q 5.10 Certainly e is non-negative and symmetric, and, if $e(x,y) = 0$, then both $|x_1 - y_1|$ and $\sqrt{(x_2 - y_2)^2 + (x_3 - y_3)^2}$ are zero, from which we get $x_1 = y_1$, $x_2 = y_2$ and $x_3 = y_3$ and therefore $x = y$. Towards the triangle inequality, for $a, b, c, d \in \mathbb{R}$, we have $(bc - ad)^2 \ge 0$, giving $b^2 c^2 + a^2 d^2 \ge 2abcd$ and so $(a^2 + b^2)(c^2 + d^2) \ge (ac + bd)^2$. It follows that $(\sqrt{a^2 + b^2} + \sqrt{c^2 + d^2})^2 \ge (a + c)^2 + (b + d)^2$. So, if we make the replacements $a = x_2 - z_2$, $b = x_3 - z_3$, $c = z_2 - y_2$ and $d = z_3 - y_3$, we get the inequality $\sqrt{(x_2 - z_2)^2 + (x_3 - z_3)^2} + \sqrt{(z_2 - y_2)^2 + (z_3 - y_3)^2} \ge \sqrt{(x_2 - y_2)^2 + (x_3 - y_3)^2}$. Since the triangle inequality for the modulus gives $|x_1 - z_1| + |z_1 - y_1| \ge |x_1 - y_1|$, the given function satisfies the triangle inequality. The ball $\flat[0\,;1]$ is a cylinder of radius 1 and length 2; its axis lies on the first axis (x-axis) of $\mathbb{R}^3$.

Q 5.13 Since every non-empty open subset of X is a non-trivial union of open balls of X, this follows easily from 4.2.1.

Q 5.14 Suppose U is open in P and $x \in U$. Then, for each $i \in \mathbb{N}_n$, there exists an open subset V_i of X_i such that $x \in \prod_{i=1}^n V_i \subseteq U$ and, by 5.2.2, there exists an open ball B_i of X_i such that $x_i \in B_i \subseteq V_i$. Then $x \in \prod_{i=1}^n B_i \subseteq \prod_{i=1}^n V_i \subseteq U$. It follows that U is the union of all products of balls that are included in U.

Q 5.16 Suppose $z \in \partial S$. Let $\epsilon \in \mathbb{R}^+$. There exist $x \in S$ and $y \in X \backslash S$ such that $\|z - x\| < \epsilon$ and $\|z - y\| < \epsilon$. Then we have $\|-z - (-x)\| = \|z - x\| < \epsilon$ and also $\|-z - (-y)\| = \|z - y\| < \epsilon$. Since $-x \in -S$ and $-y \notin -S$, it follows that $\text{dist}(-z, -S) < \epsilon$ and $\text{dist}(-z, X \backslash (-S)) < \epsilon$. Because ϵ is arbitrary in $\mathbb{R}^+$, we then get $-z \in \partial(-S)$. So $-\partial S \subseteq \partial(-S)$. The reverse inclusion follows easily by putting $-S$ in place of S, therefore $\partial(-S) = -\partial S$. Since $S^\circ = S \backslash \partial S$ and $\overline{S} = S \cup \partial S$, simple calculations yield both $\text{Int}(-S) = -\text{Int}(S)$ and $\overline{-S} = -\overline{S}$.

Chapter 6: Convergence

Q 6.1 For each $n \in \mathbb{N}$, set $x_{2n} = 0$ and $x_{2n-1} = n$.

Q 6.4 Either $\limsup \frac{x_{n+1}}{x_n} = \infty$ or there exists $s \in \mathbb{R}^+$ with $\limsup \frac{x_{n+1}}{x_n} < s$. In the latter case, there exists $m \in \mathbb{N}$ such that, for all $n \in \mathbb{N}$, we have $x_{n+m} < s x_{n+m-1}$ and hence, by induction, $x_{n+m} < s^n x_m$. So $x_{n+m}^{1/(n+m)} < s x_m^{1/(n+m)} / s^{m/(n+m)}$ and, since $x_m^{1/(n+m)} \to 1$ and $s^{m/(n+m)} \to 1$ as $n \to \infty$, it follows that $\limsup x_n^{1/n} \leq s$. Therefore, in either case, $\limsup x_n^{1/n} \leq \limsup x_{n+1}/x_n$. A similar argument shows that $\liminf x_{n+1}/x_n \leq \liminf x_n^{1/n}$. Then Q 6.3 completes the list of inequalities.

Q 6.6 If $x_n \to \infty$, then, for each $s \in \mathbb{R}$, there is a tail of (x_n) in (s, ∞), whence $\liminf x_n \geq s$. Since s is arbitrary in $\mathbb{R}$, $\liminf x_n = \infty$ and, since this is not greater than $\limsup x_n$ (Q 6.3), it follows that $\limsup x_n = \infty$ as well. For the converse, suppose that $\liminf x_n = \infty$. Then, for each $s \in \mathbb{R}$, there exists a tail of (x_n) in (s, ∞), so that $x_n \to \infty$. The other part is demonstrated similarly.

Q 6.7 If $\limsup a_n^{1/n} < 1$, then there exist $r \in (0, 1)$ and $k \in \mathbb{N}$ such that $a_n < r^n$ for all $n \in \mathbb{N}$ with $k \leq n$. Since the geometric series $\sum_{n \in \mathbb{N}} r^{n+k}$ converges, a simple argument using the comparison test establishes convergence of $\sum_{n \in \mathbb{N}} a_n$. Conversely, if $\limsup a_n^{1/n} > 1$, then (a_n) does not converge to 0 and the series cannot converge.

Q 6.8 If $\limsup a_{n+1}/a_n < 1$, then there exist $r \in (0, 1)$ and $k \in \mathbb{N}$ such that $a_{n+1} < r a_n$ for all $n \in \mathbb{N}$ with $k \leq n$. By induction, $a_{n+k} \leq a_k r^n$ for all $n \in \mathbb{N}$, so that, since the geometric series $\sum_{n \in \mathbb{N}} a_k r^n$ converges, the comparison test establishes that $\sum_{n \in \mathbb{N}} a_{n+k}$ converges and it follows that $\sum_{n \in \mathbb{N}} a_n$ converges. Conversely, if $\liminf a_{n+1}/a_n > 1$, then (a_n) does not converge to 0 and the series cannot converge.

Q 6.13 Suppose $x \in X$. There is a sequence in S that converges to x if, and only if, $x \in \overline{S}$, by 6.6.2. The result follows immediately.

Q 6.14 Let $b = \lim a_m$ and $\epsilon \in \mathbb{R}^+$. Then, for all sufficiently large $m \in \mathbb{N}$, we have $\sup\{|b_i - \pi_i(a_m)| \mid i \in \mathbb{N}\} < \epsilon$, so that, for each $i \in \mathbb{N}$, $|b_i - \pi_i(a_m)| < \epsilon$; so, because ϵ is arbitrary in $\mathbb{R}^+$, $(\pi_i(a_m))_{m \in \mathbb{N}}$ converges to b_i in X_i.

Chapter 7: Bounds

Q 7.2 Even when there is just one space in the product, boundedness need not be preserved. Endow $\mathbb{R}^+$ with the metric $(a, b) \mapsto |a^{-1} - b^{-1}|$, which produces the same open sets as the usual metric. The interval $(0, 1)$ is unbounded with respect to this metric.

Q 7.4 Call the specified metric d. For all $n \in \mathbb{N}$, we have $d(1, n) = 1 - 1/n < 1$, so that $\mathbb{N} \subseteq \flat_d[1; 1)$ and $\mathbb{N}$ is bounded. On the other hand, $d(1, 1/n) = n - 1$ for all $n \in \mathbb{N}$, so that $\text{diam}_d(\{1/n \mid n \in \mathbb{N}\}) \geq n - 1$ for all $n \in \mathbb{N}$ and is thus infinite.

Q 7.6 Since $f_n \to g$, there exists $n \in \mathbb{N}$ such that $\sup\{e(f_n(x), g(x)) \mid x \in X\} < 1$. Then $e(g(a), g(b)) \leq e(g(a), f_n(a)) + e(f_n(a), f_n(b)) + e(f_n(b), g(b)) \leq 2 + \text{diam}(f_n(X))$ for each $a, b \in X$, whence $\text{diam}(g(X)) < \infty$.

Q 7.9 Suppose that a sequence (h_n) of functions from a set X to a metric space (Y, e)

converges uniformly to a bounded function g. Let $\flat[a\,;r]$ be a ball of Y that includes $g(X)$. For sufficiently large $n \in \mathbb{N}$, we have $\sup\{e(h_n(x), g(x) \mid x \in X\} < 1$, so that $h_n(X) \subseteq \flat[a\,;r+1]$ and h_n is bounded.

Q 7.11 The constant sequence whose terms are all 1 is in $c(\mathbb{R})$ but not in $c_0(\mathbb{R})$. The sequence whose terms are alternately 0 and 1 is in $\ell_\infty(\mathbb{R})$ but is not in $c(\mathbb{R})$.

Q 7.16 For each $n \in \mathbb{N}$, let f_n be the identity function on $\mathbb{R}$ and g_n be the constant function on $\mathbb{R}$ whose value is $1/n$. Clearly, (f_n) converges uniformly to the identity function and (g_n) converges uniformly to the zero function. But, for each $x \in \mathbb{R}$, $f_n(x)g_n(x) = x/n$, and (f_ng_n) converges pointwise, but not uniformly, to the zero function. Note that (g_n) is a bounded sequence, whereas (f_n) is not.

Q 7.18 Let $\mathcal{C}$ be any infinite countable collection of bounded real sequences. Enumerate the members of $\mathcal{C}$. Construct a sequence $x = (x_n)$ of real numbers as follows: for each $n \in \mathbb{N}$, let $x_n = 1$ if the nth term of the nth member of $\mathcal{C}$ is negative and let $x_n = -1$ otherwise. Then x is a bounded sequence and, by construction, the distance from x to the nth member of $\mathcal{C}$ is at least 1. So $\text{dist}(x, \mathcal{C}) \geq 1$. Therefore $\mathcal{C}$ is not dense in ℓ_∞.

Q 7.19 Suppose X is a totally bounded metric space. For each $n \in \mathbb{N}$, choose a finite collection C_n of points of X such that $X \subseteq \bigcup\{\flat[a\,;1/n] \mid a \in C_n\}$; such C_n exist because A is totally bounded. Let $N = \bigcup\{C_n \mid n \in \mathbb{N}\}$. N is countable by B.17.4. Suppose $x \in X$ and $\epsilon \in \mathbb{R}^+$. Let $n \in \mathbb{N}$ be such that $1/n < \epsilon$. Then there exists $a \in C_n$ such that $x \in \flat[a\,;1/n]$, so that $\text{dist}(x, N) < \epsilon$. Because ϵ is arbitrary in $\mathbb{R}^+$, it follows that $\text{dist}(x, N) = 0$ and $x \in \overline{N}$ by 3.6.10. Since x is arbitrary in X, we then have $X = \overline{N}$, as required.

Q 7.21 X is certainly infinite because every finite metric space is bounded (Q 7.1). Let $w \in X$. For each $n \in \mathbb{N}$, let $A_n = \{x \in X \mid d(w, x) > n\}$. Each A_n is non-empty by hypothesis. By B.19.1, there exists a sequence (a_n) with $a_n \in A_n$ for each $n \in \mathbb{N}$. Suppose $z \in X$ is arbitrary and let $k \in \mathbb{N}$ be such that $k \geq 1 + d(w, z)$. Then $d(a_n, z) \geq d(a_n, w) - d(w, z) \geq n + 1 - k \geq 1$ for every $n \in \mathbb{N}$ with $n \geq k$, so that $\text{dist}(z, \text{tail}_k(a)) \geq 1$. So 6.7.2 ensures that (a_n) has no subsequence that converges to z. Since z is arbitrary in X, this completes the proof.

Q 7.22 (x_n) certainly has a convergent subsequence by 7.11.1. Let z be the common limit of all these subsequences. Let $r \in \mathbb{R}^+$. Suppose there is an infinite number of terms of (x_n) in $X \backslash \flat[z\,;r]$. Then there is a subsequence of (x_n) with all its terms in $X \backslash \flat[z\,;r]$. Being a subsequence of (x_n), it is bounded and, since X has the nearest-point property, has a convergent subsequence. This latter is a convergent subsequence of (x_n) whose limit is not z, contradicting our hypothesis. So $\flat[z\,;r]$ includes a tail of (x_n) and, since r is arbitrary in $\mathbb{R}^+$, (x_n) converges to z.

Q 7.23 We work with real spaces; the proof can be easily adapted for complex spaces. The trivial space $\{0\}$ certainly has the nearest-point property. Suppose that $n \in \mathbb{N}$ and that all normed linear spaces of dimension less than n have the property. Let $(X, \|\cdot\|)$ be an n-dimensional normed linear space, and let V be an $(n-1)$-dimensional subspace of X and $a \in X \backslash V$. Then every vector of X can be represented uniquely as $\lambda a + v$ for some $\lambda \in \mathbb{R}$ and some $v \in V$. By the inductive hypothesis, V has the nearest-point property and so is closed in X, yielding $\text{dist}(a, V) > 0$. Suppose (λ_n) and (v_n) are sequences in $\mathbb{R}$ and V, respectively, for which the sequence $(\lambda_n a + v_n)$ is bounded in X. Since each $-v_n \in V$, we have $\|\lambda_n a + v_n\| \geq \text{dist}(\lambda_n a, V) = |\lambda_n| \text{dist}(a, V)$, which, since $\text{dist}(a, V) > 0$, ensures that (λ_n) is a bounded sequence; it follows easily that (v_n) is also a bounded sequence. Since $\mathbb{R}$ has the nearest-point property, (λ_n) has a convergent subsequence (λ_{m_n}) (7.11.1); let $\mu \in \mathbb{R}$ be its limit. Then, since V has the nearest-point property, the bounded subsequence (v_{m_n}) of (v_n) has a convergent subsequence $(v_{p_{m_n}})$ (7.11.1); let $u \in V$ be its limit. Now $(\lambda_{p_{m_n}})$, being a subsequence of (λ_{m_n}), converges to μ (6.7.1) and

$$\left\| \lambda_{p_{m_n}} a + v_{p_{m_n}} - \mu a - u \right\| \leq \left| \lambda_{p_{m_n}} - \mu \right| \|a\| + \left\| v_{p_{m_n}} - u \right\| \to 0 \text{ as } n \to \infty,$$

so that $(\lambda_{p_{m_n}} a + v_{p_{m_n}})$ is a convergent subsequence of $(\lambda_n a + v_n)$ by 6.1.4. Therefore X satisfies the convergence criterion of 7.11.1 and hence has the nearest-point property. By the Principle of Induction, every finite-dimensional real normed linear space has the nearest-point property.

Chapter 8: Continuity

Q 8.1 With reference to 8.2.1, suppose $w, y \in X$ and $f(x) \to y$ and $f(x) \to w$ as $x \to z$. By 8.2.1, for each $\epsilon \in \mathbb{R}^+$, there exists $\delta \in \mathbb{R}^+$ such that, for each $x \in \text{dom}(f) \backslash \{z\}$ (which is a non-empty set because z is an accumulation point of $\text{dom}(f)$), we have $e(f(x), y) < \epsilon$ and $e(f(x), w) < \epsilon$. This can be true only if $e(w, y) < 2\epsilon$. Since ϵ is arbitrary, this gives $e(w, y) = 0$ and therefore $w = y$.

Q 8.3 Define f on $\mathbb{R}$ by setting $f(x) = 1$ if $x \in \mathbb{Q}$ and $f(x) = 0$ otherwise. Then $f|_{\mathbb{Q}}$ and $f|_{\mathbb{R} \backslash \mathbb{Q}}$ are both constant functions and therefore certainly continuous. But f is discontinuous at every point of $\mathbb{R}$.

Q 8.5 Suppose X is a set, d is the discrete metric on X, and e is any metric on X that does not make (X, e) into a discrete metric space. The only closed balls of (X, d) are the singleton sets and X itself. Therefore, the identity function from (X, e) to (X, d) satisfies the stated condition. It is certainly not continuous.

Q 8.8 First, if S is not closed, then the zero function from S to $\mathbb{R}$ is continuous but has graph $S \times \{0\}$, which is not closed in $\mathbb{R}^2$. For the converse, suppose that S is closed in $\mathbb{R}$ and that $f: S \to \mathbb{R}$ is continuous. Let Γ denote the graph of f. Suppose $x \in \mathbb{R}^2 \backslash \Gamma$. If $x_1 \notin S$, then $\text{dist}_{\mathbb{R}^2}(x, \Gamma) \geq \text{dist}_{\mathbb{R}}(x_1, S) > 0$, so that $x \notin \overline{\Gamma}$. If, on the other hand, $x_1 \in S$, then we have $x_2 \neq f(x_1)$ and, since f is continuous at x_1, there exists $\delta \in \mathbb{R}^+$ with $\delta < |x_2 - f(x_1)|/2$ such that, for $z \in S$, the implication $|z - x_1| < \delta \Rightarrow |f(z) - f(x_1)| < |x_2 - f(x_1)|/2$ holds, and we deduce that either $|z - x_1| \geq \delta$ or $|f(z) - x_2| \geq |x_2 - f(x_1)| - |f(x_1) - f(z)| > |x_2 - f(x_1)|/2 > \delta$, and, in either case, that $\text{dist}(x, \Gamma) \geq \delta$ and $x \notin \overline{\Gamma}$. Since x is arbitrary in $\mathbb{R}^2 \backslash \Gamma$, it follows that Γ is closed in $\mathbb{R}^2$.

Q 8.9 f^{-1} is well defined because f is injective (B.14.2). Suppose U is an open subset of X. Then, because f is an open mapping, $f(U)$ is open in Y and therefore also in $f(X)$. But $f(U) = (f^{-1})^{-1}(U)$. So f^{-1} satisfies the open set criterion for continuity.

Q 8.12 Suppose $w, z \in \mathbb{C}$ and (a_n, b_n) is a sequence in $\mathbb{C} \times \mathbb{C}$ that converges to (w, z). By 6.5.1, (a_n) converges to w and (b_n) converges to z in $\mathbb{C}$. It follows easily that $(a_n + b_n)$ converges to $w + z$ and that $(a_n b_n)$ converges to wz in $\mathbb{C}$. Since (a_n, b_n) is an arbitrary sequence converging to (w, z), this establishes that addition and multiplication satisfy the convergence criterion for continuity at (w, z). Since (w, z) is arbitrary in $\mathbb{C} \times \mathbb{C}$, these maps are continuous.

Q 8.13 Denote the graph by Γ and suppose $(a, b) \in \overline{\Gamma}$. Then there exists a sequence $(x_n, f(x_n))$ in Γ that converges to (a, b) in $X \times Y$ (6.6.2). Since the metric on $X \times Y$ is a product metric, 6.5.1 ensures that $x_n \to a$ in X and $f(x_n) \to b$ in Y. But f is continuous at a, so it follows from 8.1.1 that $b = f(a)$, whence $(a, b) \in \Gamma$.

Q 8.14 For each $i \in \mathbb{N}_n$, suppose U_i is an arbitrary open subset of X_i. Then, since π_i is continuous, $\pi_i^{-1}(U_i)$ is open in P. The finite intersection $\bigcap \{\pi_i^{-1}(U_i) \mid i \in \mathbb{N}_n\}$ is open in P by 4.3.2. But, since $\pi_i^{-1}(U_i) = \{x \in P \mid x_i \in U_i\}$ for each $i \in \mathbb{N}_n$, this intersection is $\{x \in P \mid x_i \in U_i \text{ for each } i \in \mathbb{N}_n\}$, which is precisely $\prod_{i=1}^{n} U_i$. All unions of such products are then open in P by 4.3.2. In other words, every member of the product topology is a member of the topology on P.

Q 8.18 The projection $(a_1, a_2) \mapsto a_1$ of $\mathbb{R}^2$ onto $\mathbb{R}$ is surjective and open, but the image of the closed subset $\{(x, 1/x) \mid x \in \mathbb{R}^+\}$ of $\mathbb{R}^2$ under this map is $\mathbb{R}^+$, which is not closed in $\mathbb{R}$. We have seen in 8.4.2 that the mapping f given by $x \mapsto x^3 - x^2$ on $\mathbb{R}$ is continuous, surjective and not open. If S is any closed subset of $\mathbb{R}$ and $z \in \overline{f(S)}$, then there exists a sequence (x_n) in S such that $f(x_n) \to z$. The sequence (x_n) is clearly bounded and so has a subsequence (x_{m_n}) that converges to some $w \in \mathbb{R}$. Because S is closed, $w \in S$; because f is continuous, $f(x_{m_n}) \to f(w)$; and because $f(x_n) \to z$, we have $z = f(w)$, proving that $f(S)$ is closed.

Q 8.19 Suppose f is an open mapping and F is a closed subset of X. Then $X \backslash F$ is open in X, so that $f(X \backslash F)$ is open in Y. Because f is bijective, $Y \backslash f(F) = f(X \backslash F)$. So $f(F)$, being the complement in Y of the open set $Y \backslash f(F)$, is closed in Y. Since F is an arbitrary closed subset of X, this shows that f is a closed mapping. The converse is proved similarly.

Q 8.24 Let $p = \lim f_n'$. For each $n \in \mathbb{N}$, f_n' is continuous by hypothesis; therefore p is continuous by 8.9.2. Let $z \in [a, b]$. Then p is continuous at z. Let $\delta \in \mathbb{R}^+$ be such that for all $x \in [a, b]$ with $|z - x| < \delta$, we have $|p(z) - p(x)| < \epsilon/4$. Suppose $h \in (-\delta, \delta)$ and $z + h \in [a, b]$. Then $\left| hp(z) - \int_z^{z+h} p(x)\, dx \right| \le \epsilon|h|/4$. Since (f_n) converges pointwise to g and (f_n') converges uniformly to p, there exists $m \in \mathbb{N}$ such that $|f_m(z) - g(z)| < \epsilon|h|/4$ and $|f_m(z+h) - g(z+h)| < \epsilon|h|/4$ and $\sup\{f_m'(x) - p(x) \mid x \in [a, b]\} < \epsilon/4$. Then

$$\begin{aligned}
|f_m(z+h) - f_m(z) - hp(z)| &\le \left| f_m(z+h) - f_m(z) - \int_z^{z+h} p(x)\, dx \right| + \epsilon|h|/4 \\
&= \left| \int_z^{z+h} f_m'(x) - p(x)\, dx \right| + \epsilon|h|/4 \\
&\le \epsilon|h|/2,
\end{aligned}$$

whence $|g(z+h) - f_m(z+h)| + |f_m(z) - g(z)| + |f_m(z+h) - f_m(z) - hp(z)| \le \epsilon|h|$, so that $|g(z+h) - g(z) - hp(z)| \le \epsilon|h|$. Since this is true for all $h \in (-\delta, \delta)$ for which $z + h \in [a, b]$, it follows that g is differentiable at z and that $p(z) = g'(z)$. Since z is arbitrary in $[a, b]$, the result follows.

Chapter 9: Uniform Continuity

Q 9.1 For $x \in \mathbb{R}^+$, we have $(e^x - 1)/x = \sum_{n=1}^\infty x^{n-1}/n! > 1$, so that, in particular, if $a < b$, then $(e^{b-a} - 1)/(b - a) > 1$ and, multiplying both sides of the inequality by the positive number e^a, we get $(e^b - e^a)/(b - a) > e^a$.

Q 9.4 The function $x \mapsto x^2$ defined on $\bigcup\{[2n, 2n+1] \mid n \in \mathbb{N}\}$ has this property.

Q 9.6 Consider the product metric $(a, b) \mapsto |a^{-1} - b^{-1}|$ on the trivial product $\mathbb{R}^+$. The identity map from $\mathbb{R}^+$ with this metric to $\mathbb{R}^+$ with its usual metric is not uniformly continuous because, given $\delta \in \mathbb{R}^+$, we can set $a = 1/\sqrt{\delta}$ and $b = a + 1$, yielding $|a^{-1} - b^{-1}| < \delta$ and $|a - b| = 1$.

Q 9.10 Suppose first that S is bounded and let $h \in B(X, Y)$ and $r \in \mathbb{R}^+$ be such that $S \subseteq \flat_{B(X,Y)}[h; r]$. Also, as h is a bounded function, let $z \in X$ and $t \in \mathbb{R}^+$ be such that $h(X) \subseteq \flat_Y[h(z); t]$. Then, for each $x \in X$ and $f \in S$, we have $\hat{x}(f) = f(x) \in \flat_Y[h(x); r] \subseteq \flat_Y[h(z); r + t]$. It follows that $\hat{x}$ is bounded, and, being continuous (9.4.6), is in $\mathcal{C}(S, Y)$. It follows also that, for arbitrary $a, b \in X$ and all $f \in S$, $e(\hat{a}(f), \hat{b}(f)) < 2r + 2t$, whence $\operatorname{diam}(\{\hat{x} \mid x \in X\}) \le 2r + 2t$.

For the converse, suppose that $\{\hat{x} \mid x \in X\}$ is a bounded subset of $\mathcal{C}(S, Y)$ and let $u \in B(S, Y)$ and $p \in \mathbb{R}^+$ be such that $\{\hat{x} \mid x \in X\} \subseteq \flat_{B(S,Y)}[u; p]$. Since u is bounded, let $h \in S$ and $q \in \mathbb{R}^+$ be such that $u(S) \subseteq \flat_Y[u(h); q]$. Then, for each $x \in X$ and all $f \in S$, we have $f(x) = \hat{x}(f) \in \flat_Y[u(f);p] \subseteq \flat_Y[u(h);p+q]$. It follows that, for arbitrary $f, g \in S$ and $x \in X$, $e(f(x), g(x)) < 2p + 2q$, whence $\mathrm{diam}(S) \leq 2p + 2q$.

Q 9.11 Suppose f is non-zero. Let $x \in X$ be such that $f(x) \neq 0$. For each $r \in \mathbb{R}^+$, we have $f(rx/f(x)) = r$, so that f is not a bounded function.

Q 9.13 Let $\epsilon \in \mathbb{R}^+$. Because f_n converges uniformly to g, there exists $m \in \mathbb{N}$ such that $e(f_m(x), g(x)) < \epsilon/2$ for all $x \in X$. So, for $x, z \in X$, we have

$$e(g(x), g(z)) \leq e(g(x), f_m(x)) + e(f_m(x), f_m(z)) + e(f_m(z), g(z)) < \epsilon + kd(x,z),$$

and, since ϵ is arbitrary in $\mathbb{R}^+$, we conclude that $e(g(x), g(z)) \leq kd(x,z)$ and that g is Lipschitz with Lipschitz constant k.

Chapter 10: Completeness

Q 10.3 $[1, \infty)$, being a closed subspace of $\mathbb{R}$, is complete with its usual metric. The metric $(a,b) \mapsto \left| a^{-1} - b^{-1} \right|$ is a product metric on $[1, \infty)$, but it does not make $[1, \infty)$ complete because it has the non-convergent Cauchy sequence (n).

Q 10.5 Suppose $x = (x_n)$ is a real sequence that converges to any $z \in \mathbb{R}\backslash\{0\}$, and suppose $a = (a_n)$ is a sequence in $c_0(\mathbb{R})$. Then there exists $k \in \mathbb{N}$ such that $|x_k| > 2|z|/3$ and $|a_k| < |z|/3$. Then $|x_k - a_k| \geq |z|/3$, whence $\|x - a\|_\infty \geq |z|/3$. Since a is arbitrary in $c_0(\mathbb{R})$, we then have $\mathrm{dist}(x, c_0(\mathbb{R})) \geq |z|/3$, so $x \notin \mathrm{Cl}(c_0(\mathbb{R}))$. Since x is arbitrary in $c(\mathbb{R})\backslash c_0(\mathbb{R})$, it follows that $c_0(\mathbb{R})$ is closed in $c(\mathbb{R})$ and then complete because $c(\mathbb{R})$ is complete (10.8.4).

Q 10.7 First, we consider the case when $n = 1$. Invoking 10.9.3, notice that $f - h_1 = f - f^\dagger$, whose infimum and supremum on S are $-k$ and k, and also that $h_1(X) = f^\dagger(X) \subseteq [a+k, b-k]$ and $(h_2 - h_1)(X) = (f - h_1)^\dagger(X) \subseteq [-k/3, k/3]$, as required. To check the inductive step, suppose the four statements hold when n is an arbitrary $m \in \mathbb{N}$. By 10.9.3, on S we have $f - h_{m+1} = f - h_m - (f - h_m)^\dagger$ with supremum $(\sup(f - h_m)(S) - \inf(f - h_m)(S))/3 = 2^m k/3^m$ and infimum minus this quantity. Then $(h_{m+2} - h_{m+1})(X) = (f - h_{m+1})^\dagger(X) \subseteq [-2^m k/3^{m+1}, 2^m k/3^{m+1}]$, again by 10.9.3. Last, since, by hypothesis, $h_m(X) \subseteq [a + 2^{m-1}k/3^{m-1}, b - 2^{m-1}k/3^{m-1}]$ and $(h_{m+1} - h_m)(X) \subseteq [-2^{m-1}k/3^m, 2^{m-1}k/3^m]$, we have also the required inclusion $h_{m+1}(X) \subseteq [a + 2^m k/3^m, b - 2^m k/3^m]$. So the four statements hold when $n = m + 1$ and the inductive step is justified.

Q 10.8 Suppose first that X is a Baire space. Suppose $\mathcal{C}$ is a countable collection of closed nowhere dense subsets of X. Then each member of $\{X\backslash F \mid F \in \mathcal{C}\}$ is open and dense in X by Q 4.19. By Definition 10.11.3, $\bigcap\{X\backslash F \mid F \in \mathcal{C}\}$ is dense in X. By De Morgan's Theorem (B.11.2), this intersection is the complement of $\bigcup \mathcal{C}$. So (i) implies (ii).

Next, suppose that $\mathcal{C}$ is a countable union of nowhere dense subsets of X. Then $\mathcal{D} = \{\overline{F} \mid F \in \mathcal{C}\}$ is a countable union of closed nowhere dense subsets of X. If (ii) holds, then $\bigcap \mathcal{D}$ has dense complement, and, since $\bigcap \mathcal{C} \subseteq \bigcap \mathcal{D}$, $\bigcap \mathcal{C}$ also has dense complement and therefore empty interior (Q 4.2). So (ii) implies (iii).

That (iii) implies (iv) follows from the fact that every open set is equal to its interior (4.1.1).

Last, suppose $\mathcal{C}$ is a countable collection of open dense subsets of X. For each $U \in \mathcal{C}$, $X\backslash U$ has dense complement, and therefore has empty interior (Q 4.2), so that, being closed, it is nowhere dense. Therefore $X\backslash(U \cup \overline{\bigcap\mathcal{C}})$ is also nowhere dense. But

$$\bigcup\left\{X\backslash(U \cup \overline{\bigcap\mathcal{C}}) \;\middle|\; U \in \mathcal{C}\right\} = X\backslash\bigcap\left\{U \cup \overline{\bigcap\mathcal{C}} \;\middle|\; U \in \mathcal{C}\right\} = X\backslash(\bigcap\mathcal{C} \cup \overline{\bigcap\mathcal{C}}) = X\backslash\overline{\bigcap\mathcal{C}},$$

which is open, and therefore empty if (iv) holds. In that case, $\bigcap\mathcal{C}$ is dense in X. So (iv) implies (i).

Q 10.10 The function $x \mapsto f(x) - x$ is positive at 0 and negative at 1. We then need to invoke the Intermediate Value Theorem from real analysis (see 11.3.3) to assert that there exists $z \in [0,1]$ with $f(z) - z = 0$.

Q 10.14 We write f for the extension as well. By 10.9.1, f is uniformly continuous. Let $\epsilon \in \mathbb{R}^+$. There exists $\delta \in (0, \epsilon/4)$ such that, for all $x, y \in X$, with $d(x, y) < \delta$, we have $e(f(x), f(y)) < \epsilon/4$. For $a, b \in X$, there exist $u, v \in S$ such that $d(a, u) < \delta$ and $d(b, v) < \delta$. Then $e(f(a), f(u)) < \epsilon/4$ and $e(f(b), f(v)) < \epsilon/4$, so that

$$e(f(a), f(b)) \le e(f(a), f(u)) + e(f(u), f(v)) + e(f(b), f(v)),$$

and $d(a, b) \le d(a, u) + d(u, v) + d(b, v)$, which yield $|e(f(a), f(b)) - d(a, b)| \le \epsilon$ because $d(u, v) = e(f(u), f(v))$. Because ϵ is arbitrary in $\mathbb{R}^+$, $e(f(a), f(b)) = d(a, b)$, as required.

Q 10.16 Let $s = \operatorname{dist}(w, C)$. Let $C' = \{c - w \mid c \in C\}$. Then C' is closed and convex and $\inf\{\|x\|_2 \mid x \in C'\} = s$. For each $\delta \in \mathbb{R}^+$, $S_\delta = \flat[0; s + \delta] \cap C'$ is a non-empty closed subset of $\ell_2(\mathbb{R})$. For each $a, b \in S_\delta$, we have $\|b - a\|_2^2 = \sum_{i=1}^\infty (b_i^2 - 2a_i b_i + a_i^2)$ and $\|b + a\|_2^2 = \sum_{i=1}^\infty (b_i^2 + 2a_i b_i + a_i^2)$, giving $\|b - a\|_2^2 + \|b + a\|_2^2 = 2\sum_{i=1}^\infty (b_i^2 + a_i^2)$ and then

$$\|b - a\|_2^2 = 2\|a\|_2^2 + 2\|b\|_2^2 - 4\left\|\frac{b + a}{2}\right\|_2^2.$$

Since $a, b \in S_\delta$, we have $\|a\|_2 \le s + \delta$ and $\|b\|_2 \le s + \delta$. Since C' is convex, we have $(b + a)/2 \in C'$ and therefore $\|(b + a)/2\|_2 \ge s$. So $\|b - a\|_2^2 \le 4(s + \delta)^2 - 4s^2 = 8s\delta + 4\delta^2$ and thus $\operatorname{diam}(S_\delta) \le 8s\delta + 4\delta^2$. Therefore $\inf\{\operatorname{diam}(S_\delta) \mid \delta \in \mathbb{R}^+\} = 0$. Since $\ell_2(\mathbb{R})$ is complete, the nest criterion for completeness ensures that $\bigcap\{S_\delta \mid \delta \in \mathbb{R}^+\}$ is a singleton set. In other words, there is exactly one point $z \in C'$ such that $\|z\|_2 = s$. Then $z + w$ is unique in C at distance s from w.

Chapter 11: Connectedness

Q 11.4 The components are all closed (11.5.3). Suppose C is a component. If the number of components is finite, then the union of all except C is closed (4.3.2). So C, being the complement of that union, is open.

Q 11.5 Suppose C is a connected component of $[0, 1] \setminus S$. Let $a = \inf C$ and $b = \sup C$. Since C is connected, $(a, b) \subseteq C$. If $a \in C$, then $a \ne 0$ because $0 \in S$, and, because S is closed, $\operatorname{dist}(a, S) > 0$, from which it follows that there exists $r \in \mathbb{R}^+$ such that $(a - r, a) \cap S = \varnothing$ and therefore that $(a - r, b)$ is a connected subset of $[0, 1] \setminus S$, contradicting the definition of a. So $a \notin C$. A similar argument shows that $b \notin C$. So $C = (a, b)$.

Q 11.8 It is clear that the set $\Gamma = \{(x, \sin(1/x)) : 0 < x \le 1\}$ is pathwise connected and therefore connected. So $\overline{\Gamma}$ is connected. But $\overline{\Gamma} = S$ (Q 3.4). So S is connected. Now suppose there exists a continuous path $f \colon [0, 1] \to S$ that joins $(0, 0)$ to some point of Γ. Let $r = \sup(f^{-1}(S \setminus \Gamma))$. Note that $f^{-1}(S \setminus \Gamma)$ is closed because $S \setminus \Gamma$ is closed, so that $f(r) \in S \setminus \Gamma$. So $r < 1$ because $f(1) \in \Gamma$ by hypothesis. Since f is continuous at r, there exists $\delta \in (0, 1 - r)$ such that $f([r, r + \delta])$ is included in the ball B of radius $1/2$ around $f(r)$. It is easy to check that the connected component of $B \cap S$ that contains $f(r)$ is $B \cap (S \setminus \Gamma)$, so that $f(r + \delta) \in S \setminus \Gamma$, contradicting the definition of r. So there is no such continuous path f.

Q 11.9 For each $a \in S$, the line segment joining a to (a_1, a_1^2) is in S and (a_1, a_1^2) lies on the path $\{x \in \mathbb{R}^2 \mid x_2 = x_1^2\}$, which is included in S. So S is pathwise connected. Suppose $s \in S \setminus \{(0, 0)\}$. We show that the line segment joining s to $(0, 0)$ is not included in S. We have then $0 < s_2 \le s_1^2$. Let $t = s_2/2s_1^2$. Then $t \in (0, 1)$. Set $x = ts$ and note that x is in the line segment joining $(0, 0)$ to s. But $x_1^2 = s_2^2/4s_1^2 < s_2^2/2s_1^2 = x_2$, so that $x \notin S$. It follows that no line segment in S has an endpoint at $(0,0)$ and therefore that S is not polygonally connected.

Chapter 12: Compactness

Q 12.2 Let $\mathcal{C}$ be an open cover for $\{z\} \cup \{x_n \mid n \in \mathbb{N}\}$. There exists $V \in \mathcal{C}$ with $z \in V$. Because $x_n \to z$ and V is open, there exists $k \in \mathbb{N}$ such that V includes the kth tail of (x_n). For each $i \in \mathbb{N}_k$, pick $U_i \in \mathcal{C}$ such that $x_i \in U_i$. Then $\{V\} \cup \{U_i \mid i \in \mathbb{N}_k\}$ covers $\{z\} \cup \{x_n \mid n \in \mathbb{N}\}$.

Q 12.4 By 11.3.3, $f(I)$ is an interval. By 12.3.1, it is compact and therefore closed and bounded.

Q 12.9 Since every compact metric space is totally bounded, Q 7.19 gives the result.

Q 12.11 Suppose $x \in S$. If S is open, then there is a ball $\flat_X[x;r]$ included in S. Since X is locally compact, there exists $s \in (0,r)$ such that $\flat_X[x;s]$ is compact. But $\flat_X[x;s] \subseteq \flat_X[x;r] \subseteq S$, so that $\flat_S[x;s] = \flat_X[x;s]$. So $\flat_S[x;s]$ is compact. If, alternatively, S is closed in X, let B be a compact ball of X centred at x. Then the ball $S \cap B$ of S centred at x is closed in X and therefore in B and so is compact. In either case, S satisfies the conditions for local compactness (12.7.1).

Q 12.12 Suppose that (X,d) is a locally compact metric space and that $\mathcal{U}$ is a non-empty countable collection of dense open subsets of X. If $\mathcal{U}$ is finite, we invoke 10.11.1. Suppose $\mathcal{U}$ is infinite, and let (U_n) be an enumeration of the members of $\mathcal{U}$. Let B_0 be any non-empty open subset of X, and choose a sequence (B_n) of open balls of X with radius less than $1/n$ such that $\overline{B_n}$ is compact and $\overline{B_n} \subseteq U_n \cap B_{n-1}$. This is possible for reasons similar to those listed in 10.11.4; B_n can be chosen with compact closure because X is locally compact. Then $\bigcap\{\overline{B_n} \mid n \in \mathbb{N}\} \neq \varnothing$ by 12.4.3. So $B_0 \cap \bigcap \mathcal{U} \neq \varnothing$. Since B_0 is an arbitrary non-empty open subset of X, $\bigcap \mathcal{U}$ is dense in X by 4.2.1. So X is a Baire space.

Q 12.15 Using the notation of 12.9.1, suppose $a \in \mathcal{P}_t$ is arbitrary. Then the following calculations ensue: $\sum_{i=1}^{\nu(a)-1} d(f(a_i), f(a_{i+1})) \leq \sum_{i=1}^{\nu(a)-1} k(a_{i+1} - a_i) = kt$. The result follows by taking the supremum over $a \in \mathcal{P}_t$.

Q 12.16 Suppose $r,s \in [0,1]$ with $r < s$ and $g(r) = g(s)$. Define h by $h(t) = g(t)$ if $t \in [0,r]$ and $h(t) = g(s + ((t-r)(1-s)/(1-r)))$ if $t \in (r,1]$. It is easy to check that h is a path from a to b and that $\mathrm{lth}(h) = m(1 - s + r) < m$, contradicting the definition of m.

Q 12.18 Let $v \in X \backslash S$. Since S has the nearest-point property, there exists a nearest point s of S to v. Then $\|v - s\| = \mathrm{dist}(v,S) = \mathrm{dist}(v-s,S)$, and it follows that $\mathrm{dist}((v-s)/\|v-s\|,S) = 1$. But $(v-s)/\|v-s\|$ is a vector of length 1.

Q 12.20 Let $a \in \mathbb{R}^n$ be the vector that has 1 as its first coordinate and zeroes elsewhere. Let $b \in \mathbb{R}^n$ have 1 as its second coordinate and zeroes elsewhere. Then $\|a\|_p = \|b\|_p = 1$, whereas $\|a + b\|_p = 2^{1/p}$, which exceeds 2 because $p < 1$.

Q 12.24 If $\mathbb{I} \cap \mathbb{J} = \varnothing$, then, for each $i \in \mathbb{N}_n$, either $x_i = y_i$ or $wx_i + (1-w)y_i = 0$. Then $f(w) = \sum_{i \in \mathbb{N}_n \backslash \mathbb{J}} |x_i|^p = \sum_{i \in \mathbb{N}_n \backslash \mathbb{J}} |y_i|^p$, whereas it is clear that $f(0) = \sum_{i \in \mathbb{N}_n} |y_i|^p$ and $f(1) = \sum_{i \in \mathbb{N}_n} |x_i|^p$. But, in this case, $\mathbb{J} \neq \varnothing$ because $\mathbb{I} \neq \mathbb{N}_n$ by hypothesis. For $j \in \mathbb{J}$, we have either $x_j \neq 0$ or $y_j \neq 0$, making either $f(0) > f(w)$ or $f(1) > f(w)$.

Q 12.25 Suppose $a,b \in \ell_p(\mathbb{R})$. Then the series $\sum_{n=1}^{\infty} |a_n|^p$ and $\sum_{n=1}^{\infty} |b_n|^p$ both converge; their sums are denoted, following 12.11.4, by $\|a\|_p^p$ and $\|b\|_p^p$, respectively. By 12.11.3, $\left(\sum_{n=1}^{k} |a_n + b_n|^p\right)^{1/p} \leq \left(\sum_{n=1}^{k} |a_n|^p\right)^{1/p} + \left(\sum_{n=1}^{k} |b_n|^p\right)^{1/p} \leq \|a\|_p + \|b\|_p$ for each $k \in \mathbb{N}$. Since k is arbitrary in $\mathbb{N}$, the series $\sum_{n=1}^{\infty} |a_n + b_n|^p$ converges and $\|a+b\|_p \leq \|a\|_p + \|b\|_p$. The convergence implies that $\ell_p(\mathbb{R})$ is algebraically closed under addition (B.20.5); that it is a linear space follows easily. The inequality is the triangle inequality for $\|\cdot\|_p$, and the other norm properties are clearly satisfied.

Chapter 13: Equivalence

Q 13.2 Assume, without loss of generality, that $\mathcal{U}$ is infinite, and enumerate $\mathcal{U}$ as (U_m). For each $m \in \mathbb{N}$ and $x \in U$, let $f_m(x) = 1/\text{dist}(x, U_m^c)$. For each $a, b \in \bigcap \mathcal{U}$, let

$$e(a,b) = d(a,b) + \sum_{m=1}^{\infty} \frac{1}{2^m} \frac{|f_m(a) - f_m(b)|}{1 + |f_m(a) - f_m(b)|}.$$

It is easily verified, by methods similar to those used in Q 1.11 and Q 1.16, that this is a metric on $\bigcap \mathcal{U}$. That it is topologically equivalent to d on $\bigcap \mathcal{U}$ follows from the fact that $d(a,b) \leq e(a,b) \leq d(a,b) + |f_m(a) - f_m(b)|$ for all $a, b \in \bigcap \mathcal{U}$ and the knowledge that $(a,b) \mapsto d(a,b) + |f_m(a) - f_m(b)|$ is a metric on $\bigcap \mathcal{U}$ that is topologically equivalent to d (10.3.4, 13.1.13). Suppose (x_n) is a Cauchy sequence in $(\bigcap \mathcal{U}, e)$. Because $d \leq e$, (x_n) is Cauchy in (X, d) and so there exists $z \in X$ such that $x_n \to z$ in (X, d) because (X, d) is complete. We claim that $z \in \bigcap \mathcal{U}$. If there were, on the contrary, some $m \in \mathbb{N}$ such that $z \notin U_m$, then we should have $f_m(x_n) \to \infty$ as $n \to \infty$, so that, for any $k \in \mathbb{N}$ and for all sufficiently large $l \in \mathbb{N}$, we should have $|f_m(x_l) - f_m(x_k)| > 1$ and therefore $e(x_l, x_k) > 1/2^{m+1}$, contradicting the hypothesis that (x_n) is Cauchy. So our claim is justified. That (x_n) converges to z in $(\bigcap \mathcal{U}, e)$ follows because e is topologically equivalent to d on $\bigcap \mathcal{U}$.

Q 13.3 The condition that $f^{-1}: f(X) \to X$ is continuous is that, for all $z \in f(X)$ and all $\epsilon \in \mathbb{R}^+$, there exists $\delta \in \mathbb{R}^+$ such that, for all $y \in f(X)$, if $m(z,y) < \delta$, then $d(f^{-1}(z), f^{-1}(y)) < \epsilon$. Using the fact that f is injective and writing $a = f^{-1}(z)$ and $b = f^{-1}(y)$, this becomes: for all $a \in X$ and all $\epsilon \in \mathbb{R}^+$, there exists $\delta \in \mathbb{R}^+$ such that, for all $b \in X$, $m(f(a), f(b)) < \delta \Rightarrow d(a,b) < \epsilon$; that is, $e(a,b) < \delta \Rightarrow d(a,b) < \epsilon$. But this is precisely the condition for e to be topologically stronger than d.

Q 13.4 The condition that f is uniformly continuous is that for every $\epsilon \in \mathbb{R}^+$ there exists $\delta \in \mathbb{R}^+$ such that, for all $a, b \in X$ with $d(a,b) < \delta$, we have $m(f(a), f(b)) < \epsilon$. Since $m(f(a), f(b)) = e(a,b)$, this is precisely the condition that the identity function from (X, d) to (X, e) is uniformly continuous—or that e is uniformly weaker than d.

The condition that $f^{-1}: f(X) \to X$ is uniformly continuous is that for every $\epsilon \in \mathbb{R}^+$ there exists $\delta \in \mathbb{R}^+$ such that, for all $u, v \in f(X)$ with $m(u,v) < \delta$, we have $d(f^{-1}(u), f^{-1}(v)) < \epsilon$. Writing $u = f(a)$ and $v = f(b)$, this is precisely the same as saying that, for every $\epsilon \in \mathbb{R}^+$, there exists $\delta \in \mathbb{R}^+$ such that for all $a, b \in X$ with $e(a,b) < \delta$, we have $d(a,b) < \epsilon$—in other words, d is uniformly weaker than e.

Q 13.6 f is Lipschitz if, and only if, there is $k \in \mathbb{R}^+$ with $m(f(a), f(b)) \leq kd(a,b)$ for all $a, b \in X$. Since $e(a,b) = m(f(a), f(b))$, this is precisely the condition that the identity function from (X, d) to (X, e) is Lipschitz, or that e is Lipschitz weaker than d. Also, f^{-1} is a Lipschitz function if, and only if, there exists $l \in \mathbb{R}^+$ such that $d(f^{-1}(u), f^{-1}(v)) \leq lm(u,v)$ for all $u, v \in f(X)$. Writing $u = f(a)$ and $v = f(b)$, this is the same as saying that $d(a,b) \leq le(a,b)$ or that e is Lipschitz stronger than d.

Q 13.7 Suppose $x \in S$. Since S is locally compact with respect to d, there exist an open subset U of (S, d) and a compact subset K of (S, d) such that $x \in U \subseteq K$. Now K is compact in (X, d) and there exists an open subset V of (X, d) such that $U = S \cap V$. But topological equivalence preserves both openness and compactness, so that V is open in (X, e) and K is compact in (X, e), whence $U = V \cap S$ is open in (S, e) and K, being a subset of S, is compact in (S, e). Since z is arbitrary in X, this implies that (S, e) is locally compact.

Q 13.10 Yes, by 13.1.13, because every locally compact metric space is an open subset of a complete metric space (12.7.6).

List of Symbols

Bibliography

Mathematical discoveries, small or great
are never born of spontaneous generation.
They always presuppose a soil seeded with preliminary
knowledge and well prepared by labour,
both conscious and subconscious.   *Jules Henri Poincaré, 1854–1912*

Some of these books are cited in the text. Others are listed because I have consulted them during the writing of this book.

[1] Michael Barnsley. *Fractals Everywhere*. Academic Press, Boston, 1988.

[2] Douglas S. Bridges. *Foundations of Real and Abstract Analysis*. Springer-Verlag, New York, 1998.

[3] Arlen Brown and Carl Pearcy. *An introduction to Analysis*. Springer-Verlag, New York, 1995.

[4] C.-H. Chu. *Modern Analysis*. University of London (External Programme), London, 1993.

[5] Harro G. Heuser. *Functional Analysis*. Wiley, Chichester, 1982.

[6] G. J. O. Jameson. *Topology and Normed Spaces*. Chapman and Hall, London, 1974.

[7] Jun Kigami. *Analysis on Fractals*. Cambridge University Press, Cambridge, 2001.

[8] A. N. Kolmogorov and S. V. Fomin. *Elements of the Theory of Functions and Functional Analysis (Volumes 1 and 2)*. Dover, Mineola, New York, 1999 (reprint).

[9] Elliott Mendelson. *Introduction to Mathematical Logic*. Van Nostrand, Princeton, 1964.

[10] George F. Simmons. *Introduction to Topology and Modern Analysis*. McGraw-Hill, Tokyo, 1963.

[11] W. A. Sutherland. *Introduction to Metric and Topological Spaces*. Oxford University Press, Oxford, 1975.

Index

*To those who do not know mathematics, it is difficult
to get across a real feeling as to the beauty,
the deepest beauty, of nature ...
If you want to learn about nature, to appreciate nature,
it is necessary to understand
the language that she speaks in.*   Richard Feynman, 1918–1988

Made in the USA
Lexington, KY
14 July 2012